BIOLOGICAL NITROGEN FIXATION IN FOREST ECOSYSTEMS: FOUNDATIONS AND APPLICATIONS

FORESTRY SCIENCES

Also in this series:

Baas P, ed: New Perspectives in Wood Anatomy. 1982. ISBN 90-247-2526-7
Prins CFL, ed: Production, Marketing and Use of Finger-Jointed Sawnwood. 1982. ISBN 90-247-2569-0
Oldeman RAA, *et al.* eds: Tropical Hardwood Utilization: Practice and Prospects. 1982. ISBN 90-247-2581-X
Den Ouden P and Boom BK, eds: Manual of Cultivated Conifers: Hardy in Cold and Warm-Temperate Zone. 1982. ISBN 90-247-2148-2 paperback; ISBN 90-247-2644-1 hardbound
Bonga JM and Durzan DJ, eds: Tissue Culture in Forestry. 1982. ISBN 90-247-2660-3
Satoo T and Madgwick HAI: Forest Biomass. 1982. ISBN 90-247-2710-3
Van Nao T, ed: Forest Fire Prevention and Control. 1982. ISBN 90-247-3050-3
Douglas J: A Re-appraisal of Forestry Development in Developing Countries. 1983. ISBN 90-247-2830-4

In preparation:
Németh MV: The Virus-Mycoplasma and Rickettsia Disease of Fruit Trees.

Biological nitrogen fixation in forest ecosystems: foundations and applications

Edited by

J.C. GORDON
Oregon State University, Corvallis, U.S.A.

and

C.T. WHEELER
The University of Glasgow, Glasgow, U.K.

1983 **MARTINUS NIJHOFF / Dr W. JUNK PUBLISHERS**
a member of the KLUWER ACADEMIC PUBLISHERS GROUP
THE HAGUE / BOSTON / LANCASTER

Distributors

for the United States and Canada: Kluwer Boston, Inc., 190 Old Derby Street, Hingham, MA 02043, USA
for all other countries: Kluwer Academic Publishers Group, Distribution Center, P.O.Box 322, 3300 AH Dordrecht, The Netherlands

Library of Congress Cataloging in Publication Data

Main entry under title:

Biological nitrogen fixation in forest ecosystems.

(Forestry sciences ; v. 9)
1. Nitrogen--Fixation. 2. Forest ecology. 3. Forest management. I. Gordon, J. C. (John C.), 1939- . II. Wheeler, C. T. (Christopher T.) III. Series.
QR89.7.B55 1983 581.5'2642 83-8141
ISBN 90-247-2849-5

ISBN 90-247-2849-5 (this volume)

Copyright

PRINTED IN THE NETHERLANDS

Contents

1. Silvicultural systems and biological nitrogen fixation

JOHN C. GORDON

Department of Forest Science, Oregon State University, Corvallis, OR 97331 USA

BIOLOGICAL NITROGEN FIXATION IN FORESTS AND FORESTRY

Biological nitrogen fixation occurs in most, probably all, forest ecosystems (Chap. 2), and thus is a factor in all silvicultural systems. Silvicultural systems are defined here as purposeful attempts to derive benefits for humans from forested (or afforested) lands. Such attempts vary from minimal efforts to protect wild ecosystems from catastrophic fire to the establishment of artificial stands of exotic trees which are then vigorously managed for specific products. Although 'forest products' are commonly thought of as wood in its various fuelwood and manufactured forms, of equal or greater importance in many places are clean water, stabilized soil, enhanced wildlife populations, improved grazing for domestic livestock, and enhanced recreational opportunities. Thus, this discussion of nitrogen fixation as a silvicultural tool will emphasize, but will not be limited to, silvicultural systems with wood production as their objective. Also, since all silvicultural systems probably contain some nitrogen-fixing organisms, this discussion is limited to 1) those in which naturally-occurring fixation is recognized as particularly important and is a factor in silvicultural decisions (passive systems) and 2) those in which fixation is enhanced through species introduction or silvicultural practices, or both (active systems).

SILVICULTURAL PRESCRIPTIONS

Because biological nitrogen fixation is, in most places, just now being recognized as a potential silvicultural tool, this chapter will describe the options for its use currently available to silviculturists. Also, because silvicultural prescriptions are usually made for areas defined by aggregates of trees and other vegetation that are relatively homogeneous (stands), examples of specific stand treatments will be presented and discussed where possible. More specific examples are presented in Chapters 8–12.

Increasingly, written silvicultural prescriptions are being used to document and guide silvicultural operations. The format of written prescriptions varies widely, but nearly all include these elements:

Gordon, J.C. and Wheeler, C.T. (eds.). Biological nitrogen fixation in forest ecosystems: foundations and applications

ISBN 90-247-2849-5.
Printed in The Netherlands

1. A description of the stand to be treated that includes geographic location, and information on vegetation, climate and soil,
2. a statement of the specific objectives, in terms of products, and constraints, that guide silvicultural operations on the stand and
3. a detailed set of directions for achievement of the objectives that includes a description of all silvicultural operations to be performed, and their probable consequences.

Thus, a good silvicultural prescription provides a prediction of the influence of all prescribed operations on the ecosystem being treated. Often, this is the portion of the prescription that does, or should, use information on biological nitrogen fixation. Beyond this, however, one of the commonest silvicultural objectives is to maximize production of wood of a special quality level within stated environmental and economic constraints. Because adding nitrogen in a form available to trees has proven to increase wood production in many places, fertilization with industrially-fixed nitrogen is increasingly prescribed. Knowledge of biological nitrogen fixation can aid the silviculturist prescribing nitrogen fertilizer application by 1) allowing a prediction of the impact of added nitrogen on biological fixation, 2) making possible a comparison of industrially- and biologically-fixed nitrogen in specific costbenefit terms, and 3) providing the means for either minimizing any adverse impact of added fixed nitrogen on existing biological fixation or replacing industrially- with biologically-fixed nitrogen. Specific information on identification and measurement of symbiotic and free-living nitrogen-fixing systems will be provided in Chapters 2–6.

ADVANTAGES AND DISADVANTAGES OF BIOLOGICAL NITROGEN FIXATION AS A SILVICULTURAL TOOL

All silvicultural prescriptions are, or should be, specific to an individual site and set of objectives. Some considerations are constant over a broad range of both, however, and those are treated here as background for more specific applications.

Advantages

Biological nitrogen fixation is usually compared with the application of industrially-fixed nitrogen fertilizers to forest stands. Usually, with respect to these fertilizers the introduction or maintenance of nitrogen-fixing species is:

1. Productive of a regular supply of available nitrogen over a longer time.
2. Less likely to suffer losses from volatilization and leaching.
3. Less likely to cause public concern about 'chemical' use.
4. A source of increased soil organic matter.

5. Accessible to landowners who have difficulty obtaining fertilizer nitrogen or the technology for applying it.

Disadvantages

In comparison with the application of industriallyfixed nitrogen, enhancement of biological nitrogen fixation is:
1. Currently limited by knowledge of nitrogen-fixing species' biology and silviculture.
2. Likely to be more expensive on a unit applied nitrogen basis (see Chap. 8).
3. Slower in producing fertility increases.
4. A possible source of vegetative competition for other crop species.
5. Likely to produce a managerially more complex ecosystem.

PASSIVE SYSTEMS

Silvicultural options for those systems in which naturally-occurring fixation is recognized as important, but in which no introduction of additional nitrogen-fixing systems is made, consist mainly of recognition and choice among the different ways to achieve silvicultural objectives with minimum disruptions of the naturally-occurring system. In some stands, and for some objectives, naturally-occurring systems must be thoroughly disrupted because the cost of retaining the nitrogen-fixing component exceeds the benefit derived from it. This is commonly the case in stands where host plants of nitrogen-fixing systems compete strongly with crop trees for light, moisture, and mineral nutrients.

Classically, silvicultural systems have been divided into even-age systems, those in which crop trees are all approximately the same age, and uneven-age systems, those in which crop trees present in the stand at one time vary widely in age, and can be sorted into several 'age classes'.

Naturally-occurring forest ecosystems for which silvicultural prescriptions are prepared, and which also include an important component of systems capable of biological fixation are not found in all regions of the world. One of North America's major timber-producing regions, the Pacific Northwest, contains an abundance of them. In contrast, the eastern and sourtheastern forests of Canada and the United States contain fewer nitrogen-fixing species, and they are less important ecologically and commercially. In Western Europe, and Australia and New Zealand, most forest ecosystems are highly manipulated or have as their principal tree component introduced species, often including nitrogen-fixers, and are thus classified as active systems. Tropical and sub-tropical forests, many of which have recently been subjected to intensive harvest or conversion, or both, contain perhaps the greatest number of woody nitrogen-fixing species of any major forest type. Most, but not all, of these are legumes. However, tropical silviculture is new, and little is known about the biology and silvicultural requirements of the nitrogen fixers in tropical forests. For example, there is no definitive catalog that indicates which tropical tree legumes are known to nodulate and fix nitrogen. Thus little information, derived either from research or experience, is available for inclusion here.

Even-age systems

Even-age systems are much commoner in current silvicultural practice, and are thought to be, for most usual silvicultural objectives, more economically efficient then uneven-age systems. Because most even-age systems call for the replacement of mature stands with young stands over short times (the regeneration period) the role of naturally occurring nitrogen-fixing systems is potentially quite different in these than in uneven-age systems. This is particularly true when host plants of nitrogen-fixing systems are fast-growing species that may overtop and shade out trees that are more valuable during the regeneration period.

Often, host plants of nitrogen-fixing systems can become established before harvest of the mature stand in an even-age system, particularly if the canopy is opened to facilitate regeneration, as in shelterwood systems. Although the fixed nitrogen they produce may be beneficial to the growth of the old stand, nitrogen-fixing systems may be a nuisance during the regeneration period. Most woody species capable of symbiotic nitrogen fixation are adapted to rapid growth on unoccupied sites, particularly those low in nitrogen, and thus are highly competitive, particularly with conifers that exhibit slow juvenile growth. Silvicultural prescriptions for even-age systems in which symbiotic fixation is important must then 1) determine the quantity of nitrogen-fixing vegetation likely to arise due to operations such as thinning and shelterwood harvesting, 2) assess the benefit of the quantity of nitrogen fixed to the old stand and to the regeneration that is to follow, 3) determine the effect of the vegetation carried over from the old stand on establishment and growth of the new stand, and 4) devise ways of reducing or eliminating the competitive effect of the carried-over vegetation on the new stand if this competitive effect's cost exceeds the benefit of nitrogen fixed. It may be that the choice is between competitive vegetation that is capable of symbiotic nitrogen fixation and competitive vegetation that is not. Usually, this choice would be resolved in favor of the nitrogen-fixing form, but not always. If both types of competing vegetation must be controlled in the future, it may be sensible to favor easy-to-control over hard-to-control vegetation, regardless of whether it is capable of nitrogen fixation.

In some even-aged systems that employ vigorous site preparation, including burning, nearly ideal conditions for the growth of woody species capable of symbiotic nitrogen fixation may be created. In these stands, competition between crop trees and nitrogen fixing species must be particularly carefully managed. Again, it may be preferable to have nitrogen-fixing, as opposed to non-nitrogen fixing, competition. If, however, the nitrogen-fixing species is more likely to exceed the crop trees in height growth than the non-nitrogen fixing vegetation, it may be considerably less desirable.

For some silvicultural objectives, and for some nitrogen-fixing species, retention of nitrogen-fixers as crop trees may be desirable. In other situations, it may be desirable to grow a rotation of a nitrogen-fixing species with commercial value if it becomes established after harvest, rather than incurring the expense of conversion to a higher-value species with no return. For example it was concluded that such crop rotation merited serious consideration for red alder (*Alnus rubra*) and Douglas-fir (*Pseudotsuga menziesii*) in the Pacific North-west of the United States (1). In a comparison of 6 silvicultural regimes, 3 assuming continous cropping of Douglas-fir, and 3 involving a rotation of red alder, all options were projected to be profitable. Continuous Douglas-fir, however, was more profitable than when rotated with alder. This study assumed planting costs for the alder rotation in each regime. If alder became established naturally after site preparation, and no planting costs were incurred early in the alder rotation, it might in some situations be more profitable, given adequate markets, to manage the alder stand to rotation before converting to Douglas-fir.

Another even-aged, passive, option involves leaving naturally-occurring nitrogen fixers interspersed with crop trees if the nitrogen-fixers are either carried through the regeneration period or invade the stand at a later time, often after thinning. Again, the decision must be made on cost-benefit grounds. If the value of the nitrogen fixed plus the value of the hosts, if they are merchantable or otherwise useful (browse, shade) plants, exceeds the sum of the value of growth loss on other crop trees due to competition, and the cost of removing nitrogen fixers, then they should remain unless they present an unacceptable hazard to future regeneration, as discussed earlier.

In most such silvicultural comparisons, the value of fixed nitrogen is taken to be the cost of applying industrially-fixed nitrogen. There are some situations, however, in which the application of nitrogen fertilizers is difficult for environmental, legal or operational reasons. In these situations, biologically-fixed nitrogen may be the only alternative available to silviculturists to increase crop tree growth through better nitrogen nutrition.

Uneven-age systems

Uneven-age systems, which often involve the selection of individual trees or small groups of trees for harvest, offer rich possibilities for the retention and manipulation of naturally-occurring nitrogen-fixing tree species. Theoretically it is easy in these systems to leave an optimum number of nitrogen-fixing individuals, and to control their distribution by selective harvest. However, since most unevenage systems maintain continuous tree canopy cover of most of the stand, species adapted to full-sunlight and unoccupied sites, as many nitrogen fixers are, are discriminated against. For the same reason, since quantity of nitrogen fixed is often proportional to light interception by the host, amounts of fixed nitrogen derived from plants growing in unevenage stands can be expected to be lower than amounts derived from the same species open grown.

ACTIVE SYSTEMS

Active systems are treated fully in Chapters 8–12. It will suffice here to indicate that the possibilities for using the great variety of biological materials available have rarely been explored, and probably won't be, at least in temperate regions, until the economic and social imperative for doing so is clearer. Active systems will remain important in those limited areas where they are already successful: reclamation of wasteland and fuelwood/forage systems, primarily in the tropics.

USE OF THIS BOOK

This book is intended for use by foresters and land managers wishing to use and understand biological nitrogen fixation. Thus, Chapters 2 through 6 describe the biological bases of nitrogen fixation in forests, including the recognition of organisms and the estimation and measurement of their characteristics. Chapter 7 through 12 present applications of biological nitrogen fixation in forestry and related land management activities in a variety of locations. Our intent is that a researcher or practitioner only marginally familiar with the processes and applications of biological nitrogen fixation can, through this book, gain access to its potential as a management tool.

REFERENCE

1. Atkinson WA, Bormann BT, DeBell DS: Crop rotation of Douglas-fir and red alder: a preliminary biological and economic assessment. *Botanical Gazette* 140 (Suppl): S108–S109, 1979.

2. Morphology of nitrogen fixers in forest ecosystems

A.D.L. AKKERMANS and A. HOUWERS

Laboratory of Microbiology, Agricultural University, Hesselink van Suchtelenweg 4, 6703 CT Wageningen, The Netherlands

INTRODUCTION

In nature the element nitrogen is available in large quantities in various inorganic and organic forms. Most of it is present in the stable form of atmospheric N_2; a minor fraction of the nitrogen is present either in the reduced ionic form, ammonium (NH_4^+), or in forms with different levels of oxidation: volatile N-oxides (N_2O, NO and NO_2), and the ionic forms, nitrite (NO_2^-) and nitrate (NO_3^-).

All living cells incorporate nitrogen in its reduced form into complex N-containing substances such as amino acids, proteins, nucleic acids. The forms in which cells utilise nitrogen for these synthetic processes varies between groups of organisms. Thus, animal cells usually utilise amino acids as the N-source but green plants and many fungi and bacteria can use nitrate and ammonium nitrogen. Nitrate must be reduced to ammonium before assimilation into organic combination. Certain prokaryotes can use nitrate not only as N source but also as an electron acceptor under anaerobic conditions (dissimilatory nitrate reduction or denitrification). These micro-organisms mainly produce N_2 from nitrate. The reduction of N-oxides other than nitrate occurs only in few prokaryotes.

Only a very few microbes are able to reduce N_2 and therefore these play a key role in the nitrogen cycle in nature. Microbial reduction (= fixation) of N_2 has received much attention, especially during the last decennium. Every year up to a thousand papers and several review papers and books are published on topics related to biological nitrogen fixation. Most of these publications, however, are laboratory studies of well-known microbes or are restricted to the application of N_2 fixation in agriculture. Much less attention has been paid to N_2-fixing systems in forest and natural ecosystems. In the last decennium an increasing number of N_2 fixers have been isolated, but little is known about their occurrence in nature and possible contribution to the nitrogen cycle. In the present chapter the morphology and development of various N_2-fixing systems will be described, particularly those which occur in forest ecosystems. The description will include information on physiological characteristics that are related directly to morphological features.

Gordon, J.C. and Wheeler, C.T. (eds.). Biological nitrogen fixation in forest ecosystems: foundations and applications

ISBN 90-247-2849-5.
Printed in The Netherlands

CLASSIFICATION OF N_2-FIXING ORGANISMS

All N_2-fixing organisms are prokaryotes, i.e. micro-organisms with one chromosome, without a nuclear membrane and without organelles. Prokaryotes can be classified according to different systems. The most commonly used classification system is described in 'Bergey's Manual of Determinative Bacteriology' (1). This Manual includes micro-organisms that are accepted as real prokaryotes and that can be cultivated in pure culture. Until ten years ago there was uncertainty about the taxonomic position of blue-green algae but in the last edition of Bergey's Manual in 1974 (1), blue-green algae were rehabilitated as true prokaryotes and were classified into the separate Division of Cyanobacteria. Further subdivision was not attempted but elsewhere the blue-green algae have been divided into two orders of unicellular organisms i.e. *Chroococcales* and *Pleurocapsales*, and two orders of filamentous organisms i.e. *Nostocales* and *Stigonematales*. In the literature *ca* 81 genera have been described. Nitrogen-fixing species are found in 25 of these genera (Table 1), i.e. 31%. The other prokaryotes are subdivided into 19 parts, of 1–3 orders. In Bergey's Manual (1), 245 genera are distinguished, including 39 genera with uncertain affiliation. Nitrogen-fixing bacteria are classified in 26 genera, occurring in 8 parts (Table 1), i.e. in *ca* 10% of all genera. Based on this classification system it can be concluded that the occurrence of the N_2-fixing enzyme nitrogenase, is rather exceptional in nature.

N_2-fixing prokaryotes comprise both free-living and symbiotic micro-organisms. Free-living N_2-fixers consist of C-autotrophic (Cyanobacteria) and other phototrophic bacteria and C-heterotrophic bacteria. Nitrogen fixers living in symbiosis with higher plants belong to three taxa, *Rhizobium*, *Frankia* and Cyanobacteria (mainly *Anabaena* and *Nostoc*).

Most free-living, C-heterotrophic N_2 fixers are Gram-negative rods. In some genera, e.g. *Azotobacter*, all species are able to fix N_2. In others this feature is restricted to special groups or even only to a few strains. In the genus *Clostridium*, only saccharolytic clostridia but none of the pathogenic, proteolytic ones are able to fix N_2. In the genus *Thiobacillus*, only *T. ferro-oxydans* is known to be a N_2 fixer. Within the Enterobacteria (Part 8, Table 1) most genera include only a few strains which fix N_2, e.g. in *Escherichia coli*, *Erwinia herbicola* and *Enterobacter aerogenes*. Within the Cyanobacteria a similar situation occurs. In filamentous forms with heterocysts, e.g. *Anabaena* and *Nostoc*, all members are able to fix N_2 when cultivated under proper conditions. However, in filamentous, non-heterocystous cyanobacteria, e.g. *Oscillatoria*, only certain strains fix N_2.

Carbon-heterotrophic N_2 fixers usually are present in nature under carbon-limited growth conditions. Most of these organisms utilize mono-sugars, simple alcohols and organic acids as C- and energy source. Only *Clostridium* spp. are able to utilize polysaccharides and decompose cellulose under anaerobic conditions. Since organic carbon in soil is mainly present in polymeric form, for

Table 1. Number of genera with nitrogen fixing species.

	Total	N_2-fixing	
Division 1: CYANOBACTERIA*	81?	25	
1. Chroococcales, unicellular	19	2	*Gloeocapsa*, *Synechococcus*
2. Pleurocapsales, unicellular	5	–	several strains
3. Nostocales, filamentous	38?	18	*Anabaena*, *Anabaenopsis*, *Aphanizomenon*, *Aulosira*, *Cylindrospermum*, *Nodularia*, *Nostoc*, *Scytonema*, *Tolypothrix*, *Microcheate*, *Calothrix*, *Dichothrix*, *Rivularia*, *Raphidiopsis*, *Oscillatoria*, *Lyngbya*, *Plectonema*, *Phormidium*
4. Stigonematales, filamentous	19	5	*Fischerella*, *Hapolosyphon*, *Stigonema*, *Westelliopsis*, *Mastigocladus*
*Division 2: BACTERIA***	245	26	
1. Phototrophic Bacteria	18	4	*Rhodospirillum*, *Rhodopseudomonas*, *Chromatium*, *Chlorobium*
2. Gliding Bacteria	27	–	
3. Sheathed Bacteria	7	–	
4. Budding and/or Appendaged Bacteria	17	–	
5. Spirochates	5	–	
6. Spiral and Curved Bacteria	6	2	*Azospirillum*, *Campylobacter*
7. Gram neg. Aerobic Rods and Cocci	20	8	*Pseudomonas?*, *Azotobacter*, *Azotomonas*, *Beijerinckia*, *Derxia*, *Rhizobium*, *Methylosinus*, *Methylococcus*, *Xanthobacter****
8. Gram neg. Facultatively Anaerobic Rods	26	5	*Escherichia*, *Citrobacter*, *Klebsiella*, *Enterobacter*, *Erwinia*
9. Gram neg. Anaerobic Bacteria	9	1	*Desulfovibrio*
10. Gram neg. Cocci and Coccobacilli	6	–	
11. Gram neg. Anaerobic Cocci	3	–	
12. Gram neg. Chemolithotrophic Bacteria	17	1	*Thiobacillus*
13. Methane-producing Bacteria	3	–	
14. Gram pos. Cocci	12	–	
15. Endospore-forming Rods and Coci	6	3	*Bacillus*, *Clostridium*, *Desulfotomaculum*
16. Gram pos. Asporogenous Rod-Shaped Bacteria	4	–	
17. Actinomycetes and related Organisms	37	2	*Mycobacterium****, *Frankia*
18. Ricketsias	18	–	
19. Mycoplasmas	4	–	
Total	326?	51	

* Mainly according to ref. 29
** According to ref. 1
*** N_2-fixing, H_2-utilizing Mycobacterium have recently been classified in the genus Xanthobacter (38).

example in cellulose, where O_2 usually is present as well, it is not likely that free-living N_2 fixers will grow explosively: as in all other C-heterotrophic organisms, growth is carbon-limited and not by nitrogen. By contrast, one would expect C-autotrophic N_2-fixers, viz. *Cyanobacteria*, to have great potential for growth in nature. However, in practice their distribution is restricted by availability of light, in soil to the surface and in water to the upper few meters. Higher plants have a much larger photosynthetic surface per square metre of land surface and N_2 fixers which are able to utilize specifically C-compounds from the plants are at a great advantage.

The most specialized combination of a plant and a N_2 fixer is found in the root nodule symbiosis. Root nodules are structures on the roots induced and inhabited by specific N_2-fixing organisms. The N_2 fixer is maintained by the plant in a non-growing but nitrogen-fixing stage. It utilizes C-compounds supplied by the plant as C- and energy source and excretes fixed nitrogen into the cytoplasm of the plant. The root nodule microsymbiont has the advantage over micro-organisms, living in the rhizosphere and in the soil, of a protected environment in which it does not need to compete with other microbes for food. In the nodule symbiosis between cyanobacteria and cycads, the former can even transform into a C-heterotrophic microbe.

In contrast to prokaryotes, photosynthetic higher plants usually are limited in growth by nitrogen. The symbiosis between N_2-fixing C-heterotrophic micro-organisms and higher plants therefore has become a most successful combination.

MORPHOLOGICAL AND PHYSIOLOGICAL FEATURES OF FREE-LIVING N_2-FIXERS

Cyanobacteria

As noted, cyanobacteria (= blue-green algae) are divided into two orders of unicellular and two orders of filamentous organisms (Table 1). The Botanical Code based on classical taxonomic studies does not function in practice with these organisms and Stanier *et al.* (2) suggested reclassification according to the Bacterial Code. Furthermore, Rippka *et al.* (3) have introduced a new system of classification, which now has been adopted by the American Type Culture Collection (ATCC).

It has been known for a long time that heterocystous, filamentous cyanobacteria are able to fix N_2. However, in the last decennium an increasing number of non-heterocystous cyanobacteria also have been found to fix N_2. Within the unicellular cyanobacteria N_2 fixation is not uncommon, but in most strains nitrogenase is rapidly inactivated by oxygen. Only five strains of *Gloecocapsa* sp. (= *Gloeotheca* sp.) and one strain of *Synechococcus* sp. (= *Aphanotheca pallida*) are able to fix N_2 aerobically (3, 4). Another aerobic N_2 fixer usually classified within the Chroococcales, is *Chlorogloea fritschii*. This organism can also form heterocysts and small filaments and therefore has been classified by several authors in the genus *Nostoc* (5). Rippka and Waterbury (6) have tested 55 strains of the Chroococcales and found only three strains (excluding *Gloeocapsa*) with nitrogenase activity. Within the order of Pleurocapsales, 19 of the 32 strains (6) showed nitrogenase activity when cultivated anaerobically.

A similar type of oxygen-sensitive N_2 fixation has also been found in certain filamentous, non-heterocystous species, viz. *Plectonema boryanum*, *Oscillatoria*, *Lyngbya* and *Phormidium* spp. Only a few nonheterocystous filamentous strains are able to fix N_2 aerobically, i.e. *Trichodesmium* (also classified in the genus *Oscillatoria*) and *Microcoleus* sp. In many species nitrogenase is rapidly inactivated when the cultures are not flushed with O_2-free gas and this can give significant problems during the measurement of nitrogenase activity. In *Oscillatoria* sp. C_2H_2-reducing activity could be detected only when cells are incubated in the light + 3′-(3,4-dichlorophenyl)-1′-1′-dimethylurea (DCMU) or in the dark, i.e. when O_2 evolution by the cells was prevented. (Fig. 1). How these cells protect their nitrogenase from O_2 inactivation is still unknown but several mechanisms of protection have been proposed (7).

Many non-heterocystous species grow in layers on moist rocks or are benthic in water at reduced light intensity. Nitrogenase is poorly protected from O_2 inactivation, due to the presence of a photosynthetic system and nitrogenase within one cell though some protection occurs when the cells are grown in thick layers, creating micro-aerophilic or anaerobic conditions. Due to the O_2 sensi-

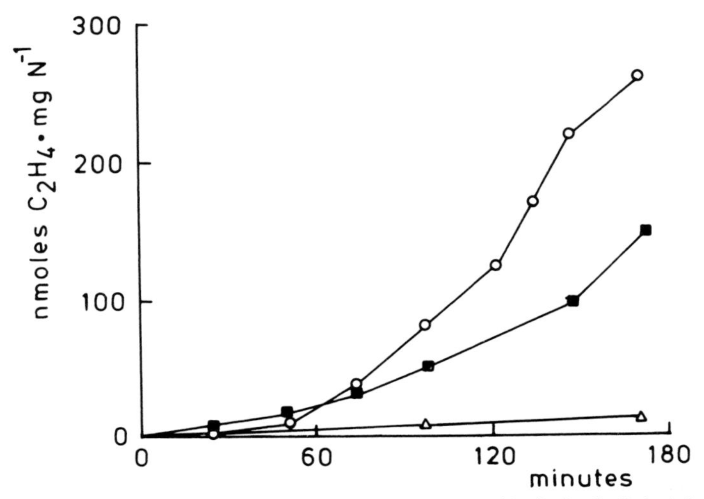

Fig. 1. Acetylene reduction of *Oscillatoria* sp. incubated anaerobically in the light (Δ), in the dark (■) and in the light in the presence of 0.03 mM DCMU (○) Remmelzwaal and Akkermans unpublished results).

tivity these N_2-fixing organisms have a restricted distribution and probably play an insignificant role in forest ecosystems.

The largest group of N_2 fixers occur in the heterocystous filamentous organisms (Table 1). These organisms have overcome the problem of O_2 sensitivity of the N_2-fixing system by localization of the nitrogenase in specialized cells, heterocysts, with an incomplete photosynthetic apparatus. These organisms form heterocysts when growing under N_2-fixing conditions. Heterocysts are non-growing, carbon-heterotrophic cells with nitrogenase. Most characteristic is the absence of photosystem 2 and consequently the inability to produce O_2 and to fix CO_2. Photosystem 1 remains intact so ATP generation in the light remains possible. Heterocysts do not multiply and they develop a thick cell wall. Their formation is associated with the induction of nitrogenase. In the presence of combined-N in the medium no heterocysts are formed and nitrogenase is absent. When cells of *Anabaena* sp. are transferred from a NH_4-medium to a N_2-medium, nitrogenase is induced when mature heterocysts are present (8). The presence of nitrogenase has also been demonstrated directly in cell-free extracts of isolated heterocysts (9). The formation and function of heterocysts have been investigated intensively and the results have been summarized and discussed in various review papers (5, 7, 10, 11).

In addition to the formation of heterocysts, filamentous cyanobacteria can also differentiate into resting cells that are specialized for survival (called akinetes or spores). Akinetes are enlarged cylindrical or spherical cells surrounded by a thick cell envelope and include granules of polyphosphate, protein (cyanophycin) and other kinds of reserve material. The formation of akinetes in e.g. *Anabaena cylindrica* usually starts adjacent to a heterocyst and spreads along the filaments (12, 13, 14, 15). The degree of sporulation varies among different strains and is apparently genetically determined. In some strains of e.g. *Anabaena* and *Nostoc* almost all vegetative cells differentiate into akinetes in the stationary phase, while in others, akinetes are formed only occasionally. The ability to form akinetes can be lost when cells are cultivated for a long time in liquid mineral medium in an exponential phase. This was found in one strain of *A. variabilis* which originally produced masses of spores. After prolonged subculture less than one percent of the cells transformed into akinetes when conditions for sporulation were optimised (Davelaar and Akkermans, unpublished results). Akinetes remain viable for a long time, even under stress conditions e.g. desiccation (16). During this resting phase these cells remain metabolically active at a low level (15) and can therefore survive conditions unfavourable for multiplication. Akinetes have a N, S and P content of respectively 8.4, 0.3 and 0.6 percent of freeze dry weight (analysis of L. van Liere, Amsterdam). Most of the nitrogen is present in protein. Akinetes, like vegetative cells, contain large quantities of the reserve proteins, cyanophycin and phycocyanin. The former is a copolymer of arginine and asparagine and is present in granules (17). The latter is a protein, coupled on the blue pigment typical of all cyanobacteria. The occurrence of proteins as reserve material is typical of cyanobacteria and not of other prokaryotes.

Germination of akinetes can be induced in the light in media containing phosphate. The outer cell wall bursts and the protoplast forms a small filament of vegetative cells (Fig. 2). When the filaments have reached a length of 2–4 cells, one polar cell enlarges and ceases to divide. This cell differentiates into a heterocyst. A thick cell wall is formed around the cell with special pores to the neighbouring vegetative cells. In a later stage of growth intercalary heterocysts are also formed. The pigment composition changes drastically and phycocyanin, the blue pigment characteristic of cyanobacteria, disappears. Photosystem 2 is lost and the cell is transformed from an autotrophic cell into a C-heterotrophic cell, dependent on carbon supplied by the host cells. Due to the absence of O_2 production from photosynthesis, the heterocyst is an excellent site of nitrogen fixation in which nitrogenase is protected. After germination of akinetes, a close correlation exists between the formation of heterocysts and the induction of nitrogenase (Fig. 3). Nitrogenase activity can be detected when the first mature heterocysts are present. A similar relationship between heterocyst formation and nitrogenase activity was found also when akinetes were incubated anaerobically in the light, under N_2/CO_2 (unpublished results) indicating that nitrogenase can not be induced in these young filaments simply by removing oxygen from the medium. It is more likely that the repression of nitrogenase is due to internal factors and there is evidence that this is due to high levels of intracellular nitrogen compounds in the young filaments. The C/N ratio of akinetes of *A. variabilis* is 5.4 and after germination this ratio increases until nitrogenase is formed. Degradation of reserve materials is probably responsible for increased levels of nitrogenous compounds in the young filaments since a gradual decrease in the number and size of the cyanophycin granules is found during the differentiation of the young filaments, indicating the consumption of reserve material for growth

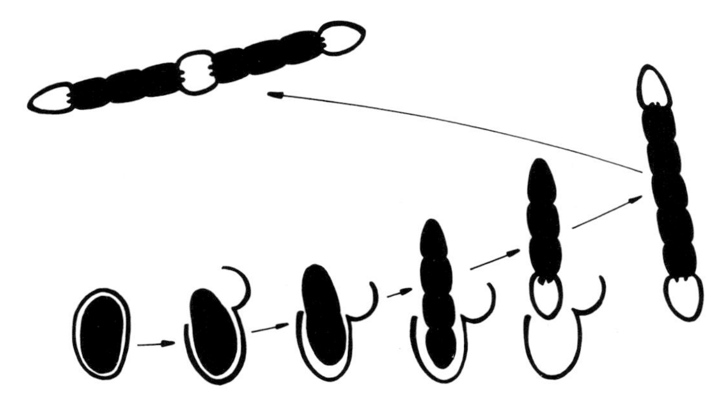

Fig. 2. Germination of akinetes of *Anabaena* sp. and formation of heterocysts.

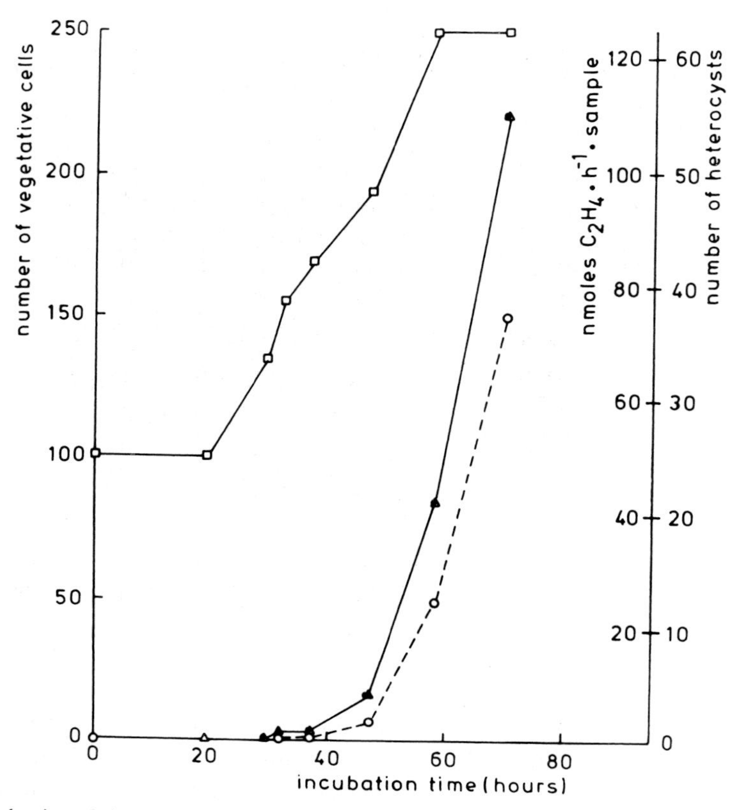

Fig. 3. Induction of nitrogenase (C_2H_2 reduction) and formation of heterocysts after germination in the light of akinetes of *Nodularia* sp. C_2H_2 reduction (○) and the relative number per 100 akinetes germinated of vegetative cells (□) and heterocysts (▲). Akkermans (unpublished results).

and differentiation. A similar phenomenon has been described previously for the differentiation of heterocysts in filaments of *A. cylindrica* after transferring cells from an ammonium-containing medium to a N-free medium (Fig. 4, ref. 8). The C/N ratio of the culture changed from 4.5 (in NH_4-grown cells) to 8 (in N_2-fixing culture).

The effect of oxygen on germination varies between species. Akinetes of *A. cylindrica* germinate faster in air than under N_2/CO_2 (18). In *A. variabilis*, however, the rate of germination is not changed by supply of external O_2 (Akkermans, unpublished results). In this latter species, when akinetes are incubated in Ar/CO_2, growth stops after formation of terminal heterocysts. At this stage the vegetative cells lack a granular structure and the arginine content, which originally is high decreases markedly (Table 2). These observations indicate that growth stops due to nitrogen limitation.

Germination of akinetes is light dependent. In the experiment described in

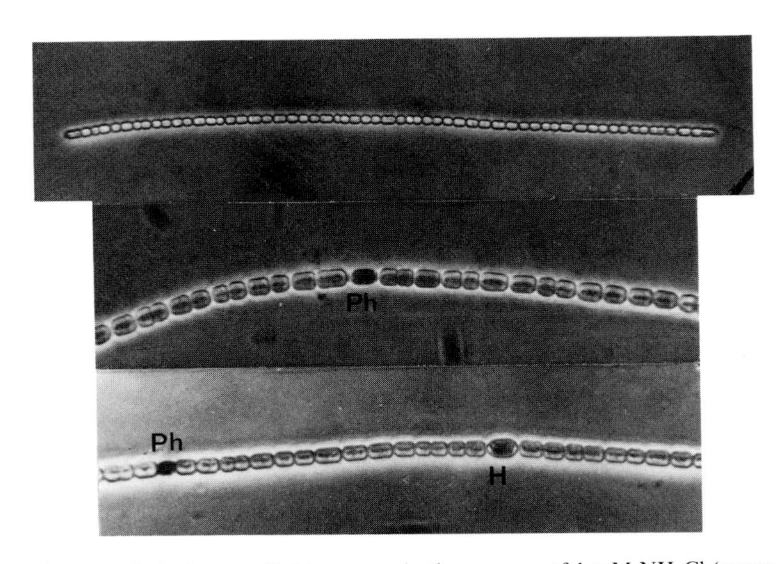

Fig. 4. Filaments of *Anabaena cylindrica* grown in the presence of 1 mM NH_4Cl (upper photo, magn. × 500), 24 hours (middle photo, magn. × 1200) and 27 hours (lower photo, magn. × 1200) after transferring to a N-free medium. Ph = proheterocyst; H = heterocyst. Nitrogenase activity was detected when mature heterocysts (H) were present. Kulasooriya *et al.* (8).

Fig. 3 up to 70 percent of the akinetes germinated. In a simultaneous experiment in the dark, under aerobic conditions only 20 percent of akinetes germinated and growth stopped at an earlier stage (unpublished results). No germination occured under anaerobic conditions in the dark. Yamamoto (18) reported that germination of akinetes of *A. cylindrica* in the dark can be stimulated by addition of acetate. Germination was enhanced with increasing light intensities, especially red light. Other species, e.g. *A. doliolum*, however, do not require red light (19).

Table 2. Arginine content of filaments of *Anabaena variabilis* A 70 after germination of akinetes.

Incubation time (h)	Gas mixture	Arginine content*	
		μmol. ml^{-1}	Percent**
0	—	0.61	14.6
20	1% CO_2 in N_2	0.64	18.7
	1% CO_2 in Argon	0.56	13.2
49	1% CO_2 in N_2	0.66	14.0
	1% CO_2 in Argon	0.29	11.6

* Whole cells were hydrolyzed in 6 N HCl for 16 h. Arginine content was determined with a Biocal 200 Amino-acid Analyzer. Values are expressed per ml incubated algal suspension.
** Percent of total content of amino acids in the hydrolysate.

The heterocyst frequency in aerobic N_2-fixing cells is ca. 5 and can be increased to 7–8 under anaerobic conditions and N-starvation (Argon instead of N_2 in gas phase) (8). This supports the hypothesis that heterocysts are the sites of N_2-fixation.

Effect of desiccation on cyanobacteria

Cyanobacteria often are found on the surface of soils and rocks which are exposed to extreme climatological conditions. Soil crusts, barks of trees and rocks frequently dry out for a period and are subsequently remoistened. Algal species found on such exposed sites must be adapted to these extreme conditions and usually differ significantly from algal species that occur in water. The difference in sensitivity of cyanobacteria to desiccation can be seen from the following experiments.

Three types of cyanobacteria from extreme environments were cultivated 'in vitro' in liquid medium: a. *Tolypothrix* sp. a sheath-forming benthic species from rocks; b. *Anabaena flos-aquae*, a typical planktonic species in fresh-water lakes and

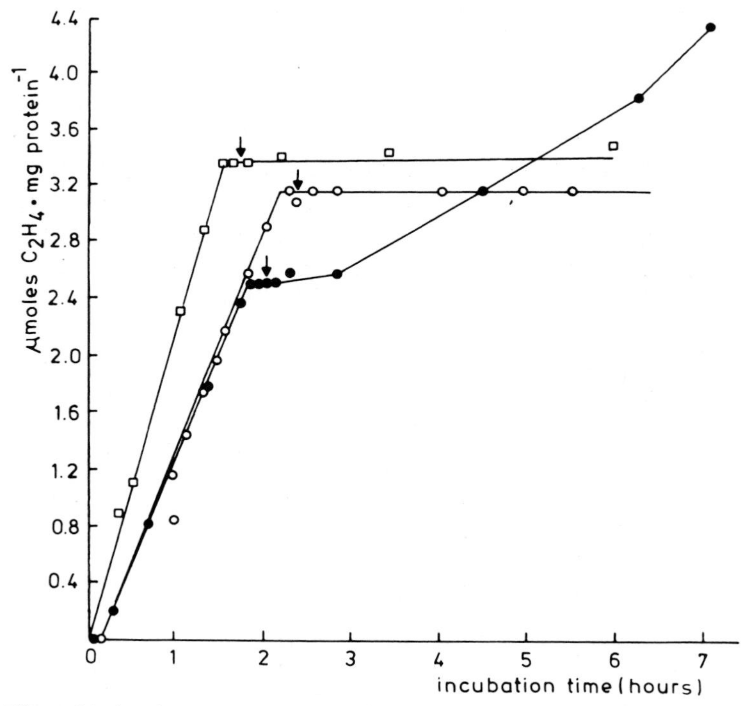

Fig. 5. Effect of desiccation and remoistening (↓) on the C_2H_2-reducing activity of *Tolypothrix* sp. (●), *Anabaena flos-aquae* (□) and *Nostoc* sp. strain 72 (○). (Stoopendaal and Akkermans, unpublished results).

c. *Nostoc* sp. Cells were harvested by filtration through a 0.8 μm filter. The algae on the filter were rinsed with water. After removing the adhering water from the filter, the algal-containing filters were incubated in vials above dry $MgCl_2$. The C_2H_2-reduction activity of the cells was measured during the gradual desiccation of the cells during incubation. As shown in Fig. 5 nitrogenase activity suddenly stopped when the water content of the cells reached a critical, low level. When the cells were remoistened by injecting water on the filters (arrows in Fig. 5), nitrogenase activity of *Nostoc* and *Anabaena* spp. was not restored, but in *Tolypothrix* sp. activity recovered slowly. A similar reaction was found for the dark respiration and the light-dependent O_2 evolution of the cells (data not shown). Microscopic observations showed that most *Nostoc* and *Anabaena* cells were lysed, in contrast to *Tolypothrix* cells, and after lysis of the algae, phycocyanin, a blue, water-soluble pigment, was visible by eye on the filters. These experiments clearly demonstrate the higher resistance to desiccation of *Tolypothrix* compared to both fresh water micro-organisms.

The effect of desiccation on the nitrogenase activity of *Tolypothrix* can be demonstrated directly in the field by incubating rocks or paving-stones which are

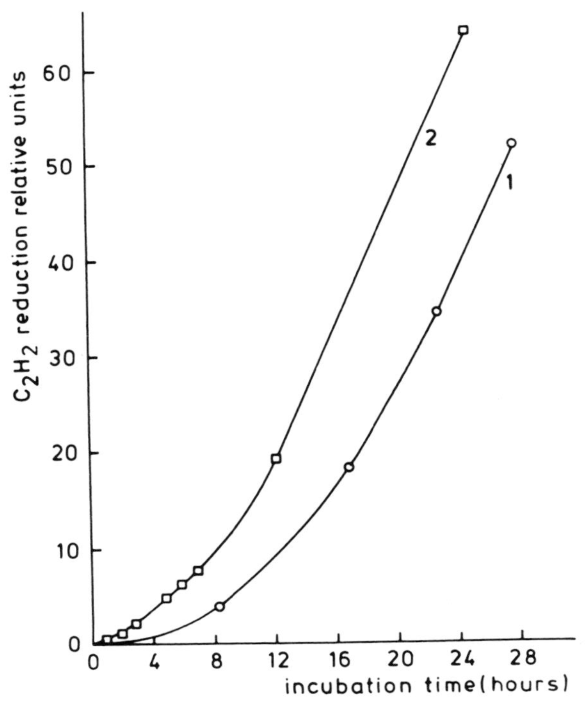

Fig. 6. Effect of moistening on the C_2H_2-reducing activity of cyanobacteria (mainly *Tolypothrix* sp.) grown on paving-stones (30 × 30 cm) after air drying for 22 days (0) and drying for a further 2 days before remoistening (□). (Stoopendaal and Akkermans, unpublished results).

covered by a layer of algae, containing *Tolypothrix* spp. The effect of moistening paving-stones which were air-dried for 22 days is shown in Fig. 6. Nitrogenase activity was induced within 8 hours of moistening (line 1). When the tiles subsequently were air-dried for 2 days and then remoistened, nitrogenase activity recovered within a few hours (line 2). The high resistance to desiccation of *Tolypothrix* spp. is possibly related to the possession of a thick sheath and extracellular slime (Fig. 7). Even more pronounced resistance to desiccation is found in cyanobacteria which grow in symbiosis with fungi, i.e. lichens. The recovery of nitrogenase activity after drying of the lichens *Collema tuniforme*, *Peltigera* sp. (20) and *Stereocaulon paschale* (21) has been investigated extensively. In the latter species nitrogenase activity recovered within 12 hours of moistening even after a previous long period of dryness.

Phototrophic bacteria

Phototrophic bacteria, excluding cyanobacteria, can grow anaerobically in the light with CO_2 as C-source and reduced inorganic compounds as electron donors. Their physiology has been reviewed recently (7). Cells are spherical, rod, vibrio or spiral shaped. Gram negative and colored purple, green, brown or red, depending on the pigment composition. The group is divided into purple sulfur

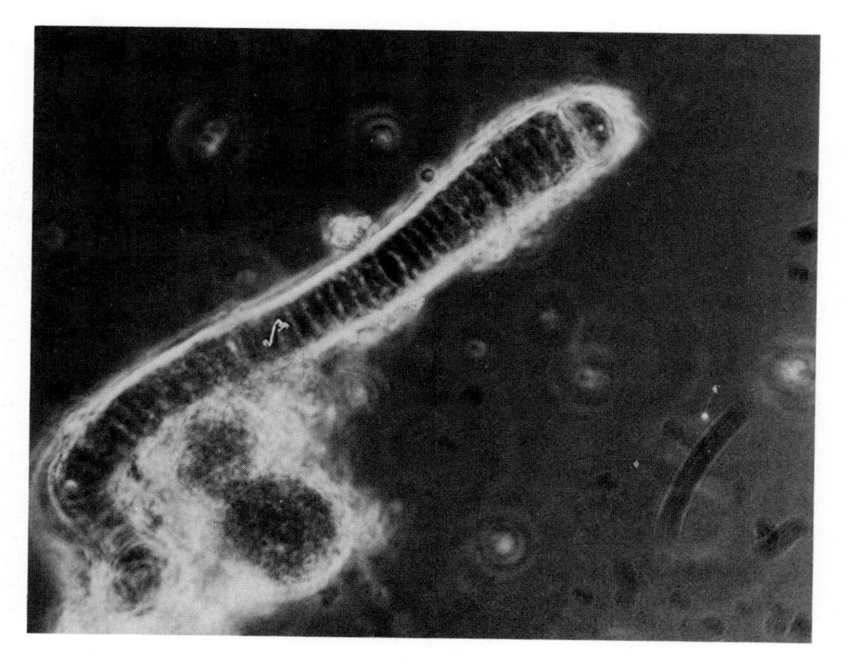

Fig. 7. Detail of filament of *Tolypothrix* sp. on paving-stone. Note the terminal heterocyst.

bacteria and purple non-sulfur bacteria. The former consists of strict anaerobes with a photo-autotrophic metabolism and utilize H_2S as electron donor. Most characteristic of these organisms is the deposition of inorganic sulfur in or outside the cells (e.g. *Chromatium*).

Purple non-sulfur bacteria have a photoheterotrophic type of metabolism. These bacteria typically occur in fresh water with large amounts of organic matter and low concentrations of sulfide. Nitrogen-fixing strains frequently occur within the genera *Rhodospirillum* and *Rhodopseudomonas*. A second group of photosynthetic bacteria are the Green bacteria: strictly anaerobic photo-autotrophic organisms which utilize reduced inorganic sulfur compounds or H_2 as electron donors. Nitrogen fixers are found within the genus *Chlorobium*.

From this description of their metabolism it will be evident that these N_2-fixing phototrophic bacteria play a role only in anaerobic environments, with high concentrations of organic compounds, e.g. in water and in muddy soils. In forest ecosystems they play an insignificant role.

Spiral- and curved bacteria

There are two genera with N_2-fixing species, viz. *Azospirillum* and *Campylobacter*. These are Gram-negative, aerobic chemoheterotrophic, helical cells which are motile by a polar flagellum (Fig. 8). So far N_2-fixing strains of *Campylobacter* have been found only on the roots of the halophytic plant *Spartina alternaria*. *Azospirillum* spp., however, have been isolated from roots

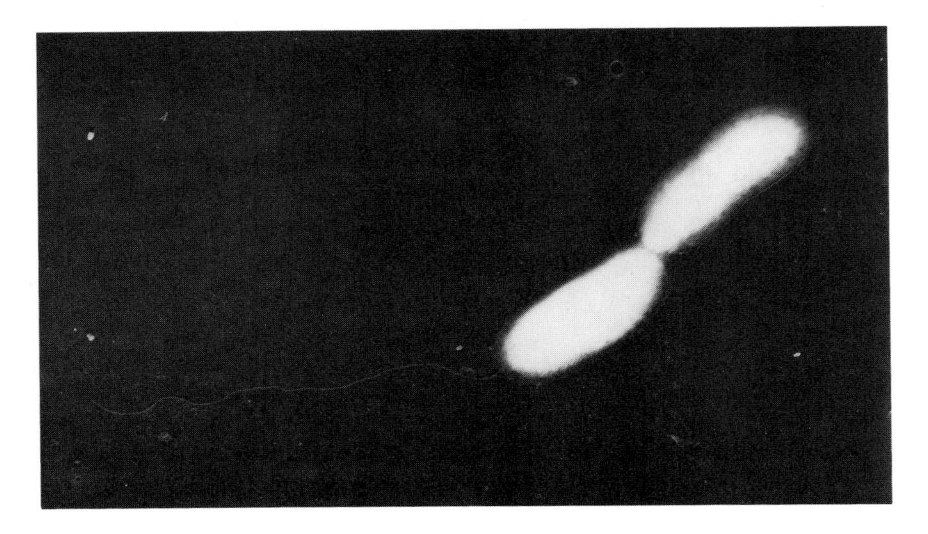

Fig. 8. Transmission electron micrograph of *Azospirillum lipoferum* (×7,500) grown on a yeast extract glucose nutrient solution. After Becking (30).

of various plants growing in different ecosystems. These organisms grow preferentially in the rhizosphere of plants and utilize organic acids excreted by the plant.

It is still debatable whether or not these free-living microbes can contribute significantly to the nitrogen economy of a soil. It has been reported that up to 20 kg N per hectare can be fixed annually but usually their contribution is much lower. The stimulatory effect of *Azospirillum* on plant growth, as has been reported in the literature, may be due in part to its production of plant growth substances.

Gram-negative aerobic rods and cocci

Comprises a large group of N_2 fixers, namely *Azotobacter*, and related genera, *Rhizobium*, one single strain of *Pseudomonas* with uncertain affiliation, methane-oxidizing bacteria (*Methylosinus*, *Methylococcus*) and H_2-utilizing *Xanthobacter* sp. (38), previously classified as *Mycobacterium* sp. (see p. 29). All are Gram-negative rods and usually are motile.

Azotobacter and methane-oxidizing bacteria produce resting stages, cysts (Fig. 9), while the latter group also forms exospores. Exospores are small spher-

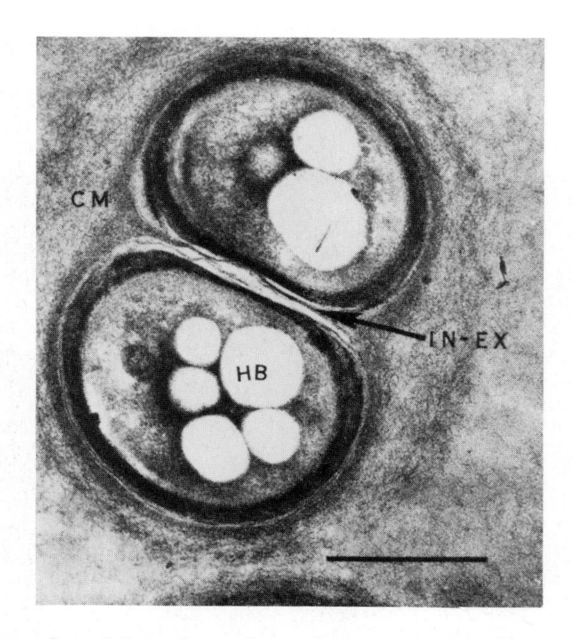

Fig. 9. Mature cysts from 5-day culture of *Azotobacter vinelandii* on Burk's butanol medium. Division is complete as evidenced by intine (IN) and exine (EX) complex between the two cysts. Capsular material (CM) adhering to cysts may be reason for failure to separate. Also shown are sites of poly-β-hydroxybutyric acid deposits (HB). Marker represents 1 μm. After Cagle and Vela (61).

ical cells, attached to one pole of the mother cell (Fig. 10). Cysts are large cells with an extra cell-wall layer. Both cysts and exospores are resistant to desiccation.

Nitrogen-fixing methane-oxidizing organisms are obligate methylotrophs and utilize only C_1-compounds (methane, methanol) and a few other compounds with C-atoms that are not directly linked to one another. Nitrogenase activity of these organisms cannot be demonstrated simply by the acetylene-reduction method, unless methanol instead of methane is supplied as energy- and C-source (22, 23). Acetylene was found to be a potent inhibitor of the first step in the oxidation of methane.

Methane-oxidizing bacteria are wide-spread in the aerobic layer of soil and water. In stagnant water they grow in the upper layer and utilize methane that is produced under anaerobic conditions in the mud. In aerobic soils little methane will be available (except gas leakages from pipelines) and the methylotrophs probably utilize methanol, which is formed upon degradation of pectin by pectin methylesterase (pectinase). Pectin is an important cell-wall component of plants and a significant production of methanol can be expected even in forest soils.

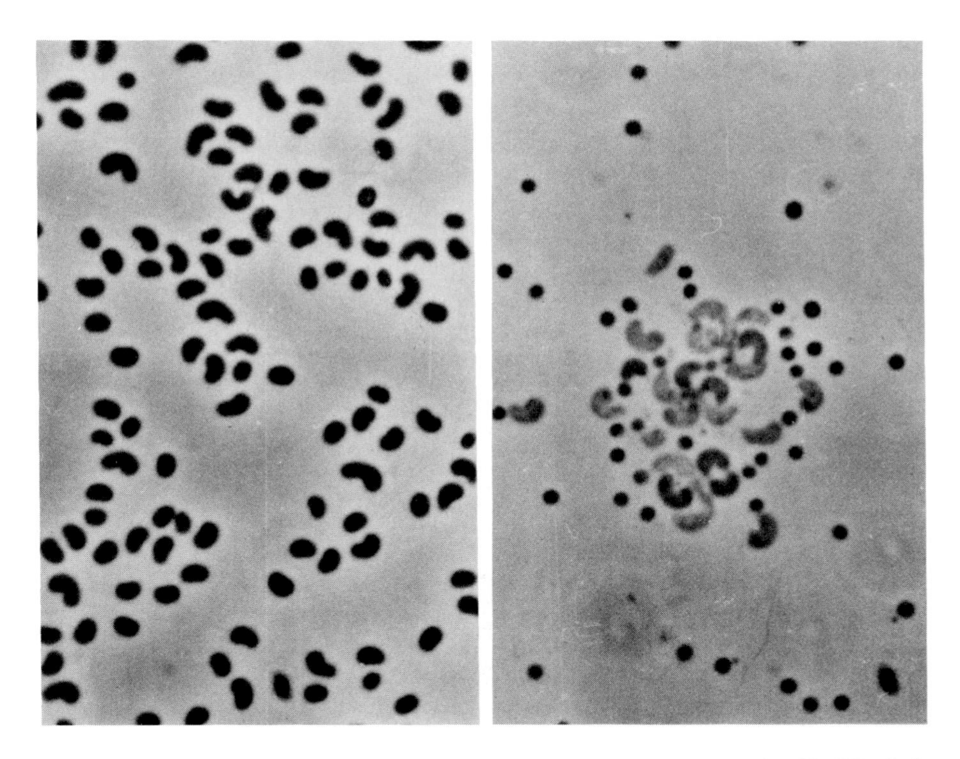

Fig. 10. *Methylosinus* sp. strain 41, grown on slants of mineral salts solution under 10% CH_4. *Left*, a five-day culture and *right*, a three-week culture, with exospores and lysed cells. From de Bont (23).

Since there are many micro-organisms other than methylotrophs which can utilize methanol, it is unlikely that methylotrophs will contribute significantly to the N-cycle in forest ecosystems.

Azotobacter spp. have been the subject of intensive investigation since their first discovery in 1901 by Beijerinck. Azotobacters are rod-shaped Gram-negative, aerobic bacteria, which usually grow as duplococs. They form large amounts of extracellular polysaccharides (EPS) and intracellular poly-β-hydroxybutyrate (PHB). The structure of the cell is dependent on the N-source in the medium (Fig. 11). These organisms occur in low numbers ($10^2 - 10^4 \cdot g^{-1}$ soil) in many neutral and alkaline soils. Soil-inhabiting azotobacters produce large amounts of EPS. *Azotobacter* spp., in contrast to other members of the Azotobacteriaceae, viz. *Beijerinkia*, *Derxia* and *Azotomonas*, produce cysts under growth-limiting conditions. Cysts are resting cells with a thick extra cellular wall consisting of two parts: an outer layer containing a lipoprotein-lipopolysaccharide-lipid complex with calcium as the dominant mineral component and an inner layer with relatively more polysaccharide, free lipids and calcium and less protein (24). Stevenson and Socolofsky (25) have shown that the formation of cysts in *A. vinelandii* is related to the accumulation of PHB which is consumed and utilized for cell wall synthesis (Fig. 12). Cysts can germinate when an appropriate C-source is available. In the absence of combined nitrogen, nitrogenase is induced rapidly in the newly formed vegetative cells.

Azotobacters have a very high rate of respiration which protects nitrogenase from damage by free O_2 (reviewed in 26). Nitrogenase activity has its optimum when cells are cultivated at low O_2 concentration. When free O_2 is present in the medium nitrogenase activity decreases. However, when cells are cultivated in aggregates or in flocks instead of in homogeneous suspensions of single cells, a part of the population inside the flocks will be growing under O_2 limitation because of O_2 consumption by cells in the superficial layer. The advantage of this form of growth for nitrogen fixation by a part of the colony can be demonstrated by measuring simultaneously O_2 consumption and nitrogenase activity of *Azotobacter* culture by methods similar to those employed for studies on Rhizobium (27). Uptake of O_2 was measured with an O_2 electrode in a closed vessel which was saturated with air initially. Acetylene reduction was measured by dissolving C_2H_2 in the medium and measuring the amount of dissolved C_2H_4 produced (Fig. 13). When the growth medium of cells precultivated homogeneously in suspension was saturated with air for a short time, most nitrogenase activity was lost (Fig. 14). However, cells grown in flocks remained active and nitrogenase activity increased rapidly after 1 h incubation. In this latter growth form, consumption of O_2 by cells in the outer layer of the flocks protects nitrogenase of cells inside the flocks from O_2 damage. Low oxygen concentrations due to respiration of neighbouring cells has profound effects on the metabolism of bacteria other than *Azotobacter*. For example, Krul (28) has

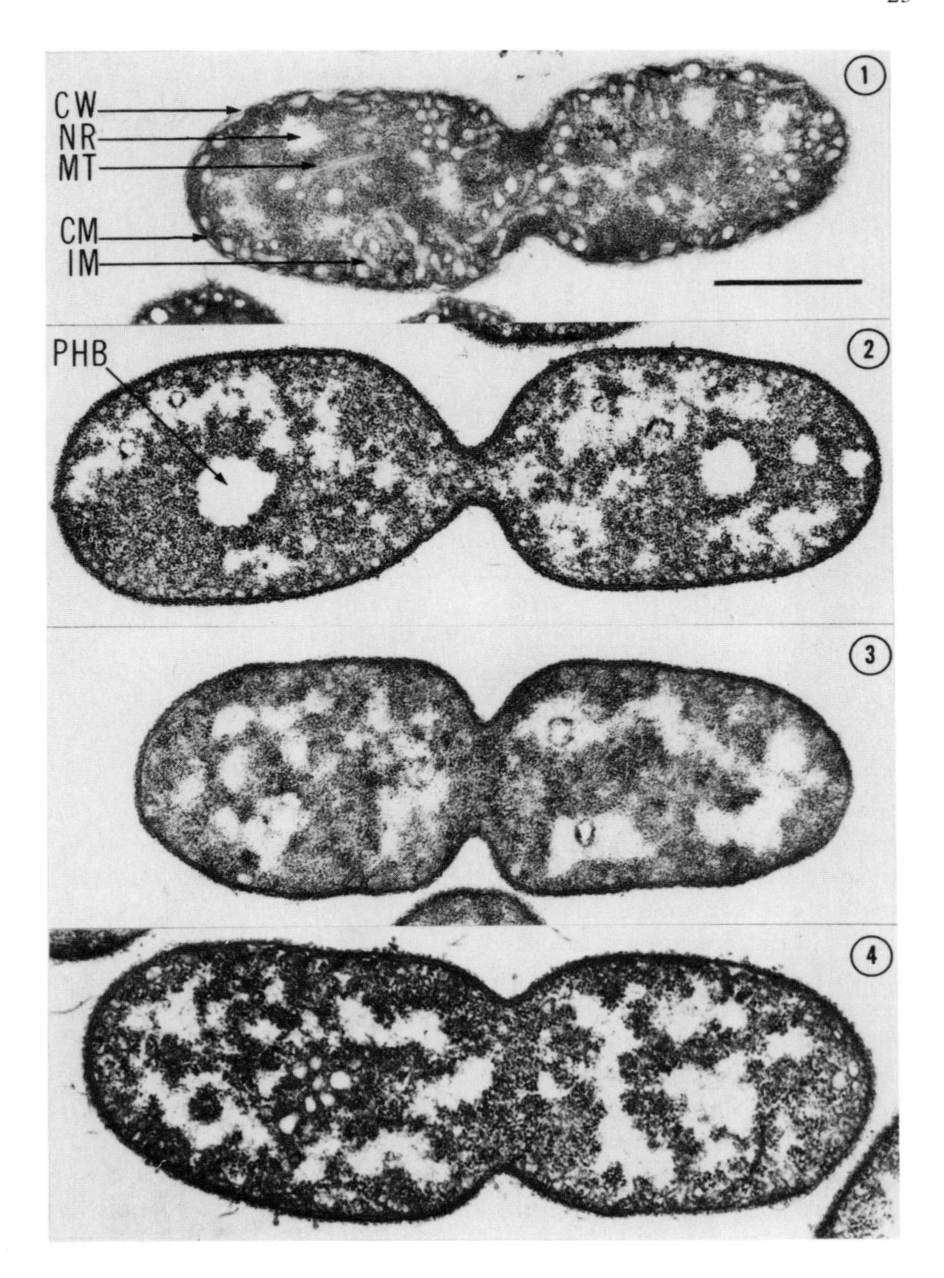

Fig. 11. Sections of *Azotobacter vinelandii* growing exponentially in **1** (N_2), **2** 0.25% NH_4Cl, **3** 0.25% $NaNO_3$ and **4** 0.5% casamino acids. Visible are the cell wall (CW), the internal membrane system (IM), the nuclear region (NR), a microtubule (MT) and poly-β-hydroxybutyrate (PHB) granules. The bar represents 1 μm. After Oppenheim and Marcus (62).

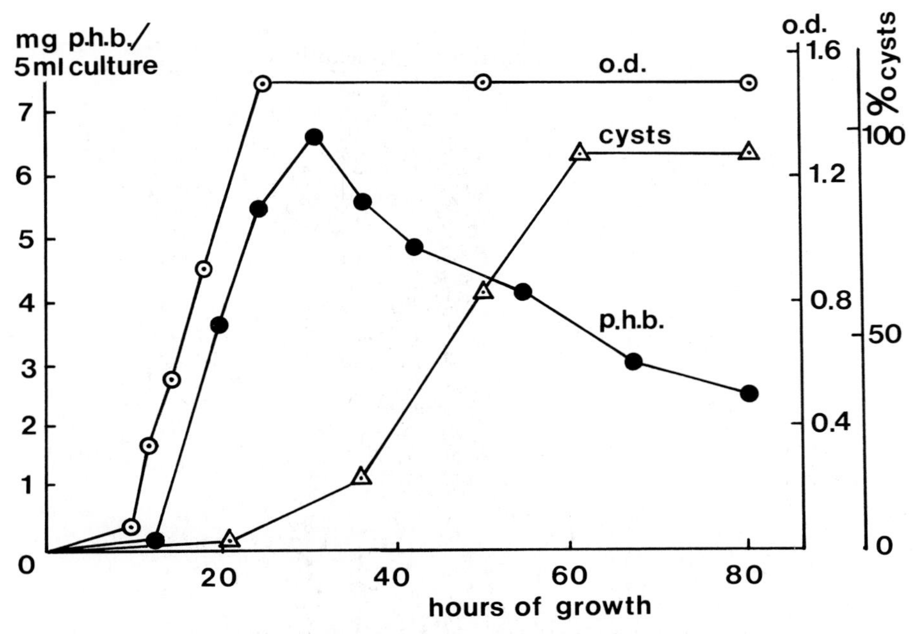

Fig. 12. Relationships between growth, PHB accumulation, encystment and viscosity when cells of *Azotobacter vinelandii* 12837 were cultivated in a liquid medium supplemented with $CaCO_3$. After Stevenson and Socolofsky (25).

demonstrated that dissimilatory nitrate reduction (denitrification) by *Alcaligenes* sp. may occur in the presence of low concentrations of O_2 in the solution, provided cells are grown in flocks. No denitrification occurs under these conditions when cells are cultivated in a homogenous suspension.

Soil bacteria never grow in suspension and all are attached to a surface, usually in small colonies. The examples described above indicate that growth conditions in compact colonies may be quite different from the conditions usually prevailing in laboratory studies on microbes.

The last main group of N_2-fixing Gram-negative aerobic rods are classified in the genus *Rhizobium*. These organisms can grow and fix N_2 in root nodules on leguminous plants and a few others. Outside the plant the free-living rhizobia are short rods which are motile by means of flagella (Fig. 15, ref. 104). Rhizobia usually are subdivided into slow-growing and fast-growing types. A number of slow-growing strains are able to form nitrogenase 'in vitro'. Nitrogenase activity, however, is low and insufficient to supply N for growth. Until now no fast-growing strains have been found that can fix N_2 in the free-living stage.

Within the root nodule rhizobia are transformed into a non-growing stage: bacteroids. The induction of nitrogenase seems to be related to the formation of a non-growing stage, as is also the case in heterocysts of cyanobacteria and in

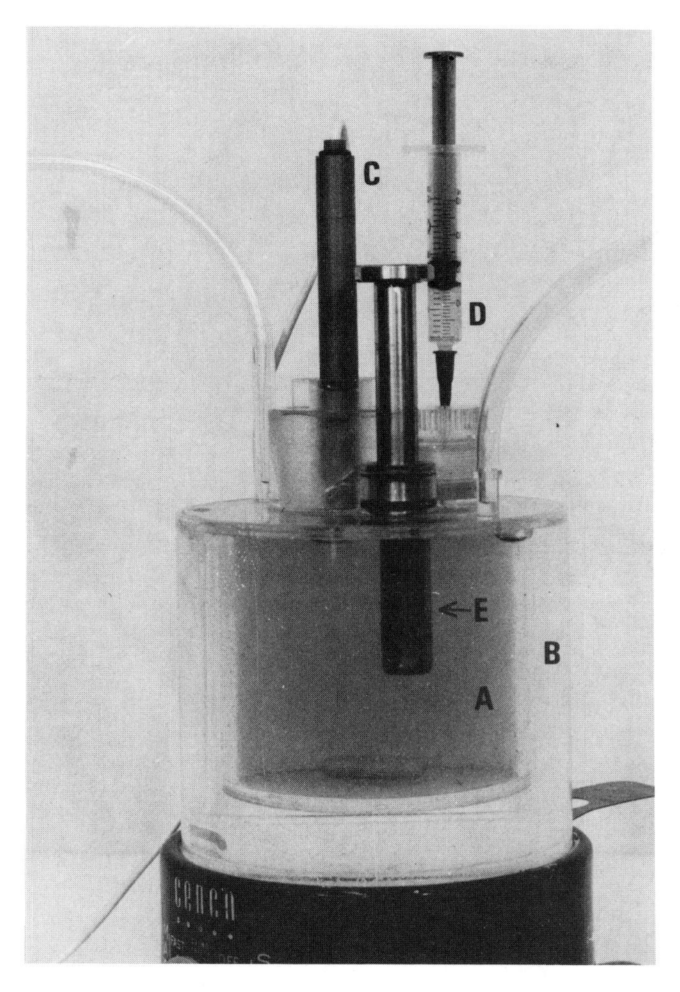

Fig. 13. Simultaneous measurement of the respiration (O_2 uptake) and C_2H_2-reduction in a closed system. Sample chamber of 100 ml (A); waterjacket (B) connected to circulation thermostat; oxygen electrode (C) connected to respirometer (Yellow Springs); sampler for assay of soluble C_2H_4 (D) and compensation plunger (E). The suspension is flushed with air containing 10% of C_2H_2 and subsequently closed in the absence of gas bubbles. O_2 concentration is recorded continuously. The amount of C_2H_4 produced is measured discontinously by taking 1 ml liquid samples. The volume in the sample chamber is adjusted with plunger (E). Akkermans, unpublished.

Frankia (See Fig. 16). A more detailed description of the symbiotic stage of *Rhizobium* spp. is given below.

Gram-negative facultatively anaerobic rods

This group includes large numbers of members of the Enterobacteriaceae, native

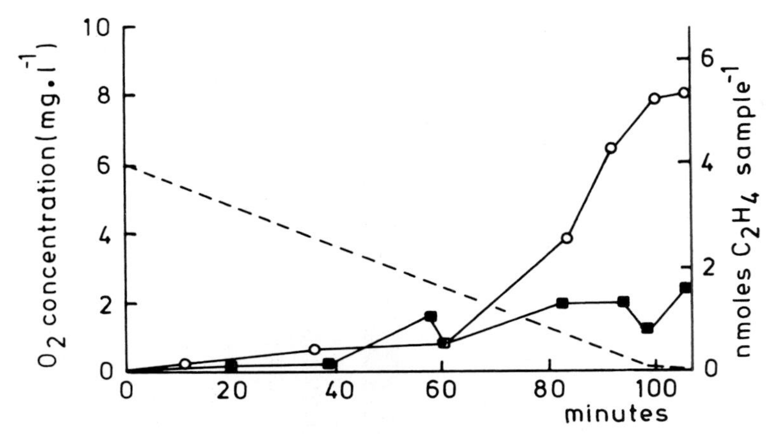

Fig. 14. Effect of aeration on the respiration (dotted line) and the C_2H_2-reducing activities of *Azotobacter vinelandii* A66 grown in flocks (○) and in suspension (■). Incubation method see Fig. 13. The oxygen concentration is recorded continuously. Van Beusichem and Akkermans, unpublished results.

Fig. 15. Electron micrograph of *Rhizobium lupini* 'var. *densiflora*' with subpolar flagellum. After Abdel-Ghaffar and Jensen (104).

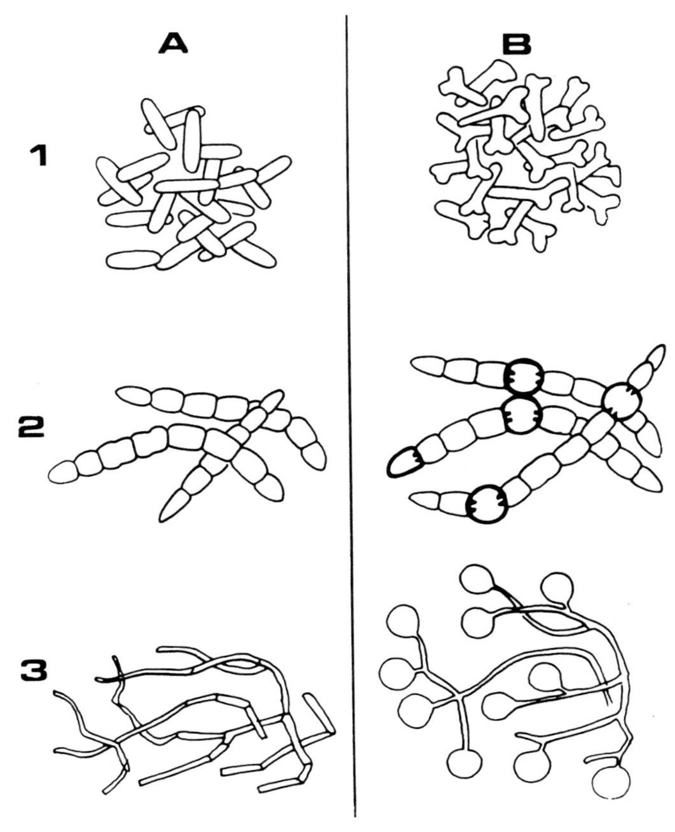

Fig. 16. Schematic representation of the differentiation of *Rhizobium* (1), *Anabaena* (2) and *Frankia* (3) in the presence (A) or absence (B) of combined nitrogen. B-1 and B-2 represent the symbiotic stages in root nodules.

to the intestines of animals (e.g. *Escherichia coli*) and various pathogens on plants (e.g. *Erwinia* spp.). With the exception of one *Klebsiella* species, very few strains from other genera fix nitrogen under aerobic conditions. Usually nitrogen fixation takes only place under anaerobic conditions or at low O_2 concentration. N_2-fixing, fermentative bacteria have been isolated from decaying tissue in living white fir trees in Oregon (31). These micro-organisms can occur in large quantities (ca. $10^5 \cdot ml^{-1}$) in the centre of the tree together with cellulose decomposing fungi. This is an example of nutritional interaction: fungi decompose the cellulose and the N_2-fixers utilize a part of the sugars and acids released. It is not known to what extent and in what form the fixed nitrogen in turn becomes available to fungi, or to other organisms.

Citrobacter has been isolated from the gut of termites and utilizes carbohydrates released by decomposition of wood. Other enterobacters have been found in paper mill effluent and in other substrates with a high C/N ratio.

All N_2-fixing species recorded in this taxon ferment sugars through the Embden-Meyerhof pathway. The mode of degradation of pyruvate may vary, but one pathway has not been found in other bacterial groups: the formation of formate and acetyl CoA from pyruvate and CoA. Many species, e.g. *E. coli*, degrade formate to CO_2 and H_2 in equimolar quantities. In this they differ significantly from the production of H_2 and CO_2 by spore-forming *Clostridium* and *Bacillus* spp.

Gram-negative anaerobic bacteria

Only one genus includes N_2-fixing species viz. *Desulfovibrio*. These are motile, curved rods with one or more polar flagella. They are characterized by the ability to reduce sulphate to sulphide: sulphate serves as the terminal electron acceptor. Sulphate-reducing, like nitrate-reducing micro-organisms, contain haem pigments (cytochrome c), which participate in the anaerobic electron-transport system. In contrast to denitrifiers, the sulphate-reducing bacteria are strictly anaerobic and occur in anaerobic soils and water (e.g. muds and marine sediments), rich in organic substrates and sulphate. They do not play a role in aerobic forest soils. Lactate, malate and pyruvate are utilized as carbon sources. The end product of the oxidation is always acetate, since *Desulfovibrio* sp. does not contain a functional tricarboxylic acid cycle.

Gram-negative chemolithotrophic bacteria

This group comprises 17 genera. Recently Mackintosh (32) reported nitrogenase activity in *Thiobacillus ferrooxidans*. This is an autotrophic, filamentous aerobic iron-oxidizing bacterium, characterized by iron deposits around the cells, which can grow at extremely low pH, *ca* 2. Moreover this organism can oxidize the S-component from pyrite by a direct-contact mechanism and can utilize the energy produced for its growth (33, 34, 35). Nitrogen is fixed only at low O_2 concentrations.

T. ferrooxidans is an extremely slow-growing microbe. Recently Arkesteyn (35) showed that the strain, used in previous studies, was not a pure culture but was contaminated with the heterotroph, *T. acidophilus*. It is likely that these organisms occur and act together in the soil.

Pyrite-oxidizing bacteria play an important role only in acid sulphate soils in coastal regions and will be absent in ordinary aerobic forest top soils.

Endospore-forming rods and cocci

These organisms are distinguished from all other prokaryotes by the ability to form endospores inside the cells. Endospores remain viable for long periods and are highly resistant to heat, desiccation and toxic chemicals. The vegetative cells usually are rod shaped (except in one genus) and are Gram-positive. Older cells, however, become Gram-negative.

Nitrogen fixation occurs in one species of the genus *Bacillus* (a facultative anaerobe), several *Clostridium* (strict anaerobes) and *Desulfotomaculatum* spp. Nitrogenase activity in *Bacillus polymyxa* has been observed only when cells are cultivated anaerobically i.e. under fermentative conditions.

Nitrogen-fixing clostridia are all of the saccharolytic type e.g. *C. pasteurianum*, *C. butyricum*. So far, nitrogen fixation has not been detected in the proteolytic and pathogenic clostridia.

Actinomycetes and related organisms

This taxon includes various types of filamentous and pleomorphic bacteria. Nitrogen-fixing micro-organisms are found within a number of *Mycobacterium* and *Frankia* species.

Mycobacterium spp. are gram variable, non-motile, coryneform organism with snapping division. The first nitrogen-fixer, *M. flavum* 301, was isolated by Fedorov and Kalininskaya (36). More recently a number of strains related to this organism have been isolated (37, 38). All are autotrophic, H_2-utilizing bacteria which have little similarity to other members of the Corynebacteria. Some strains are now reclassified as *Xanthobacter autotrophicus* (38) and it is suggested that these strains be included within the family Azotobacteriaceae (see p. 20).

The genus *Frankia* consists of slow-growing actinomycetes which can grow and fix nitrogen in symbiosis with a number of non-leguminous plants in so-called root nodules (actinorrhizas). First strains were isolated from *Comptonia* sp. nodules by Callaham *et al.* (39) and subsequently an increasing number of *Frankia* strains have been isolated (see p. 39).

Pure cultures of *Frankia* form clusters of hyphae and large sporangia, from which spores are released (Fig. 17). Under nitrogen-limiting conditions, vesicles (Fig. 18) are formed and nitrogenase is induced (40, 41 and 42). Within the nodule the hyphae grow from cell to cell and many vesicles are formed intracellularly. This stage is called the vesicle-cluster stage (Fig. 19, 20) and it is assumed to be the stage where N_2 is fixed (43, 44, 45, 46).

The growth experiments summarized above suggest that at least some strains of *Frankia*, in contrast to *Rhizobium* spp., are free-living N_2-fixers with much similarity to heterocystous cyanobacteria. According to this hypothesis the vesicles of *Frankia* may have the same function as heterocysts: both represent the non-growing N_2-fixing fraction of the population (Fig. 16). This hypothesis needs further experimental evidences.

All *Frankia* strains which have been investigated so far are able to grow on the

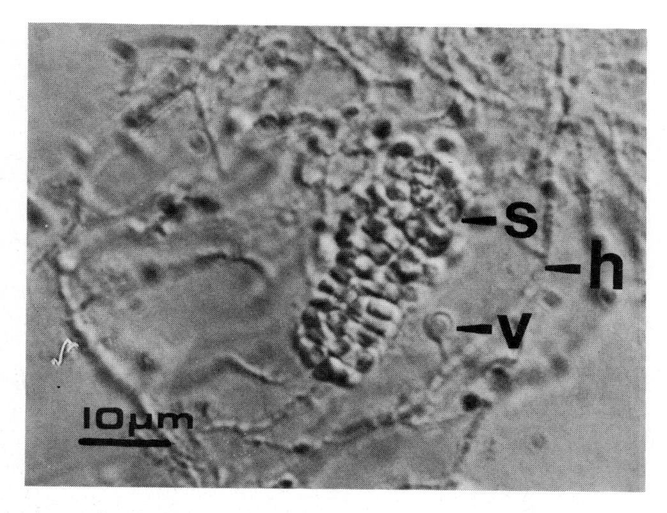

Fig. 17. Pure culture of *Frankia* Cc.1, derived from root nodules of *Colletia cruciata* and grown in propionate + NH_4^+ medium. Sporangium (s) with spores; hypha (h) and vesicle (v). Full description of the features of the strain to be published elsewhere (Baas and Akkermans). Nomarski interference optics at Plant Protection Service (P.D.), Wageningen. Courtesy of Fauzia Hafeez.

C_3 – fatty acid, propionic acid (42). The growth requirements of *Frankia* spp. have not been investigated in detail, mainly because only a few strains are yet available. A recent study by Blom of *Frankia* AvcI1, isolated by Baker and Torrey from *Alnus viridis* ssp *crispa*, indicated that this strain (AvcI1) only utilizes Tweens or free fatty acids as C-sources (47, 48). This hypothesis has been supported by a metabolic study of the enzymes involved in the carbon metabolism of *Frankia* in pure culture (49, 50, 51) and in root nodules (52, 53).

More recently, other strains of *Frankia* have been isolated which also can consume sugars and organic dicarboxylic acid, e.g. succinate (105) and it is possible that *Frankia* strains may be subdivided into groups based on their C-utilization. More detailed information on the nutritional requirements of different *Frankia* strains will be published in 1983 (see 54).

MORPHOLOGY OF N_2 FIXING SYMBIOSES

Cyanobacterium symbioses

A number of heterocystous, filamentous cyanobacteria are able to fix nitrogen both as free-living organisms and in symbiosis with eukaryotes. There are two physiologically distinct types of symbiosis: endo-symbiosis (cyanobacteria lives inside the host cells) and ecto-symbiosis. Most N_2-fixing cyanobacteria live as ecto-symbionts inside multicellular eukaryotes (58, 59, 60).

Simple forms of ecto-symbiosis occur in thalloid liverworts, *Anthoceros*, *Bla-*

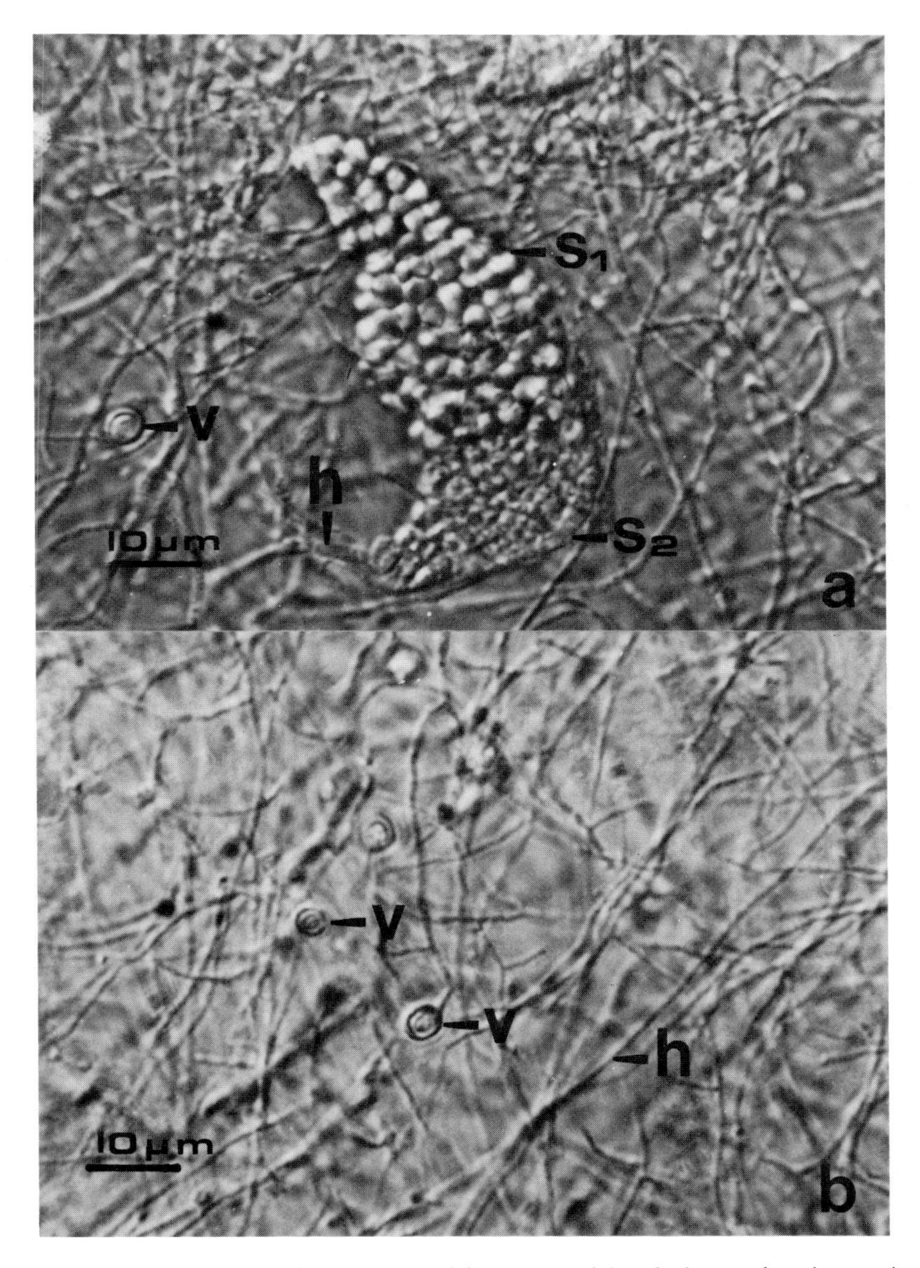

Fig. 18. Pure culture of *Frankia* An.1, derived from root nodules of *Alnus nitida* and grown in N-free propionate medium. a. Detail of a terminal mature sporangium (s), attached on hypha (h). Immature spores are visible in the basal part of the sporangium (S1). Mature spores are released from the older apical part of the sporangium (S2). b. Detail of culture showing branched hypha (h) and terminal vesicles (v) on short side branches.

Full description of the features of the strain to the published elsewhere (Hafeez, Akkermans, Chaudhary). Nomarski interference optics at Plant Protection Service (P.D.), Wageningen. Courtesy of Fauzia Hafeez.

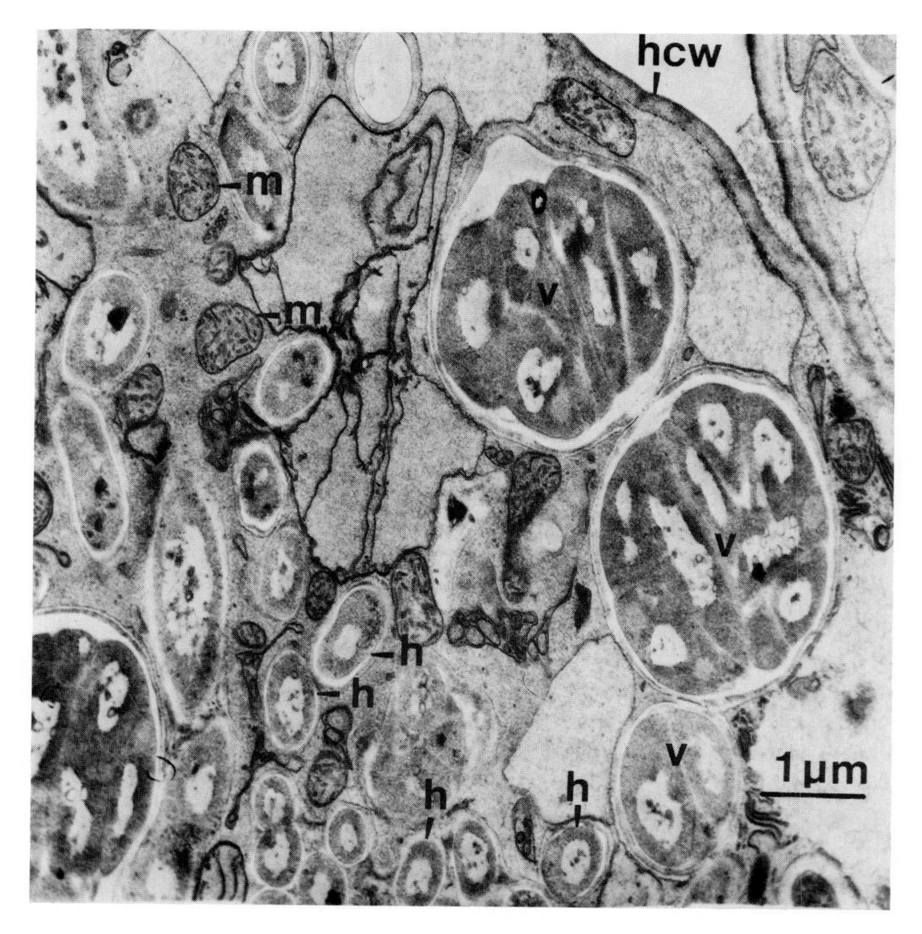

Fig. 19. Transmission electron microscopic picture of root nodule cell of *Alnus nitida* infected by *Frankia* sp. Septate vesicles (v) of the endophyte are located at the periphery of the hypha cluster, near the host cell wall (hcw). Cross section of hypha (h) and plant mitochondria (m). Courtesy of Miss E. Bouw, Technical and Physical Engineering Research Service (TFDL), Wageningen and Fauzia Hafeez, Quaid-i-Azam University, Islamabad, Pakistan.

sia and *Cavicularia* spp. (Fig. 21). These liverworts fix nitrogen in symbiosis with a *Nostoc* species. The *Nostoc* enters the liverwort in different ways e.g. it enters *Anthoceros* via pores and comes into mucilaginous cavities on the ventral side of the leaves (55). The *Nostoc* cells multiply inside the cavities. The number of heterocysts in the filament increases from ca. 5% in the free-living stage up to 40% in the symbiotic stage. The plant cells surrounding these cavities swell and produce branched multicellular filaments between the cyanobacteria from a region opposite the cavity pore. *Nostoc* fixes more nitrogen per cell in the symbiotic stage than when free-living, mainly due to the increased heterocyst fre-

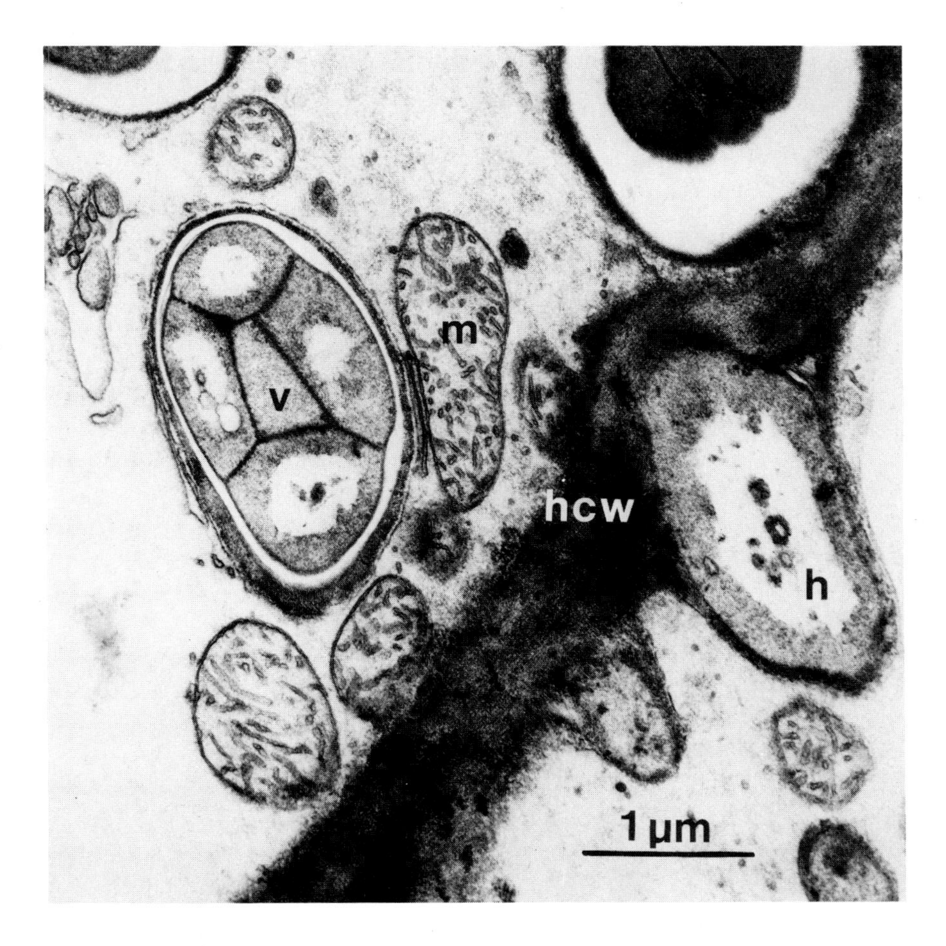

Fig. 20. Transmission electron microscopic picture of a septate vesicle (v) surrounded by a capsule, and part of a hypha (h) passing the host cell wall (hcw) in the root nodule of *Alnus nitida*. Note the concentration of mitochondria (m) surrounding the endophyte. Courtesy of Miss E. Bouw, Technical and Physical Engineering Research Service (TFDL), Wageningen and Fauzia Hafeez.

quency in the former state. The excess of ammonium is excreted and is taken up by the host. Nitrogen-fixing liverworts usually occur in low numbers on moist sites. Although quantitative data are still lacking, it is likely that their impact on the nitrogen cycle is restricted to local sites in nature.

A second form of ecto-symbiosis is the association with the waterfern *Azolla* (Fig. 22, 23), which has a wide distribution in stagnant fresh water in tropical as well as in temperate regions. The N_2-fixing symbiont, *Anabaena azollae*, lives in cavities in the leaf bases (Fig. 24). As in liverworts, the surrounding plant cells extend between the cyanobacteria (56). The heterocyst frequency also is increas-

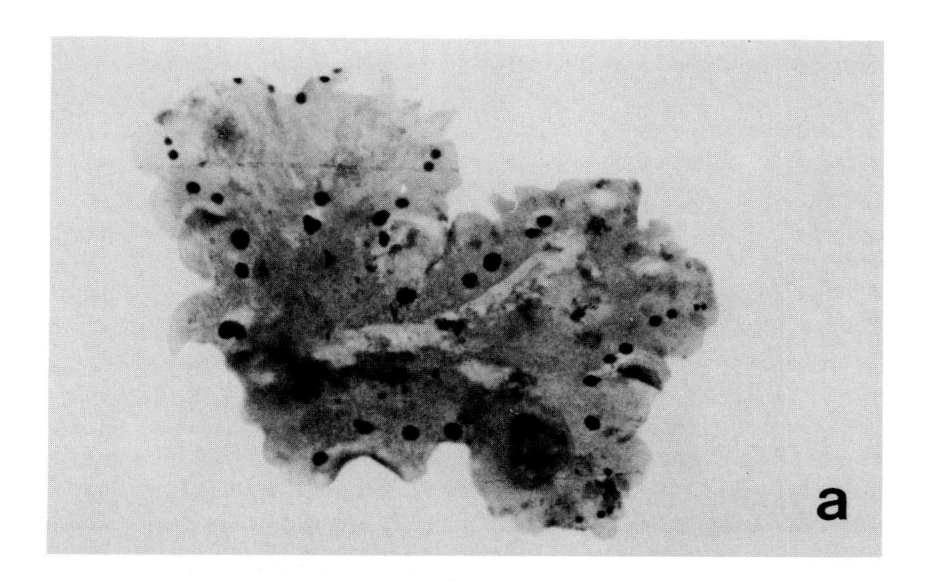

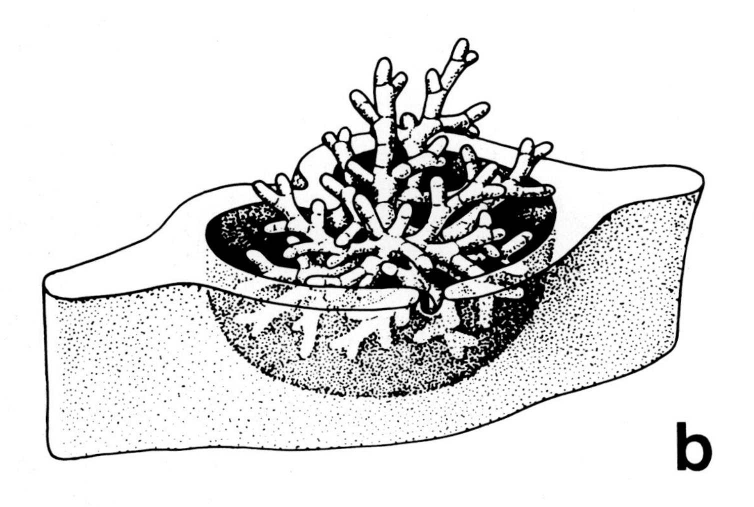

Fig. 21. a. Thallose gametophyte of *Blasia pusilla*. *Nostoc* colonies (dark spots) are present in cavities on the undersurface (× 6). b. Diagrammic representation of algal cavities, 6 weeks after infection with the *Nostoc* phycobiont (omitted from diagram) showing the development of branched septate filamentous protrusions arising from the cavity wall opposite the cavity pore (× 100). c. Filamentous protrusions extruded from an algal-containing cavity. (× 400). d. Algal cells extruded from cavities showing the high frequency of heterocysts (× 700). After Rodgers and Stewart (55).

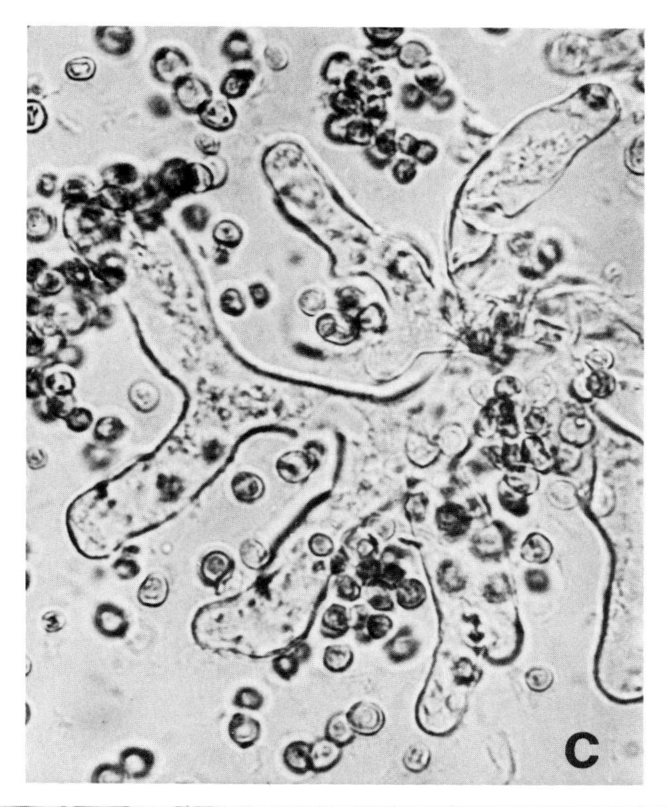
c

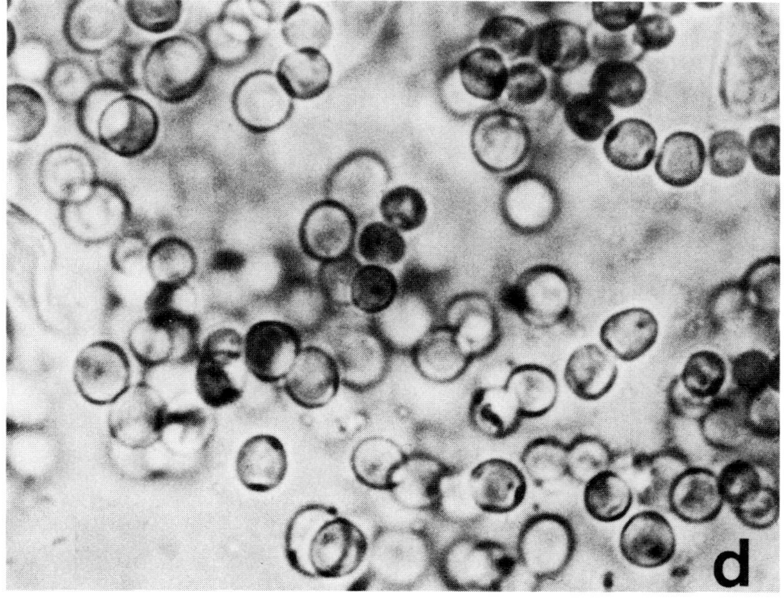
d

Fig. 22. *Azolla* vegetation floating on stagnant fresh water in a ditch near Utrecht (The Netherlands).

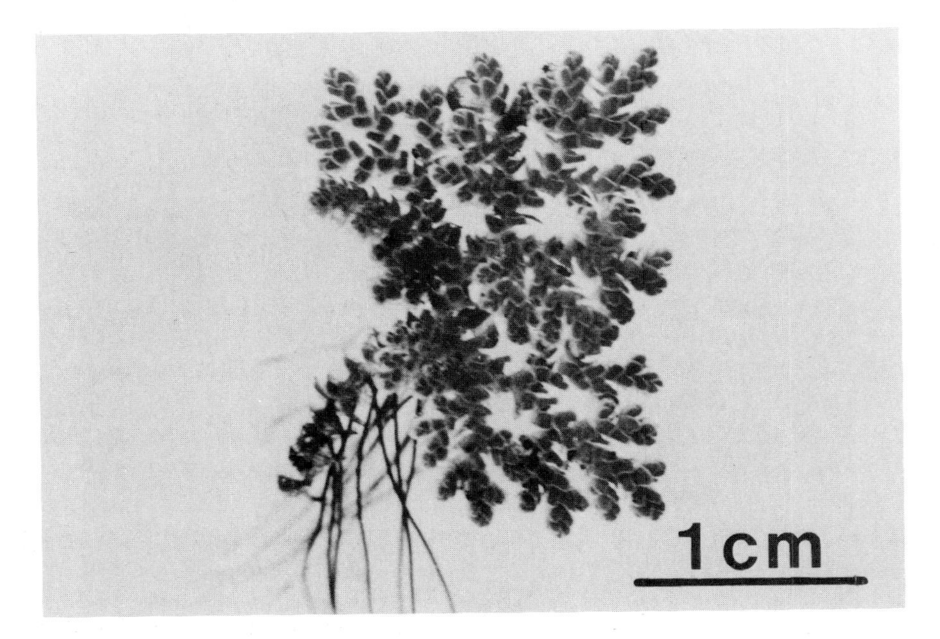

Fig. 23. *Azolla* plant.

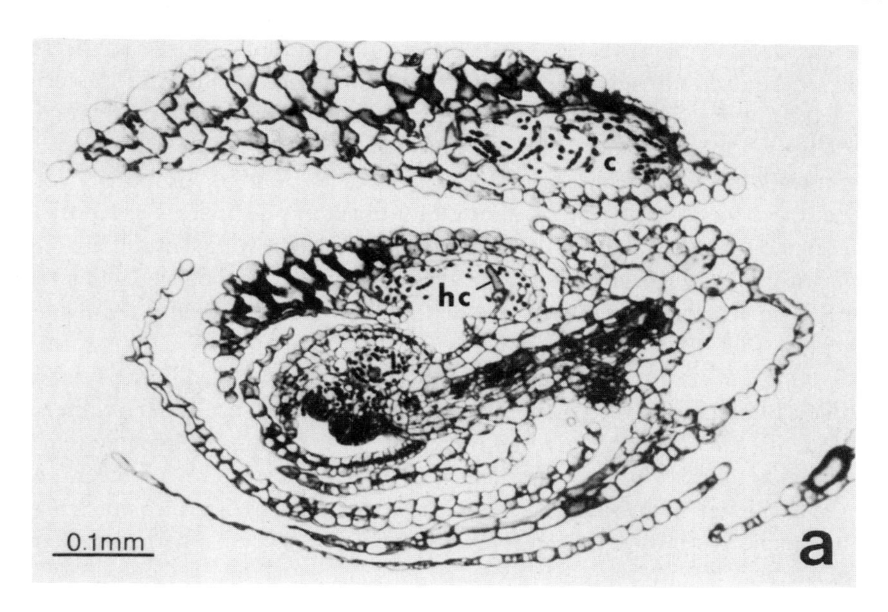

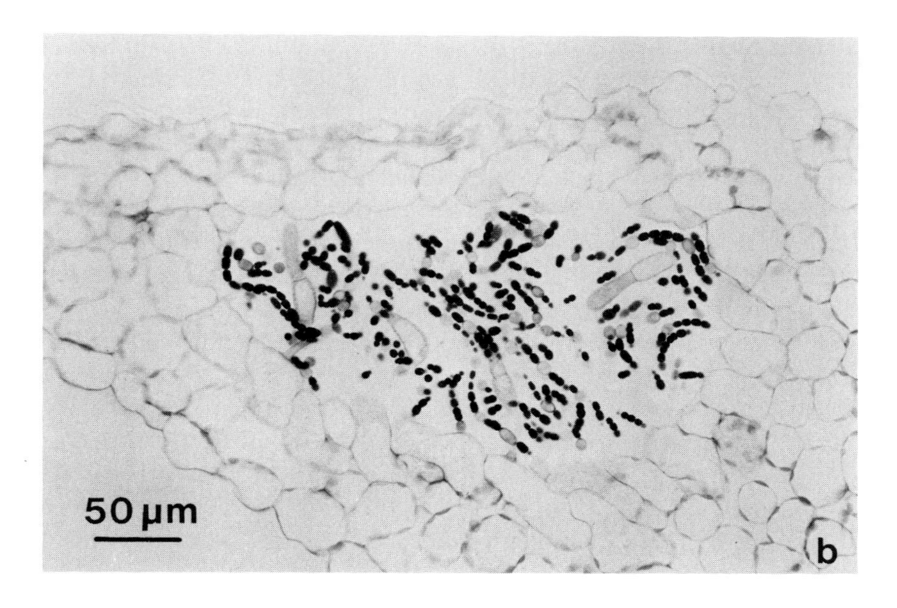

Fig. 24. a. Light micrograph of a section through the second dorsal leaf lobe of *Azolla caroliniana*. Filaments of the algal symbiont and hair cells (hc) are shown within the leaf chamber (c). After Peters *et al.* (73). b. Detail leaf chamber (Calvert, Kettering Research Lab., Yellow Springs, Ohio). 400 ×.

ed from 5 to 40 percent after transition into the symbiotic phase and the major part of the fixed nitrogen is excreted as ammonia and is utilized by the fern.

A third type of ecto-symbiosis with cyanobacteria is found within the group of lichens. Lichens are associations between fungi and photosynthetic micro-organisms, usually green algae or cyanobacteria. A small proportion of the lichens, 5–10 percent, have N_2-fixing cyanobacteria as micro symbionts (57). The cyanobacteria sometimes are intermingled with the fungal mycelium (e.g. in *Collema*) or are restricted to a distinct layer covered by the fungus (e.g. *Peltigera*). The heterocyst frequency usually is as low as in free-living alga, viz. *ca* 4 percent.

In some lichens there are more than two partners involved in the symbiosis. Triple-symbioses between a fungus, a green-alga and a cyanobacterium occur e.g. in *Stereocaulon*, *Peltigera* and *Lobaria* spp. In this complex system, the cyanobacterium has an increased heterocyst frequency (up to 30 percent) compared to the free-living stage. As a result nitrogenase activity is also much higher. In the triple symbioses, the N_2 fixer is localized in special spherical structures, cephalodia.

Lichens are slow-growing organisms which can survive extremes of temperature and dryness. In low productivity ecosystems, such as deserts, Arctic tundras and Northern forests, lichens can provide a significant N-input to the soil (21).

The remaining group of ectosymbiosis are found in the Cycadaceae (seed-ferns). The plants form typical deformed roots, which are infected by *Nostoc*. This cyanobacterium multiplies within the intercellular spaces of a defined concentric layer in the root cortex. The infected roots divide dichotomously and form large clusters (root nodules, rhizothamnia) which are similar to the *Alnus*-type of actinorrhizas (Fig. 25). The *Nostoc* is visible in cross sections as a dark-green layer (Fig. 25). *Nostoc* cells remain as an ectosymbiont in the intercellular spaces (63). Occasionally *Nostoc* cells have been reported as an intracellular endosymbiont in nodules of *Macrozamia communis* (64).

Endosymbiotic growth of N_2-fixing cyanobacteria is known only for a few *Nostoc* species living inside the cells of a few eukaryotes. Of these, only the *Gunnera-Nostoc* symbioses are of ecological importance in forestry. These plants are native to tropical and subtropical regions in the Southern Hemisphere (65), e.g. *G. macrophylla* in volcanic soils in Indonesia (Fig. 26). The *Nostoc* is located in special glands at the basis of the petioles (Fig. 27). The anatomy of glands of *G. albocarpa* has been described in detail by Silvester (66). As in the algal symbioses described bove, the heterocyst frequency of the microsymbiont is high compared to that of the free-living culture: in free-living *Nostoc ca.* 5% and in apical *Gunnera* glands 23–41%, depending on the age of the glands. The CO_2-fixing ability of the cyanobacterium inside the cell has been lost and micro-symbiont behaves as a carbon-heterotrophic organism, like *Frankia* and *Rhizobium* in root nodules. It is not known whether this change in the photosynthetic

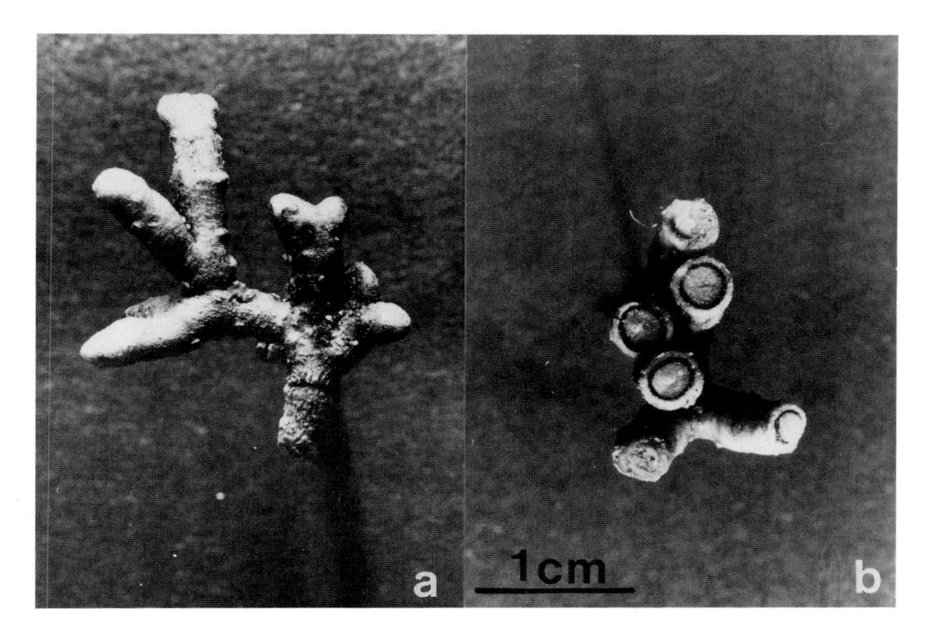

Fig. 25. Root nodules of *Cycas rhumphii* (a). *Anabaena* sp. is localized in the concentric layer in the cortex (b).

system is typical of the algal symbioses. In most of the symbiotic systems described above, the microsymbiont is a *Nostoc* species, which is morphologically very similar to the free-living *Nostoc* spp. Little is known about the taxonomic position of these organisms. It has been shown that liverworts can only be infected by certain *Nostoc* strains (55) indicating the occurrence of host-specificity as in other types of N_2-fixing symbioses.

Frankia symbioses

About 200 plant species, classified in 20 genera and 8 families are able to form root nodules (actinorrhizas) with actinomycetes (*Frankia*) as N_2-fixing endosymbionts (see Ch. 3). The endophyte enters the roots through deformed root hairs and multiplies within the living cortex cells. The infection process and the morphogenesis of the actinorrhizas have been studied mainly in *Alnus* spp. (67, 68), but recently detailed descriptions of the anatomy of root nodules of e.g. *Casuarina* (69), *Comptonia* (70) and *Ceanothus* (71) have been published. In many other examples, however, especially in the Rhamnaceae, Rosaceae and Datiscaceae the description of the nodule symbioses is still incomplete.

In most actinorrhizas the tips of the hyphae of the endophyte soon swell to form vesicles which may be spherical (typical of the endophyte of *Alnus* and

Fig. 26. *Gunnera macrophylla*, Cibodas, Indonesia.

Hippophae, Fig. 19, 20) pear- or club-shaped (typical of *Myrica*, *Coriaria* and *Datisca* spp., Fig. 28). In the majority of host-plant genera these vesicles are aggregated round the periphery of the host cell, close to the cell wall. In *Coriaria* and also in *Datisca* spp. the vesicles are swollen hypha oriented inwards in the plant cell and the endophyte is visible in tires in the plant cells (Fig. 29).

An important question is to what extent the morphology of *Frankia* is genetically controlled or is determined by its host plant. Evidence exists for both possibilities. In pure cultures the shape and size of the hyphae and sporangia varies among the strains. In a few cases it has been demonstrated that the form of the vesicles is determined by the host (72). Strain Cp, derived from actinorrhizas of *Comptonia peregrina* (39) is able to form spherical vesicles under N-limited conditions in pure culture (40, 41). Similar structures are produced after infec-

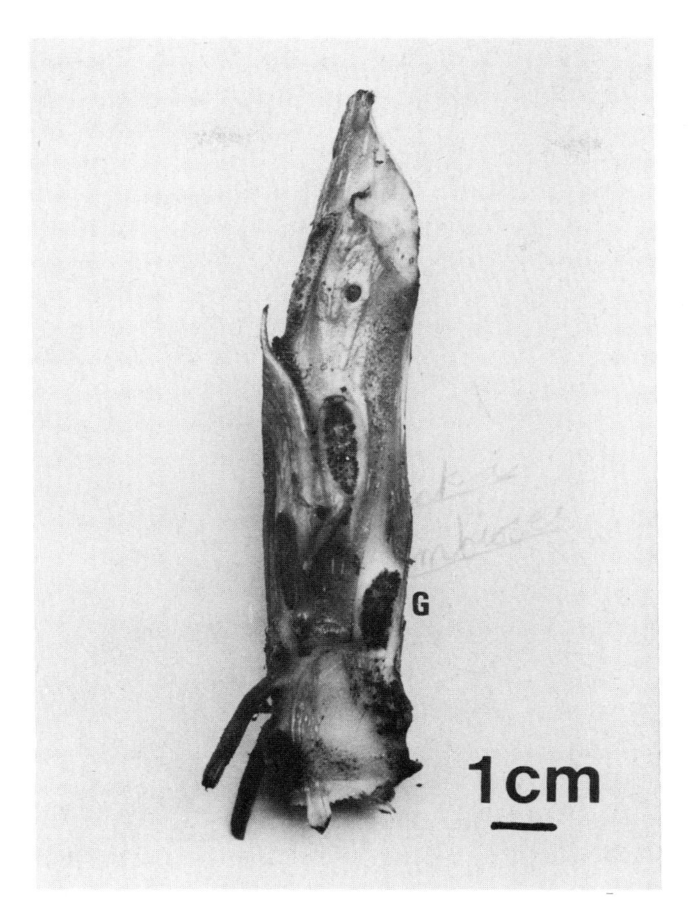

Fig. 27. Shoot of *Gunnera macrophylla* from which the leaves and petioles are removed. Algal symbionts are localized in stem glands (G) at the basis of the petioles.

tion of *Alnus* nodules. However, in the nodules of *Comptonia* this same strain forms the club-shaped structures typical of Myricaceae. Additional information is needed to characterize the possible pleomorphic character of other *Frankia* strains.

All *Frankia* strains are able to produce sporangia in pure cultures (Fig. 18, 19) although certain carbon substrates are known to suppress spore formation (Hafeez, Akkermans, Chaudhary, in prep.). Some strains are also able to produce sporangia within the nodules. The structure of these sporangia and their morphogenesis inside the nodule (74) is similar to that in pure culture (39).

Homogenates of nodules of *Alnus glutinosa* with a spore-positive *Frankia* strain are 100–1000 times more infective than nodules with a spore-negative strain (75, 76, 77). The high infectivity of the spore positive nodule homogenates

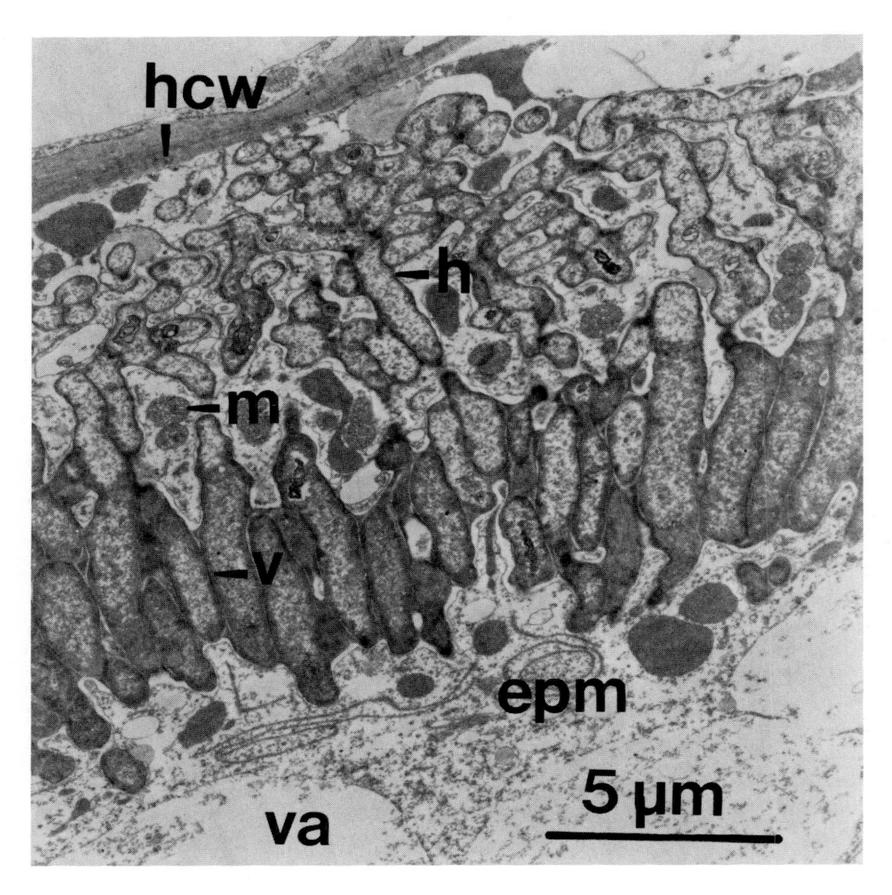

Fig. 28. Detail of infected root nodule cell of *Datisca cannabina* showing ordinary hypha of *Frankia* (h) and enlarged, electron-dense vesicle-like cells (v) orientated to the centre of the host cell. Mitochondria (m) are visible in between the hypha. The centre of the host cell is filled with vacuoles (va) and endoplasmatic reticulum (epm). hcw: host cell wall. Transmission electron micrograph prepared by E. Bouw, Technical and Physical Engineering Research Service (TFDL), Wageningen. Courtesy of Fauzia Hafeez.

may be due to the presence of large numbers of viable spores which are easily disseminated. These results indicate that there are large differences in infectivity among *Frankia* strains. Preliminary experiments indicate that the infectivity of *Frankia* strains when grown in pure culture may vary as well. Infectivity of a strain, i.e. the number of nodules that can be formed per unit of inoculum, is determined by the number of infective particles. The infectivity doubtless is dependent on the physiological condition of the vegetative cells and its ability to disseminate in the rhizosphere. A future program for selection of *Frankia* strains as inoculum in forestry therefore must include determination of the

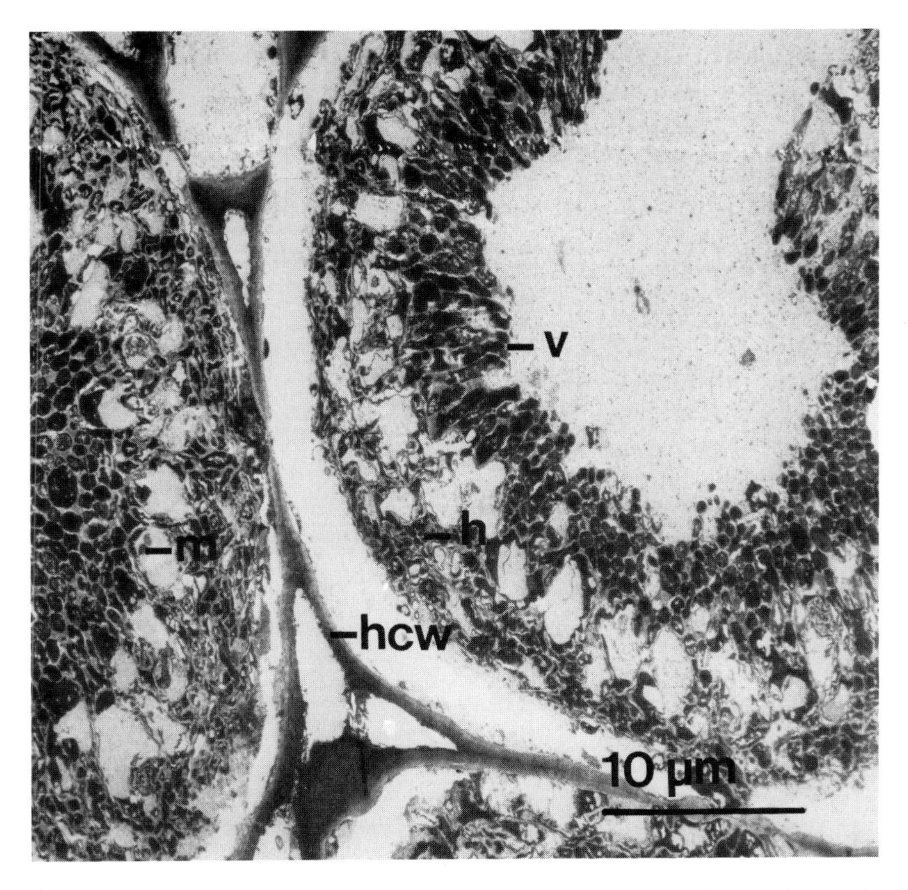

Fig. 29. Cross section of infected root-nodule cell of *Coriaria nepalensis*, showing electron dense vesicles (v), orientated to the centre of the host cell. Between the vesicles (v) and the host cell wall (hcw), *Frankia* hypha (h), mitochondria (m) and plant membranes are visible. Transmission electron micrograph prepared by E. Bouw, Technical and Physical Engeneering Research Centre (TFDL), Wageningen. Courtesy of Fauzia Hafeez.

infectivity of both the free-living and symbiotic growth phases.

All actinorrhizas are perennial structures (Fig. 30) that may survive for several years (75). From an anatomical point of view nodules of *Alnus glutinosa* can be described as adventitious lateral roots, induced by an appropriate *Frankia* strain, between the normal points of lateral root production (67). After decay of the nodules viable endophyte particles are released into the soil and are disseminated in the soil. The factors affecting distribution of *Frankia* in soil have recently been described by van Dijk and the present authors (75, 76, 77).

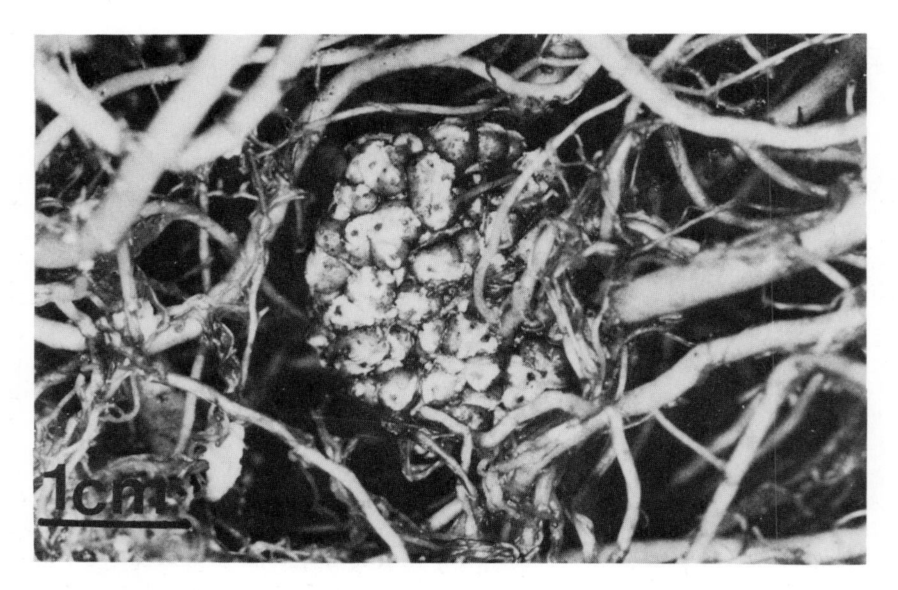

Fig. 30. Root nodule of *Datisca cannabina*

Rhizobium symbioses

Many leguminous plants, e.g. lupins, are pioneers on N-poor soils, due to the presence of N_2-fixing nodules with a *Rhizobium* as microsymbiont. Most of these plants form nodules on the roots (Fig. 31) or, occasionally on the stems, e.g. in *Aeschynomene indica* (78, 79, 80) and *Sesbania rostrata* (81), or on aerial adventitious water roots (82).

For a long time it was assumed that this kind of *Rhizobium* symbiosis was confined to leguminous plants. However, a few other plants have now been found to bear nodules with a *Rhizobium* as symbiont. Non-legume-*Rhizobium* symbioses have definitely been proven in several species of *Parasponia* (Ulmaceae) (83, 84, 85, 86). *Rhizobium* symbioses have also been reported in a number of Zygophyllaceae (87, 88), but little detailed information is yet available on this group. Since only relatively few plant species have been searched for nodulation it is likely that other kinds of plants with *Rhizobium* symbioses will be discovered in the near future.

Rhizobium symbioses have been investigated intensively for several decades. Biological studies, especially on fundamental processes, have been restricted mainly to a limited number of agriculturally important annual crops, e.g. peas and beans. More recently, however, there has been increasing interest in nodulated leguminous trees and shrubs because of their promising economic importance in the tropics (see e.g. ref. 89, 90).

In this section the present knowledge of legume symbioses will be reviewed only very briefly. For further details we refer to recent reviews on this subject (91, 92, 93).

The infection process is preceded by growth of *Rhizobium* in the rhizosphere of the host and by specific recognition of the plant cells. *Rhizobium* strains are characterized by a specific range of hosts which can be infected. Some strains have very restricted host specificity while others are more promiscuous (e.g. rhizobia from the cowpea group). Usually rhizobia enter the plant through deformed root hairs. Inside the plant cells the bacteria are encapsulated by an infection thread, i.e. cell wall and membrane material of plant origin. The bacteria are suggested to produce growth hormones which stimulates local multiplication of cortex cells. The infection process continues in cells located in lower layers of the cortex (Fig. 32). Subsequently a root primordium is induced inside the pericycle. After extension of this primordium rhizobia infect the newly formed cells, but still remain inside the infection threads. At a later stage bacteria are released from the infection threads and are encapsulated by plant membrane envelopes. Each host cell may contain up to several hundreds envelopes. Inside

Fig. 31. Root nodule of *Leuceana* sp. Courtesy of M.J. Trinick, C.S.I.R.O., Wembley, W.A.

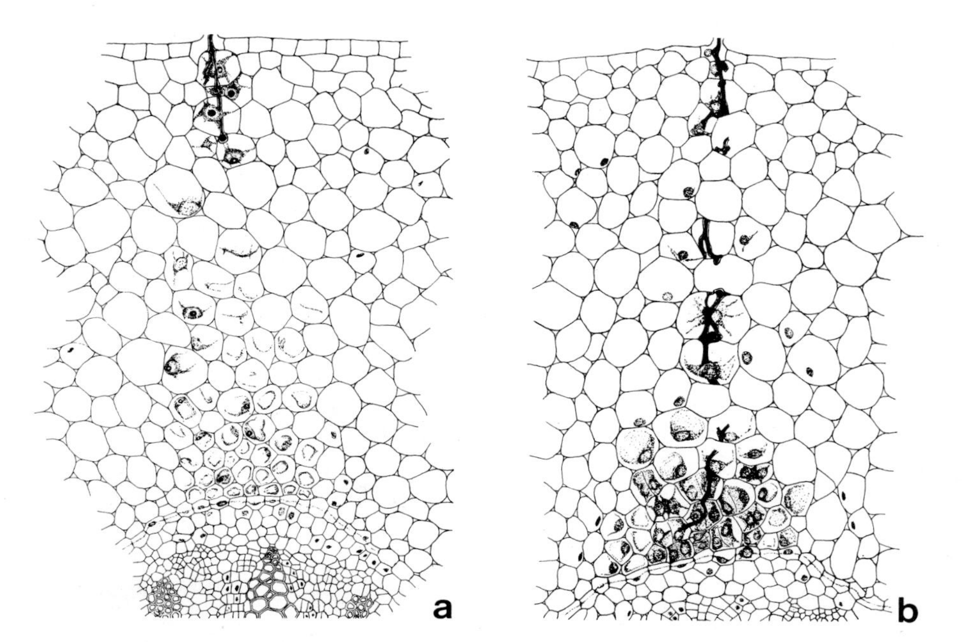

Fig. 32. Early development of root nodules of *Pisum sativum* after infection with *Rhizobium leguminosarum* PRE. a. Early stage of infection. Infection thread in upper cortical cells. Stimulation of cell activity in cortex near endodermis (hypertrophy: enlargement of nuclei). b. Infection thread (containing rhizobia) has reached endodermis. multiplication of inner cortex cells. c–e. Further expansion of inner cortex cells. Induction of cell division in pericycle, and finally also in middle cortex. After stage e. the nodule primordium expands mainly by cell enlargement and passes the outer cortex and epidermis. Courtesy of K.R. Libbenga (97).

the envelopes division of the rhizobium cells is inhibited by unknown plant factors and after one or several divisions, depending on the plant species, growth stops. The bacterial volume as well as the DNA content (94) increases, and the cells become branched. This non-growing phase of the rhizobia is called the 'bacteroid stage'. During the transition of a vegetative cell into a bacteroid, significant changes in the cell metabolism occur, especially in nitrogen metabolism. Nitrogenase is induced and the activities of enzymes involved in ammonia assimilation are reduced. The fixed nitrogen is excreted into the host cytoplasm (95) and assimilated into amino acids (96) and transported to the shoots and leaves.

The shape of the nodules and the form of the bacteroids is not only characteristic of certain strains of rhizobia (98) but also is determined by the host (99, 100, 101). Staphorst and Strijdom (100) and Jansen van Rensburg (101) have described the presence of spherical rhizobium cells as a general property of

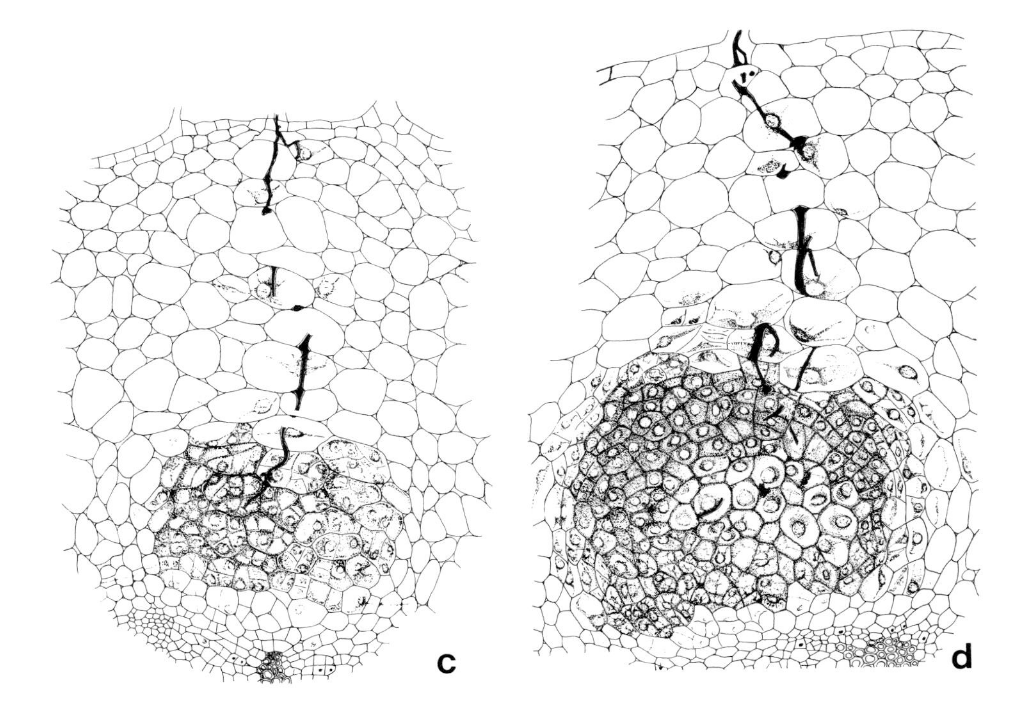
c
d

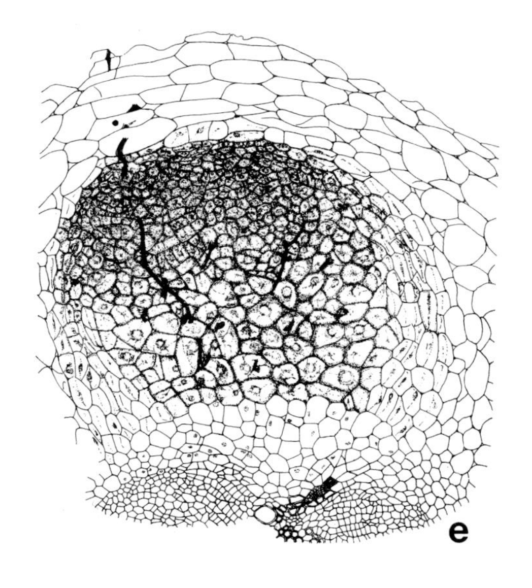
e

the rhizobium strains of the cowpea group of peanut (*Arachis* spp.). However, these strains formed the regular type of branched bacteroids on cowpea (*Vigna sinensis*).

Spherical bacteroid forms of bacteroids also have been reported in a number of fast-growing strains in nodules from peas (99) and red clover (102). *R. leguminosarum* strain PF2 forms effective (N_2-fixing) nodules on peas (*Pisum sativum*) but almost ineffective nodules on broadbean (*Vicia faba*) (99). In early stages of nodule formation, PF2 bacteroids on pea are branched also. At a later stage these cells transform into spherical structures, without losing the ability to fix nitrogen (Fig. 33). Most of the PF2 bacteroids on broadbean remain in the

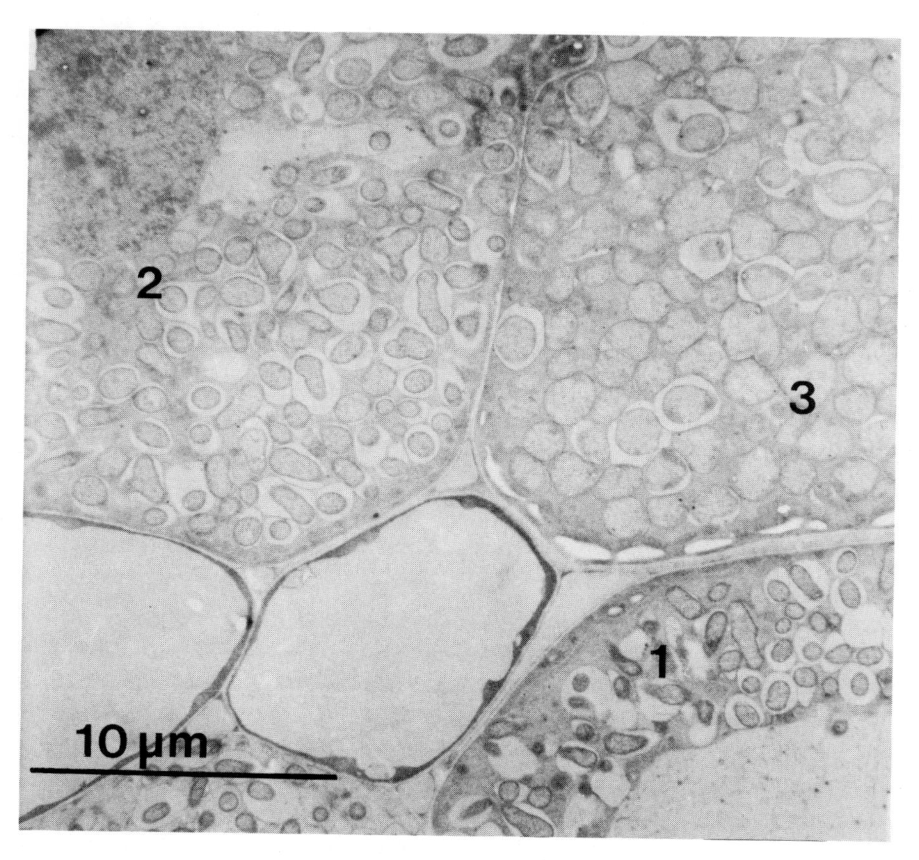

Fig. 33. Transmission electron micrograph of bacteroid tissue in root nodules of *Pisum sativum* var. Rondo, induced by *Rhizobium leguminosarum* PF2, 2½ weeks after inoculation. Early infected cell with branched bacteroids (1). Increase in the number of bacteroids in the cell (2) and transformation of bacteroids into spheroplasts (3). The latter stage is predominant. Earlier stages are visible only in young nodules. Sections prepared by Jan Kijne, Univ. of Leiden. Unpublished result of the late Erik van den Berg, Lab. of Microbiology, Univ. of Wageningen.

rod form and only a few branched bacteroids are found. The presence of a very small number of bacteroids is correlated with very low nitrogenase activity in these 'ineffective' nodules. These examples clearly demonstrate that the host plant exerts considerable influence on the morphogenesis and physiology of the bacteria. As discussed above this phenomenon also is found in other N_2-fixing symbioses and it is likely that this is a general feature of all symbioses.

REFERENCES

1. Buchanan RE, Gibbons NE (eds): *Bergey's Manual of determinative bcteriology*, Baltimore, The Williams and Wilkins Comp. 1974.
2. Stanier RY, Sistrom WR, Hansen TA, Whitton BA, Castenholz RW, Pfennig N, Gorlenko VN, Kondratieva EN, Eimhjellen KE, Whittenbury R, Gherna RL, Truper HG: Proposal to place the nomenclature of the cyanobacteria (Blue-green Algae) under the rules of the international code of nomenclature of bacteria. *Intern J System Bacteriol* 28(2): 335–336, 1978.
3. Rippka R, Deruelles J, Waterbury JB, Herdman M, Stanier RY: Generic assignments, strain histoies and properties of pure cultures of cyanobacteria. *J Gen Microbiol* 111:1–61, 1979.
4. Wyat JT, Silvey JKG: Nitrogen fixation by Gloeocapsa. *Science* 165: 908–909, 1969.
5. Fogg GE, Stewart WDP, Fay P, Walsby AE: *The Blue-green Algae*, London, New York, Academic Press, 1973.
6. Rippka R, Waterbury JB: The synthesis of nitrogenase by non-heterocystous cyanobacteria. *FEMS Microbiol Lett* 2: 83–86, 1977.
7. Gallon JR: Nitrogen fixation by phototrophs. In: *Nitrogen Fixation. Proc. Phytochem Soc Europe. Symposium Sussex, September 1979.* Stewart WDP, Gallon JR (eds) Academic Press, London, 1980, pp 197–238.
8. Kulasooriya SA, Lang NJ, Fay P: The heterocysts of blue-green algae III. Differentiation and nitrogenase activity. *Proc Roy Soc Lond* B 181: 199–209, 1972.
9. Tell-Or E, Stewart WDP: Photosynthetic electron transport, ATP synthesis and nitrogenase activity in isolated heterocysts of *Anabaena cylindrica. Biochem Biophys Acta* 423: 189–195, 1976.
10. Stewart WDP: *Anabaena cylindrica*, a model nitrogen-fixing alga. In: *Perspectives in experimental Biology*. Sunderland N (ed), Pergamon, Oxford 2: 235–246, 1976.
11. Stewart WDP: Systems involving Blue-green Algae (Cyanobacteria). In: *Methods for evaluating Biological Nitrogen Fixation*. Bergersen FJ (ed), John Wiley and Sons, Ltd New York 1980, p 583–635.
12. Wolk CP: Control of sporulation in a blue-green alga. *Develop Biol* 12: 15–35, 1965.
13. Frisch FE: *The Structure and Reproduction of the Algae*, vol. 2. Cambridge University Press, Cambridge, 1945.
14. Wolk CP: Evidence of a role of heterocysts in the sporulation of a blue-green alga. *Am. J. Bot.* 53: 60–262, 1966.
15. Fay P: Metabolic activities of isolated spores of *Anabaena cylindrica. J Exp Bot* 20: 100–109, 1969.
16. Yamamoto Y: Effect of desiccation on the germination o akinetes of *Anabaena cylindrica. Plant cell Physiol* 16: 749–752, 1975.
17. Simon RD: Cyanophycin granules from the blue-green alga *Anabaena cylindrica:* a reserve material consisting of copolymers of aspartic acid and arginine. *Proc Nat Acad Sci USA* 68: 265–267, 1971.
18. Yamamoto Y: Effect of some physical and chemical factors on the germination of akinetes of *Anabaena cylindrica. J Gen Appl Microbiol* 22: 311–323, 1976.
19. Kaushik M, Kumar HD: The effect of light on growth and development of two nitrogen fixing blue-green algae. *Arch Mikrobiol* 74: 52–57, 1970.

20. Hendriksson E, Simi B: Nitrogen fixation by lichens. *Oikos* 22: 119–121, 1971.
21. Huss-Danell K: *Nitrogen fixation in the lichen Stereocaulon paschale*. Thesis University Umeä, Sweden, 1979.
22. Bont JAM de, Mulder EG: Nitrogen fixation and co-oxidation of ethylene by a methane-utilizing bacterium *J Gen Microbiol* 83: 113–121, 1974.
23. Bont JAM de: *Nitrogen-fixing methane utilizing bacteria*. Thesis Agricultural University, Wageningen, The Netherlands, 1976.
24. Lin LP, Sadoff HL: Chemical composition of *Azotobacter vinelandii* cysts. *J Bacteriol* 100: 480–486, 1969.
25. Stevenson LH, Socolofski MD: Role of poly-β-hydrxybutyric acid in cyst formation by Azotobacter. *Antonie van Leeuwenhoek*, J Microb Serol 39: 341–350, 1973.
26. Yates MG, Partridge CCP, Walker CC, Werf AN van der, Campbell TO, Postgate JR: Recent research in the physiology of heterotrophic, non-symbiotic nitrogen-fixing bacteria. In: *Nitrogen Fixation. Proc. Phytochem. Soc. Europe Symposium. Sussex September 1979*. Stewart WDP, Gallon JR (eds), Academic Press, London, 1980, p. 161–176.
27. Bergersen FJ, Turner GL: Leghaemoglobin and the supply of O_2 to nitrogen-fixing root nodule bacteroids: studies of an experimental system with no gas phase. *J Gen Microbiol* 89: 31–47, 1975.
28. Krul JM: *Denitrification, activity of bacterial flocs and growth of a filamentous bacterium in relation with the bulking of activated sludge*. Thesis Agricultural University, Wageningen, The Netherlands, 1978.
29. Desikachary TV: *Cyanophyta*, New Delhi, Indian Council of Agricultural Research, 1959.
30. Becking JH: The Family Azotobacteriaceae. In: *The Prokaryotes, A Handbook on Habitats, Isolation and Identification of Bacteria* Vol. I. Starr MP, Stolp H, Trüper HG, Balows A, Schlegel HG (eds), Springer Verlag, Berlin Heidelberg, 1981, p 795–817.
31. Ao PE, Seidler RJ, Evans HJ, Nelson AD: Association of nitrogen fixing bacteria with decay in white fir. In: *Proceedings of the 1st International Symposium on Nitrogen Fixation* vol. 2, Newton WE, Nyman CJ (eds), Washington State University Press, 1976, p 629–640.
32. Mackintosh ME: Nitrogen fixation by *Thiobacillus ferrooxidans*. *J Gen Microbiol* 105: 215–218, 1978.
33. Beck JV, Brown DG: Direct sulfide oxidation in the solubilization of sulfide ores by *Thiobacillus ferrooxidans*. *J Bacteriol* 96: 1433–1434, 1968.
34. Dncan DW, Landesman J, Walden CC: Role of *Thiobacillus ferrooxidans* in the oxidation of sulfide minerals. *Can J Microbiol* 13: 397–403, 1967.
35. Arkesteyn GJMW: *Contribution of microorganisms to the oxidation of pyrite*. Thesis Agricultural University, Wageningen, The Netherlands, 1980.
36. Fedorov MN, Kalininskaya TA: A new species of nitrogen-fixing *Mycobacterium* and its physiological properties. *Mikrobiologiya* 30: 9–14, 1961 (English translation).
37. Bont JAM de, Leyten MWM: Nitrogen fixation by hydrogen-utilizing bacteria. *Arch Microbiol* 107: 235–240, 1976.
38. Wiegel J, Wilke D, Baumgarten J, Opitz R, Schlegel HG: Transfer of the nitrogen-fixing hydrogen bacterium *Corynebacterium autotrophicum*, Baumgarten *et al.* to *Xanthobacter* gen. nov. *Intern J System Bacteriol* 28: 573–581, 1978.
39. Callaham D, Torrey JG, Tredici P del: Isolation and cultivation *in vitro* of the actinomycete causing root nodulation in *Comptonia*. *Science* 189: 899–902, 1978.
40. Tjepkema JD, Ormerod W, Torrey JG: Vesicle formation and acetylene reduction activity in *Frankia* sp. CpI1 cultured in defined nutrient media. *Nature London* 287: 633–635, 1980.
41. Tjepkema JD, Ormerod W, Torrey JG: Factors affecting vesicle formation and acetylene reduction (nitrogenase activity) in *Frankia* sp CpI1. *Can J Microbiol* 27: 815–823, 1981.
42. Burggraaf AJP, Shipton WA: Estimates of *Frankia* growth under various pH and temperature regimes. *Plant and Soil* 69: 135–147, 1982.
43. Akkermans ADL, Straten J van, Roelofsen W: Nitrogenase activity of nodule homogenates of *Alnus glutinosa:* a comparison with the *Rhizobium*-pea system. In: *Recent Developments in Nitrogen Fixation*, Newton W, Postgate JR, Rodriguez-Barrueco C (eds). Academic Press, London, 1977, p 591–603.

44. Straten J van, Akkermans ADL, Roelofsen W: Nitrogenase activity of endophyte suspensions derived from root nodules of *Alnus*, *Hippophaë*, *Shepherdia* and *Myrica* spp. *Nature London* 266: 257–258, 1977.
45. Akkermans ADL, Roelofsen W, Blom J: Dinitrogen fixation and ammonia assimilation in actinomycetous root nodules of *Alnus glutinosa*. In: *Symbiotic Nitrogen Fixation in the Management of temperate Forests*, Gordon JC, Wheeler CT, Perry DA (eds), Forest Research Laboratory, Oregon State University, Corvallis 1979, p 160–174.
46. Akkermans ADL, Roelofsen W: Symbitic nitrogen fixation by actinomycetes in *Alnus* type root nodules. In: *Nitrogen Fixation. Proc Phytochem Soc Europe Symposium Sussex*, September 1979. Stewart WDP, Gallon JR (eds), London, Academic Press 1980, p. 279–299.
47. Blom J, Roelofsen W, Akkermans ADL: Growth of *Frankia* AvcI1 on media containing Tween 80 as C-source. *FEMS Microbiol Lett* 9: 131–135, 1980.
48. Blom J: Utilization of fatty acids and NH_4^+ by *Frankia* AvcI1. *FEMS Microbiol Lett* 10: 143–145, 1981.
49. Blom J, Harkink R: Metabolic pathways for gluconeogenesis and energy generation in *Frankia* AvcI1. *FEMS Microbiol Lett* 11: 221–224, 1981.
50. Blom J: Carbon and nitrogen source requirements of *Frankia* strains. *FEMS Microbiol Lett* 13: 51–55, 1982.
51. Blom J: *Carbon and nitrogen metabolism of free-living Frankia spp. and of Frankia-Alnus symbioses.* Thesis Agricultural University, Wageningen, The Netherlands, 1982.
52. Akkermans ADL, Huss-Danell K, Roelofsen W: Enzymes of the tricarboxylic acid cycle and the malate-aspartate shuttle in the N_2-fixing endophyte of *Alnus glutinosa*. *Physiol Plant* 53: 289–294, 1981.
53. Huss-Danell K, Roelofsen W, Akkermans ADL, Meyer P: Carbon metabolism of *Frankia* spp. in root nodules of *Alnus glutinosa* and *Hippophaë rhamnoides*. *Physiol Plant* 54: 461–466.
54. Torrey JG and Tjepkema JD (eds): The Biology of *Frankia* and its association with higher plants. *Can J Bot* to be published.
55. Rodgers GA, Stewart WDP: The cyanophyte-hepatic symbiosis. I Morphology and physiology. *New Phytol* 78: 441–458, 1977.
56. Duckett JG, Toth R, Soni SL: An Ultrastructural study of the *Azolla*, *Anabaena azollae* relationship. *New Phytol* 75: 111–118, 1975.
57. Millbank JW: Lower plant associations. In: *A Treatise on Dinitrogen Fixation. Section III Biology*. Hardy RWF, Silver WS (eds), John Wiley and Sons Inc., New York 1977, p 25–151.
58. Akkermans ADL: Root-nodule symbioses in non-leguminous N_2-fixing plants. In: *Interactions between non-pathogenic soil microorganisms and plants*. Dommergues YR, Krupa SV (eds), Developments in Agricultural and managed Forest Ecology 4. Elsevier Scientific Publ Comp Amsterdam 1978, p 335–372.
59. Stewart WDP, Rowell P, Rai AN: Symbiotic nitrogen-fixing cyanobacteria. In: *Nitrogen Fixation, Proc Phytochem Soc Europe Symposium Sussex, September 1979*. Stewart WDP, Gallon JR (eds), Academic Press, London, 1980, p. 239–277.
60. Allen EK, Allen ON: Non-leguminous plant symbiosis. In: *Microbiology and Soil Fertility*. Gilmore CM, Allen ON (eds), Oregon State University Press, Corvallis 1965, p. 77–106.
61. Cagle GD, Vela GR: Giant cysts and cysts with multiple central bodies in *Azotobacter vinelandii*. *J Bacteriol* 107: 315–319, 1971.
62. Oppenheim J, Marcus L: Correlation of ultrastructure in *Azotobacter vinelandii* with nitrogen source for growth. *J Bacteriol* 101: 286–291, 1970.
63. Milandasuta, BE: Developmental anatomy of corralloid roots in cycads. *Am J Bot* 62: 468–472, 1975.
64. Nathanielz CP, Staff IA: On the occurrence of intracellular blue-green algae in cortical cells of the apogeotrophic roots of *Macrozamia communis*. L. Johnson. *Ann Bot* 39: 363–368, 1975.
65. Silvester WB: Dinitrogen fixation by plant associations excluding legumes. In: *A Treatise on Dinitrogen Fixation. Section IV Agronomy and Ecology*. Hardy RWF, Gibson AH (eds). John Wiley and Sons Inc., New York 1977, p. 141–190.
66. Silvester WB: Endophyte adaptation in Gunnera-Nostoc symbiosis. In: *Symbiotic nitrogen fixation in plants*. Nutman PS (ed.). Cambridge University Press, Cambridge, London, New York, Melbourne. 1976, p. 521–538.

67. Angulo Carmona AF: La formation des nodules fixateurs d'azote chez *Alnus glutinosa*. *Acta Bot Neerl* 23: 257–303, 1974.
68. Lalonde M: Techniques and observations of the nitrogen fixing Alnus root nodule symbiosis. In: *Recent Advances in Biological Nitrogen Fixation*. Subba Rao NS (ed.) Edward Arnold, London, 1980, p. 421–434.
69. Torrey JG: Initiation and development of root nodules of Casuarina (Casuarinaceae). *Am J Bot* 63: 335–344, 1976.
70. Callaham D, Torrey JG: Prenodule formation and primary nodule development in roots of Comptonia (Myricaceae). *Can J Bot* 55: 2306–2318, 1977.
71. Strand R, LaetschWM: Cell and Endophyte structure of the nitrogen-fixing root nodules of *Ceanothus integerrimus* H. and A. I. fine structure of the nodule and its endosymbiont, II. progress of the endophyte into young cells of the growing nodule. *Protoplasma* 93: 165–178 (I), 179–190 (II), 1977.
72. Lalonde M: Immunological and ultrastructural demonstration of nodulation of the European *Alnus glutinosa* (L.) Gaertn. host plant by an actinomycetal isolate from the North American *Comptonia peregrina* (L.) Coult. root nodule. *Bot Gaz* 140 (Suppl): 35–43, 1979.
73. Peters GA, Ray TB, Mayne BC and Toia RE: Azolla-Anabaena Association: Morphological and Physiological Studies. In: *Nitrogen Fixation*, Vol. II. Newton WE, Orme-Johnson WH (eds.). University Park Press, Baltimore, p. 293–309, 1980.
74. Dijk C van, Merkus E: A microscopic study of the development of a spore like stage in the life cycle of the root-nodule endophyte of *Alnus glutinosa* L. Gaertn. *New Phytol.* 77: 73–91, 1976.
75. Akkermans ADL, van Dijk C: Non-leguminous root-nodule symbioses with actinomycetes and Rhizobium. In: *Nitrogen Fixation. Vol. I. Ecology*. Broughton WJ (ed). Oxford University Press, p. 57–103, 1981.
76. van Dijk C: Endophyte distribution in the soil. In: *Symbiotic Nitrogen Fixation in the Management of temperate Forests*. Gordon JC, Wheeler CT, Perry DA (eds). Oregon State University Press. P. 84–94, 1979.
77. Houwers A, Akkermans ADL: Influence of inoculation on yield of *Alnus glutinosa* in the Netherlands. *Plant and Soil* 61: 189–202, 1981.
78. Arora N: Morphological development of the root and stem nodules of *Aeschynomene indica* L. *Phytomorphology* 4: 211–216, 1954.
79. Yatazawa M, Yoshida S: Stem nodules in *Aeschynomene indica* and their capacity of nitrogen fixation. *Physiol Plant* 45: 293–295, 1979.
80. Subba Rao NS, Tilak KVBR, Singh CS: Root nodulation studies in *Aeschynomene aspera*. *Plant and Soil* 56: 491–494, 1980.
81. Dreyfus BL, Dommergues YR: Nitrogen-fixing nodules induced by Rhizobium on the stem of the tropical *Sesbania rostrata*. *FEMS Microbiol Lett* 10: 313–317, 1981.
82. Schaede R: Die Knöllchen der adventiven Wasserwurzeln von *Neptunia oleracea* und ihre Bakterien-symbiose. *Planta* 31: 1–21, 1940.
83. Trinick MJ: Rhizobium symbiosis with a non-legume. In: *Proc 1st Intern Symp Nitrogen Fixation* Vol. 2. Newton WE, Nijman CJ (eds). Washinton state University, p. 507–517, 1976.
84. Trinick MJ: Structure of nitrogen-fixing nodules formed by Rhizobium on roots of *Parasponia andersonii* Planch. *Can j Microbiol* 25: 565–578, 1979.
85. Akkermans ADL, Abdulkadir S, Trinick MJ: N_2-fixing root nodules in Ulmacea: *Parasponia* or (and) *Trema* spp. *Plant and Soil* 47: 711–715, 1978.
86. Trinick MJ: Symbiosis between Rhizobium and the non-legume *Trema aspera*. *Nature London* 244: 459–460, 1973.
87. Sabet YS: Bacterial root nodules in Zygophyllaceae. *Nature London* 157: 656, 1946.
88. Athar M, Mahmood A: Root nodules in some members of Zygophyllaceae growing at Karachi, University Campus. *Pak J Bot* 4: 209–210, 1972.
89. Anon.: *Leucaena. Promising Forage and Tree Crop for the Tropics*. Report 26, National Academy of Sciences, Washington DC, 118 p., 1977.
90. Anon.: *Firewood crops. Shrubs and species for energy production*. Report 27. National Academy of Sciences, Washington DC, 237 p., 1980.
91. Sprent JI: *The Biology of Nitrogen Fixing Organisms*. McGraw-Hill Book Company (UK) Limited. 196 p., 1979.

92. Stewart WDP and Gallon JR (eds): *Nitrogen Fixation. Proc Phytochem Soc Europe Symposium Sussex* 1979. Academic Press, London. 451 p., 1980.
93. Gibson AH, Newton WE: *Current Perspectives in Nitrogen Fixation. Proc 4th Intern Symp on nitrogen fixation.* Canberra 1980. Australian Academy of Science, Canberra, 534 p., 1981.
94. Bisseling T, Bos RC van den, Kammen A van, Ploeg M van der, Duyn P van, Houwers A: Cytofluorometrical determination of the DNA contents of bacteroids and corresponding broth-cultured Rhizobium bacteria. *J Gen Microbiol* 101: 79–84, 1977.
95. Haaker H, Laane C, Veeger C: Dinitrogen fixation and proton-motive force. In: *Nitrogen Fixation.* Stewart WDP, Gallon JR (eds.). Academic Press, London. p. 113–138, 1980.
96. Scott DB: Ammonia assimilation in nitrogen-fixing systems. In: *Limitations and potentials for biological nitrogen fixation in the tropics.* Dobereiner J, Burris RH, Hollander A (eds). Plenum Press, New York, London. p. 223–235, 1978.
97. Libbenga KR: *Nodulatie bij Leguminosae.* Thesis University Leiden, 99 p., 1970.
98. Jordan DC: The bacteroids of the genus Rhizobium. *Bacteriol Rev* 26: 119–141, 1962.
99. Berg EHR: The effectiveness of the symbiosis of *Rhizobium leguminosarum* on pea and broad bean. *Plant and Soil* 48: 629–639, 1977.
100. Staphorst JL, Strijdom BW: Some observations on the bacteroids in nodules of *Arachis* spp. and the isolation of Rhizobia from these nodules. *Phytophylactica* 4: 87–92, 1972.
101. Jansen van Rensburg H, Hahn JS, Strijdom BW: Morphological development of Rhizobium bacteroids in nodules of *Arachis hypogea* L. *Phytophylactica* 5: 119–122, 1973.
102. Chen HK, Thornton HG: The structure of ineffective nodules and its influence on N_2-fixation. *Proc Roy Soc* B 129: 208–229, 1940.
103. Trinick MJ: Relationship among the fast-growing rhizobia of *Lablab purpureus, Leuceana leucocephala, Mimosa* spp., *Acacia farnesiana* and *Sesbania grandiflora* and their affinities with other rhizobial groups. *J Appl Bacteriol* 49: 39–53, 1980.
104. Abdel-Ghaffar, Jensen HL: The Rhizobia of *Lupinus densiflorus* Benth., with some remarks on the classification of root nodule bacteria. *Arch Mikrobiol* 54: 393–405, 1966.
105. Burggraaf AJP, Shipton WA: Studies on the growth of *Frankia* isolates in relation with infectivity and nitrogen fixation (acetylene reduction). *Can J Bot,* 1983.

3. Taxonomy and distribution of non-legume nitrogen-fixing systems

GEORGE BOND*

Department of Botany, University of Glasgow, Scotland, U.K.

In this chapter more attention will be paid to plants bearing *Alnus*-type nodules than to other non-legume nitrogen-fixing systems, partly because it has been with the former that the author's main research interest has lain, but also for the reason that of the various systems to be considered in the chapter, the nitrogen fixation associated with *Alnus*-type nodules seems to offer the best prospects for exploitation.

ALNUS-TYPE (ACTINORRHIZAL) NODULE SYSTEMS

Historical introduction

Reports of the presence of swollen, nodular structures on the roots of non-leguminous plants, not attributable to invasion by insect pests, began to appear during the 19th century, firstly in respect of alder in 1829, and by the end of the century there were records of nodulation in a total of some 12 species belonging to the genera *Alnus*, *Hippophaë*, *Elaeagnus*, *Myrica*, *Ceanothus* and *Casuarina*. Also around that time it was demonstrated that nodulated plants of *Alnus* (1, 2), *Elaeagnus* and *Shepherdia* (3) could grow satisfactorily in a rooting medium free of combined nitrogen, suggesting that the nodules were nitrogen-fixing. Efforts to identify the micro-organism detected within the nodule cells were handicapped by a lack of adequate optical equipment, but it was correctly observed (4) that the endophyte was finely hyphal in nature, and that it resembled an actinomycete (5).

During the first half of the present century relatively few workers were attracted to the study of these non-legume nodules, though from the taxonomic standpoint the detection of nodulation in *Coriaria* (6) was of much interest, since that genus had already attracted the attention of plant geographers because of evidence that in former eras it was extremely widespread (see also p. 59). The publication in 1932 of the book 'Root Nodule Bacteria and Leguminous Plants' (7) which included a detailed review of such information as was available at the time concerning non-legume nodules, served to remind readers of the

* Now retired

Gordon, J.C. and Wheeler, C.T. (eds.). Biological nitrogen fixation in forest ecosystems: foundations and applications

ISBN 90-247-2849-5.
Printed in The Netherlands

existence of such nodules and of the opportunities for their further study, as did also the book 'Die Pflanzlichen Symbiosen' (8), by an author who was personally engaged in the study of the nodules, which in its successive editions provided up-dated reviews of the subject. Also noteworthy was a substantial compendium (9) on woody plants of the U.S.A. in which 60 species of the nodule-bearing genera *Alnus*, *Myrica*, *Shepherdia*, *Elaeagnus* and *Ceanothus* that occur in that country were listed and their distribution described. Of much interest was the publication in 1953 of the first record of nodulation in the Rosaceae, namely in *Dryas* (10).

At the time of the commencement of the International Biological Programme, which continued from 1967–73, nodulation had been recorded in some 118 species of non-legumes, distributed among 12 genera (11), the record then included for *Arctostaphylos uva-ursi* being later deleted for reasons provided elsewhere (12). As part of the I.B.P. a search for further examples of nodulation was organised, and as a result of the efforts of some 50 collaborators records of nodulation in 36 additional species including two further genera, *Colletia* and *Rubus*, were secured (11). At international meetings organised under the I.B.P. sessions were for the first time devoted to non-legume nodules, with the result that many more biologists became aware of the existence of such nodules and of their importance as providers of fixed nitrogen for the biosphere, some being influenced to take up work on them. In the post-I.B.P. period further records of nodulation have appeared, as will be made clear below, while recent successes in the isolation of the nodule endophytes (Chapter 4) have provided a further stimulus to interest in these nodules.

Non-legume plants now known to be capable of bearing Alnus-type nodules

So far 20 genera have been shown to include nodulating species, and they are listed together with their families and orders in Table 1. It will be noted that some of the genera are closely related; thus the Rosaceae and the Rhamnaceae each include six nodulating genera, and in both families five of the genera fall within a particular tribe. Three other genera are in the Elaeagnaceae. Over-all, however, the impression is one of taxonomic unrelatedness. The monogenic families Casuarinaceae and Coriariaceae, especially the latter, have long been recognised by taxonomists as showing little or no affinity to any other families, while it should also be noted that in Table 1 the eight families to which the nodulating belong are all assigned to different orders, confirming that at least in Engler's classification little affinity between the families is recognised. Although other classifications may suggest slight relationships between certain of the families, it remains that over-all the disparities in respect of floral and foliar characteristics are too great for any attempt to explain the nodulating habit by a common descent to be contemplated. The only obvious features that the genera have in common are that they are all of tree or shrub habit (with the possible exception

Table 1. Plant genera which include species bearing *Alnus*-type nodules, and their classification according to Engler (13).

Genus	Family	Order
Casuarina Adans.	Casuarinaceae	Verticillatae
Myrica L.*	Myricaceae	Juglandales
Alnus Mill.	Betulaceae	Fagales
Dryas L.	Rosaceae (tribe *Dryadeae*)	Rosales
Cercocarpus Kunth	Rosaceae (tribe *Dryadeae*)	Rosales
Purshia DC.	Rosaceae (tribe *Dryadeae*)	Rosales
Chamaebatia Benth.	Rosaceae (tribe *Dryadeae*)	Rosales
Cowania D. Don	Rosaceae (tribe *Dryadeae*)	Rosales
Rubus L.	Rosaceae (tribe *Rubeae*)	Rosales
Coriaria Hook.	Coriariaceae	Sapindales
Colletia Comm.	Rhamnaceae (tribe *Colletieae*)	Rhamnales
Discaria Hook.	Rhamnaceae (tribe *Colletieae*)	Rhamnales
Trevoa Miers ex Hook.	Rhamnaceae (tribe *Colletieae*)	Rhamnales
Talguenea Miers	Rhamnaceae (tribe *Colletieae*)	Rhamnales
Kentrothamnus Suess. & Overkott	Rhamnaceae (tribe *Colletieae*)	Rhamnales
Ceanothus L.	Rhamnaceae (tribe *Rhamneae*)	Rhamnales
Elaeagnus L.	Elaeagnaceae	Thymelaeales
Hippophaë L.	Elaeagnaceae	Thymelaeales
Shepherdia Nutt.	Elaeagnaceae	Thymelaeales
Datisca L.	Datiscaceae	Violales

* Includes *Comptonia* L'Herit. ex Ait.

of *Datisca*, whose species are classed by some authorities as shrubs but by others as perennial herbs), and all, together with their families, are dicotyledonous and members of the Engler series Archichlamydeae, though a further presumed common feature is that their root hairs have special properties which assist in the entry of the appropriate actinomycete into their roots.

It is hardly necessary to point out that the nodulating genera so far detected in the Rosaceae and Rhamnaceae form only a small proportion of the total genera complements of the families, generally reckoned as about 100 and 58 respectively. Indeed, apart from the monogenic families – two of which are mentioned above, the third being the Myricaceae (setting aside a recently proposed new genus, *Canacomyrica*, with one species) – it is only in the Elaeagnaceae that nodulation is a family characteristic. The Betulaceae is regarded by some taxonomists as including six genera, by others only two, but in either event only *Alnus* is nodulated. Again in respect of the Datiscaceae there is a measure of disagreement, since besides *Datisca* two other small genera, *Tetrameles* and *Octomeles*, which are trees found in Malaysia, are sometimes included in the family, though other taxonomists place them elsewhere.

A possible explanation of the taxonomic scatter among the plants that bear *Alnus*-type nodules will be advanced below. That they should be distributed through eight families stands in marked contrast to the near restriction, at least on present knowledge, of rhizobial nodules to one family of plants, the Leguminosae (see p. 72). In passing it may be noted that in the Engler classification that family is placed in the order Rosales.

The heading to Table 1 includes the phrase 'genera which include species bearing *Alnus*-type nodules'. Such caution is necessary, since at present it is only in respect of certain genera that the ability to form nodules can be said with full confidence to be a generic character. Data bearing on this aspect are provided in Table 2, where it will be noted that although in *Alnus* and *Coriaria* the number of species recorded to bear nodules actually exceeds the number of species that the genera are shown as comprising – a situation which reflects some disagreement among taxonomists as to what is a good species – in most of the other genera the disparity is in the opposite direction. This does not mean that some species have been found to be devoid of the capacity to form nodules, for to the author's knowledge this has never happened in the listed genera except in the case of

Table 2. Further details of the plant genera listed in Table 1.

Genus	Number of species recognised in genus*	Present distribution	Number of species recorded to bear nodules
Casuarina	45	Australia, tropical Asia, Pacific islands	24
Myrica	35	Many tropical, subtropical and temperate regions	26
Alnus	35	Europe, Siberia, N. America, China, Japan, Andes	39
Dryas	4	Arctic, mountains of north temperate zone	3
Cercocarpus	20	West and south-west U.S.A., Mexico	4
Purshia	2	N. America	2
Chamaebatia	2	California	1
Cowania	5	South-west U.S.A., Mexico	1
Rubus	250**	Many regions, especially north temperate	2
Coriara	15	Mediterranean, Japan, China, New Zealand, Chile, Mexico	16
Colletia	17	Temperate and subtropical S. America	3
Discaria	10	S. America, New Zealand, Australia	5
Trevoa	6	Andes	2
Talguenea	1	Chile	1
Kentrothamnus	2	Bolivia, Argentina	1
Ceanothus	55	N. America	31
Elaeagnus	45	Asia, Europe, N. America	28
Hippophaë	3	Europe, temperate Asia to Kamchatka, Japan, Himalaya	1
Shepherdia	3	N. America	2
Datisca	2	Mediterranean to Himalaya and central Asia, south-west U.S.A.	2

* According to Willis (14).
** In addition many sub-species have been recognised.

Rubus. Rather it is either because some species have not yet been examined for nodulation – often for the reason that they grow in areas not yet penetrated by workers interested in actinorrhizal nodules – or, though they may have been examined and nodules found, the results have not been circulated internationally. It is obvious that nodulation is a generic character in *Alnus*, *Coriaria* and the genera with a single species, while there is fairly strong evidence for that in *Myrica*, *Dryas*, *Hippophaë* and *Shepherdia* also. However it is already clear that nodulation is not a generic character in *Rubus* (11), though some additional nodulating species may still await discovery (see below).

Table 2 also provides information on the world distribution of the nodulating genera. Some of them, such as *Myrica*, *Alnus*, *Elaeagnus*, *Hippophaë* and *Datisca*, have a very widespread occurrence, while the genus *Myrica* also shows a remarkable range in habit, varying from semi-prostrate shrubs to massive trees. As noted earlier, the genus *Coriaria* has long attracted the interest of plant geographers (15) because it is now represented in four widely separated parts of the world (Table 2), so providing an extreme example of discontinuous distribution which suggests that the genus is very ancient and now in a senile state, but that in an earlier period – the Tertiary, according to the fossil evidence – it had a wide distribution in Eurasia and the southern hemisphere. The typical habitats of these non-legumes will be indicated in p. 66.

Table 3 lists all the species that have been recorded to bear *Alnus*-type nodules, according to information available to the author by the end of 1981. Compared with a previous list (12) the present one includes four additional genera and 27 newly-recorded species, and it is only for these that the sources of the reports of nodulation are now cited; citations for the remaining species, or indications of where these can be found, are provided in reference (12). The lack of an authority for some specific names in Table 3 indicates that none was provided in the original report.

The total number of species listed in Table 3 is 193. It is hoped that its inclusion will stimulate the search for further examples of nodulation, and will be of assistance to those concerned with the exploitation of non-legume nodule fixation.

The further study of nodulation in the genus *Rubus* may be aided by the following information. Of the two species in which nodulation has already been observed (Table 3), *R. ellipticus* has much the wider distribution since it occurs (18), usually at altitudes over 1000 m, in the Himalayas, India, Thailand, China (provinces Yunnan, Szechuan and Kweichow) and the Philippines, and has been introduced into Java, Australia, Jamaica, and according to other records (21) also into Florida and California where, at least for a period, it was grown as a crop plant under the name Golden Evergreen raspberry. The existing record of nodulation in this species relates to plants growing in Java (12), and it would be of interest to know whether nodules are also present in other regions where the

Table 3. Plant species known to bear *Alnus*-type nodules.

CASUARINA

C. cunninghamiana Miq., *C. descussata* Benth., *C. deplanchei* Miq., C. distyla Vent., *C. equisetifolia*, *C. fraseriana*, *C. glauca*, *C. huegeliana*, *C. lepidophloia*, *C. littoralis* Salisb., *C. montana*, *C. muellerana* Miq., *C. muricata*, *C. nana* Sieb. ex Spreng., *C. nodiflora*, *C. papuana* S. Moore, *C. pusilla* Macklin, *C. quadrivalvis*, *C. rigida* Miq., *C. rumphiana* Miq., *C. stricta*, *C. sumatrana*, *C. tenuissima*, *C. torulosa* Ait.

MYRICA

M. adenophora Hance, *M. asplenifolia* (= *Comptonia peregrina*), *M. brevifolia* E. May ex C.DC., *M. californica* Cham. & Schlecht., *M. carolinensis*, *M. cerifera* L., *M. cordifolia* L., *M. diversifolia* Adamson, *M. esculenta* Ham., *M. faya* Ait., *M. gale*. L., *M. humilis* Cham. & Schlecht., *M. integra* (Chev.) Killick, *M. javanica* Blume, *M. kandtiana* Engl., *M. kraussiana* Buching. ex Meisn., *M. microbracteata* Weim., *M. nana** Cheval., *M. pensylvanica*, *M. pilulifera* Rendle, *M. pubescens* Willd., *M. quercifolia* L., *M. rubra* Sieb. et Zucc., *M. salicifolia* Hochst., *M. sapida* var. *longifolia*, *M. serrata* Lam.

ALNUS

A. cordata (Lois.) Desf., *A. cremastogyne** Burk., *A. crispa* (Ait.) Pursh., *A. fauriei* Lev. et Vnt., *A. ferdinandi** Makio, *A. firma* Sieb. et Zucc., *A. formosana* Mak., *A. fruticosa*, *A. glutinosa*, *A. hirsuta*, *A. incana*, *A. inokumai* Murai & Kusaka, *A. japonica* Sieb. et Zucc., *A. jorullensis* Kunth., *A. mandshurica** (Callier) Hand.-Mzt., *A. maritima* Nutt., *A. matsumurae* Callier, *A. maximowiczii* Callier, *A. mayrii* Callier, *A. mollis*, *A. multinervis* Matsum., *A. nepalensis*, *A. nitida*, *A. oblongifolia* Torr., *A. orientalis*, *A. rhombifolia* Nutt., *A. rubra* Bong., *A. serrulata*, *A. serrulatoides* Callier, *A. sibirica** Fisch., *A. sieboldiana* Matsum., *A. sinuata* Rydb., *A. tenuifolia* Sarg. var. *glabra* Call., *A. tinctoria* Sarg., *A. trabeculosa* Hand.-Mzt., *A. undulata*, *A. viridis* Regel, *A. rugosa*

DRYAS

D. drummondii Richards., *D. integrifolia* Vahl., *D. octopetala* L.

CERCOCARPUS

C. betuloides, *C. ledifolius* Nutt., *C. montana* Raf., *C. paucidentatus* Britt.

PURSHIA

P. glandulosa Curran, *P. tridentata* (Pursh.) DC.

CHAMAEBATIA

*C. foliolosa*** Benth.

COWANIA

C. mexicana var. *stansburiana*⁺ (Torr.) Jeps.

RUBUS

R. ellipticus J.E. Smith, *R. ferdinandi-muelleri*⁺⁺ Focke

CORIARIA

C. angustissima, *C. arborea* Lindsay, *C. intermedia* Matsum., *C. japonica* A. Gray, *C. kingiana*, *C. lurida*, *C. myrtifolia* L., *C. nepalensis* Wall., *C. plumosa*, *C. pottsiana*, *C. pteriodoides*, *C. ruscifolia*ˣ L., *C. sarmentosa*, *C. sinica** Mixim., *C. terminalis** Hemsl., *C. thymifolia* Humb. & Bonpl.

COLLETIA

C. armata Miers (= *C. spinosa* Lam.), *C. paradoxa* (Spreng.) Escal., *C. spinosissima* Gmel.

DISCARIA

D. americana Gill. et Hook., *D. nana* (Clos.) Weberb., *D. serratifolia* (Vent.) Benth. et Hook., *D. toumatou* Raoul., *D. trinervis* (Gill.) Reiche

TREVOA

T. trinervis Miers, *T. patagonica*ˣ Speg.

TALGUENEA

*T. quinquenervis*ˣˣ (Gill. & Hook.) Johnston

KENTROTHAMNUS

*K. weddellianus*ˣ (Miers) Johnston

Table 3 (continued).

CEANOTHUS

C. americanus L., *C. azureus*, *C. cordulatus* Kell., *C. crassifolius* Torr., *C. cuneatus* (Hook.) Nutt., *C. delilianus*, *C. divaricatus* Nutt., *C. diversifolius* Kell., *C. fendleri*, *C. foliosus* Parry, *C. fresnensis* Dudley, *C. glabra*, *C. gloriosus* Howell var. *exaltatus*, *C. greggii* Gray, *C. griseus* (Trel.) McMinn, *C. impressus* Trel., *C. incana* T. & G., *C. integerrimus* H. & A., *C. intermedius*, *C. jepsonii* Greene, *C. leucodermis* Greene, *C. microphyllus*, *C. oliganthus* Nutt., *C. ovatus*, *C. parvifolius* Trel., *C. prostatus* Benth., *C. rigidus* Nutt., *C. sanguineus* Pursh., *C. sorediatus* H. & A., *C. thyrsiflorus* Esch., *C. velutinus* Dougl.

ELAEAGNUS

E. angustifolia L., *E. argentea* Pursh., *E. bockii** Diels, *E. commutata* Bernh., *E. conferta* Roxb., *E. crispa** Thunb., *E. edulis*, *E. formosana* Nakai, *E. glabra* Thunb., *E. gonyanthes** Benth., *E. henryi** Warb., *E. latifolia* L., *E. longipes* A. Gray, *E. macrophylla* Thunb., *E. matsunoana* Makino, *E. mollis** Diels, *E. moorcroftii** Wall., *E. multiflora*, *E. murakamiana* Thunb., *E. oldhami** Maxim., *E. oxycarpa** Sclecht., *E. pungens*, *E. rhamnoides*, *E. stellipila** Rehd., *E. triflora* Roxb., *E. umbellata* Thunb., *E. viridis* var. *delavayi** Lec., *E. yoshinoi* Makino

HIPPOPHAË

H. rhamnoides L., *H. rhamnoides* var. *porcera** Rehd.

SHEPHERDIA

S. canadensis Nutt., *S. argentea*

DATISCA

D. cannabina L., *D. glomerata* Baill.

* Nodulation in these species was first observed by Professor Huang Jia-bin and his colleagues at the Institute of Forestry and Pedology, Academia Sinica, Shenyang, China. The observations are included here with his permission and are expected to be published in a Chinese journal in 1983.

** nodulation reported in reference (16).

+ Nodulation reported in reference (17).

++ Nodulation reported in reference (18).

× Nodulation reported in reference (19).

×× Detection of nodules on this species growing in Chile was reported to the present author by Dr. J. Kummerow of San Diego University and included here with his permission.

species occurs. *R. ferdinandi-muelleri* on the other hand is confined to New Guinea, where the record of nodulation was obtained, and to New Britain (18).

On the question of where in this very large and taxonomically very difficult genus the search for further nodulating species should be concentrated, it may be noted that in the standard but rather old taxonomic treatment of the genus (20) the two species named above are both placed in the subgenus *Idaeobatus* (the raspberries), though in different sections. A species *R. pinfaënsis* Lévl. & Van., occurring in China, is held to be closely related to *R. ellipticus* and might prove to be nodule-bearing, as could also be true of some of the nearly 30 species that are classed alongside *R. ferdinandi-muelleri* in the section *Pungentes*, including *R. eucalyptus* Focke, *R. lutescens* Franchet, *R. pileatus* Focke, *R. sikkimensis* Hook. and *R. horridulus* Hook., of which the first three occur in China and the others in the Himalayas. In a more recent classification (18) confined chiefly to species occurring in the taxonomic region Malesia, the only species regarded as close to *R. ellipticus* is *R. wallichianus* W. & A., occurring in Vietnam, the Himalayas and

Taiwan. *R. papuana* Schlect. and *R. montis-wilhelmi* van Royen, both occurring in New Guinea, are held to be closely related to *R. ferdinandi-muelleri.*

Evidence that the nodules of the listed species are of Alnus-type

Recapitulating what may have been said in Chapter 2, nodules of *Alnus*-type have three chief characteristics, as follows.

(1) The nodules commence as simple, swollen structures on the bearing root, but repeated branching soon begins, leading to the production of coralloid nodule clusters; these are usually perennial and by continued growth over a period of some years the clusters, often roughly spherical in shape, may attain a diameter of several cm.

(2) The original simple nodule, or a lobe of a cluster, shows a general internal structure akin to that of a root, though there is now a superficial periderm and an enlarged cortex in which some cells are filled with a dense growth of the nodule endophyte, which exhibits the characters of an actinomycete.

(3) The nodules are nitrogen-fixing and provide a source of combined nitrogen which enables the plant bearing them to grow vigorously in a nitrogen-free rooting medium.

In relatively few of the species listed in Table 3 is the statement that their nodules are of *Alnus*-type based on direct evidence apart from the external form of the nodules. However, except for *Cowania*, *Trevoa*, *Talguenea* and *Kentrothamnus*, evidence on all three counts listed above is available in respect of the nodules of at least one, or in some instances several species from each of the genera listed in Table 3, and it is reasonable to assume that the nodules of other species of a given genus will be of similar structure and properties. There is a slight qualification in respect of *Rubus*, for although the occurrence of fixation in the nodules has been demonstrated (21), evidence of the actinomycetal nature of the endophyte rests only on a light microscope examination.

Since from the economic standpoint the fixation of nitrogen associated with *Alnus*-type nodules is their most important characteristic, it will be appropriate to refer briefly to the evidence for its occurrence in the various genera. Until relatively recently the only method available was to test the ability of nodulated plants to grow in a nitrogen-free rooting medium as compared with non-nodulated plants, with total nitrogen estimations being made on the harvested plants. As indicated in an earlier review (22), long-term tests of this kind have been carried out with species of *Casuarina*, *Myrica*, *Alnus*, *Dryas*, *Cercocarpus*, *Coriaria*, *Ceanothus*, *Discaria*, *Elaeagnus*, *Hippophaë* and *Shepherdia*, and later of *Purshia* (23) and *Colletia* (24). In all species the nodulated plants grew better than the controls, and usually grew vigorously, implying that the nodules are nitrogen-fixing. Under the conditions in the author's own glasshouse, nodulated plants of *Alnus glutinosa* accumulated up to 490 mg nitrogen per plant after six months' growth from seed, compared with less than 1 mg by a non-nodulated plant.

With the advent of the ^{15}N technique about 1940 and of the acetylene assay in 1967, it became possible to make short-term tests for fixation on nodules detached from field or laboratory-grown plants or, as is preferable, on nodules still attached to short lengths of root, since fixation usually falls off rapidly after complete detachment. Roots alone are used as controls. In these ways evidence of fixation has been obtained for a wider range of species, though the extent to which the nodules are satisfying the nitrogen-needs of the plant cannot be determined from such data alone.

It is well-known that some rhizobial nodules are of 'ineffective' type, showing little or no fixation of nitrogen. Ineffective actinorrhizal nodules have been obtained experimentally (25, 26) by inoculating seedlings of a foreign species with crushed nodules of a species of the same plant genus native to Britain, e.g. the inoculation of the North American *Myrica cerifera*, or of *Myrica faya* from the Canary Islands, with crushed *Myrica gale* nodules. It seems possible that the first cross could occur naturally, since the distributions of *M. cerifera* and of *M. gale* in the U.S.A. have been reported to overlap (9). The ineffective nodules obtained experimentally were also characterised by being formed in very large numbers and scattered all over the root system, but a different situation was reported in work on *Purshia* (27) in which the nodules borne on a batch of plants raised from seed collected at a particular site showed no nitrogenase activity, although the nodules were of quite normal appearance externally. Elsewhere it was reported (28) that under glasshouse conditions the formation of short-lived pseudo-nodules containing the fungi *Penicillium albidum* or in other cases *Cylindrocarpon radicicola* had been observed.

Regularity of nodulation under field conditions

It is obvious that before any claim of a significant role for these plants in the maintenance of biosphere nitrogen can be entertained, it must be established how regularly they are nodulated under natural conditions of growth. The information to be presented here will be based partly on the data obtained during a survey conducted as a part of the International Biological Programme (see also p. 56), and also on normal papers published before or after that survey. Statements made below that are not accompanied by a reference will be of I.B.P. origin, and if desired the names of the collaborators responsible can be learned from the survey report (11). The genera will be considered in the order followed in the preceding tables, though information is not available for every genus.

Casuarina. One collaborator examined trees of *C. cunninghamiana*, *C. glauca* and *C. cristata* at 100 sites in Australia for each species, and found nodules on the first two species at practically every site, but at very few on the third. Another collaborator in a different region of that country was, however, often unable to find nodules on *C. glauca*. It was suggested that under Australian conditions the nodules of casuarinas are sometimes situated in deeper, moister levels of the soil which are difficult to reach by hand excavation. Possibly the same applies to Indonesia, where nodules could be found on only three out of 83 trees of *C. equisetifolia* that were inspected, though they were easily located on all 72 trees of *C. sumatrana* that were examined. Species of *Casuarina* growing as introductions in Pakistan, Sénégal and Morocco were all nodulated, though not those in Portugal. It had previously been reported (29) that nodulation is widespread in casuarinas introduced into Florida and the Caribbean region, while there are many reports of nodulation in species growing in India, e.g. (30).

Myrica. Regular nodulation was reported for *M. pensylvanica* and *M. asplenifolia* at various sites in Canada and the U.S.A., though there were some negative results for *M. gale* in the former country. All examinations of *M. javanica* in Indonesia and of *M. rubra* in Japan gave positive results, while inspection of *M. gale* at some 40 sites in France and further sites in Scandinavia and Britain showed a majority of the plants to be nodulated, with some indication that nodulation was best where the soil was permanently wet.

Alnus. The inspection of close on 1000 trees of *A. glutinosa* at sites in France, Scandinavia, Britain, Yugoslavia, Portugal and Turkey showed nodulation to be almost unfailing, the same appearing to be true of *A. incana* and *A. viridis*, though inspections were fewer. In Canada *A. rugosa* was regularly nodulated at sites in the Quebec and Montreal regions, though findings for *A. crispa* were more variable.

Dryas. The original discovery of nodules in this genus was made, as noted already, in respect of *D. drummondii* growing at Glacier Bay, Alaska (10) and was later reported also for *D. octopetala* and *D. integrifolia* growing in the same area (31). During the I.B.P. survey nodules were found on *D. drummondii* in Canada, though none could be detected on *D. integrifolia*. A sustained search was made for nodules on *D. octopetala* outside of North America, namely at sites in Greenland, Scotland, Ireland, Sweden, the Pyrenees, the Alps and Japan, all with negative results, even though the roots were excavated to a depth of 1.5 m at one site. It seems possible that some climatic or other barrier has prevented the migration of the endophyte out of North America.

Purshia. The original report (32) of the detection of nodules on *P. tridentata*, a species of widespread occurrence on open ranges or as an understory in woodland in many western states of the U.S.A., indicated that nodulation is not unfailing in the field. More recently it was recorded (27) that at a site in the pumice region of central Oregon, out of several hundred plants examined only 46 percent were nodulated, though seedlings grown in the glasshouse in soil brought in from the site nodulated freely. It was suggested that nodulation in the field was often adversely affected by low temperatures and by drought.

Chamaebatia. In the report (16) announcing the discovery of nodules on *C. foliolosa* growing at a site in the Sierra Nevada of California, it is mentioned that no nodules could be found at a second site which differed from the first in that the mineral soil was covered by a substantial layer of organic debris; however it is admitted that the lower roots growing in the mineral stratum were not examined for nodules.

Coriaria. It is reported (33) that nodulation can be detected regularly on younger plants of *C. myrtifolia* growing in the provinces of Barcelona and Lérida, Spain, but that nodules are located less readily in older plants.

Discaria. Plants of *D. toumatou* growing in several widely separated localities in South Island, New Zealand, were all nodulated (34), the same applying to *D. americana*, *D. nana*, *D. trinervis* and *D. serratifolia* growing in Argentina (35).

Ceanothus. A study (36) was made of nodulation in *C. velutinus* at two recently burned forest sites in central Oregon. At the first, which previously had been dominated by ponderosa pine, over 80 percent of the *Ceanothus* seedlings became nodulated during their first year, but at the second site, which previously had carried fir and western hemlock, no nodules were found on the seedlings during their first two years of growth, while only 7 percent formed nodules in the third year. Other workers (37) selected stands of *C. velutinus*, 12 in all, in the Cascade Mountains, Oregon, which provided plants 3–15 years of age, growing in a soil which was rich in pumice, stony and poor in organic matter. Nodules were present on 28 percent of 3-year old plants, on 65 percent of 5-year old plants, while after seven years of growth practically all plants bore nodules.

Elaeagnus. I.B.P. reports showed that different species were regularly nodulated in Japan and Pakistan, but that the situation was more variable in Canada. Nodules were present on most of the plants growing as introductions in France, Spain and Britain. Many inspections (38) of the particular species *E. commutata* in Alberta, Canada, showed unfailing nodulation.

Hippophaë. The I.B.P. survey showed that nodulation is the rule for *H. rhamnoides* at numerous sites in Scandinavia, France, Britain, Spain, Austria, Turkey and Pakistan.

Shepherdia. Most of the plants of *S. canadensis* that were inspected under the I.B.P. at various sites in Canada and at one in the U.S.A. were nodulated, which was also true of the smaller number of plants of *S. argentea* examined. In Alberta *S. canadensis* nodulates most freely in poor sandy soils, and less well if there is a humus layer on the surface (38).

Datisca. Profuse nodulation was found on a large number of plants of *D. cannabina*, both male and female, examined at a site in Pakistan (39). Nodulation is also recorded for that species in Italy (40) and for *D. glomerata* in California (41).

Conclusions emerging from the above review are that uncertainties concerning nodulation in *Casuarina* await resolution, and that nodulation in the rosaceous genera seems to be subject to greater hazards than in most other genera.

When non-legumes with a nodule-forming capacity are to be employed in forestry, it is obviously important to know beforehand whether good nodulation will occur without human aid, and even if the transplants are already nodulated from nursery beds it is useful to have a good reserve of organisms in the forest soil to provide for continued nodule formation. Although some of the information provided above may be helpful in a general way, a preliminary investigation of a proposed planting site will obviously be desirable, either by planting test plants on the site, or more quickly by arranging nodulation tests in a glasshouse, using soil brought from the site.

Habitats to which species bearing Alnus-type nodules are adapted

The possession of an internal source of combined nitrogen endows these plants with the ability to colonise substrates that are low in, or even devoid of nitrogenous nutrients, such as deposits resulting from volcanic action or the leached substrate left by melting glaciers.

One report (42) illustrating that ability referred to *Coriaria arborea* in New Zealand, where the species (noted to be regularly nodulated) was observed to flourish on raw, unconsolidated substrates, and, in particular, on areas of barren pumice in the Kaingaroa plains, the contrast between the luxuriant green clumps of *C. arborea* with its associated plants, and the starved appearance of other vegetation was said to be remarkable.

Another example is provided by detailed studies on the colonisation of recently deglaciated areas in Alaska (10, 31, 43) which, as noted already, resulted among other things in the first detection of nodules on species of *Dryas*. The latter, especially *D. drummondii*, are in some areas the dominant early colonisers of the nitrogen-deficient matrix left by the melting ice, and these *Dryas* plants eventually form a lush green carpet covering most of the area. After 20–30 years the *Dryas* gives way to the more rapidly growing *Alnus crispa*, which by its soil-building properties enables Sitka spruce to become established later. In other areas the *Dryas* phase is absent, perhaps owing to the lack of a seed source. Thus of the various higher plants whose seed is present, these two genera bearing actinorrhizal nodules are the only ones that thrive on the newly-glaciated areas. Analyses showed a marked increase in soil nitrogen under both *Dryas* and *Alnus*, though greater under the latter.

The above findings prompt the question of whether nitrogen fixation was a factor in the prominence, revealed by fossil evidence, of *Dryas octopetala* in the flora of north-central Europe during the waning stages of the Pleistocene ice sheet. A doubt hangs over this because, as noted already, in modern times *Dryas* appears to be nodulated only in North America. However this uncertainty need not be extended to *Hippophaë rhamnoides* which, as shown by evidence reviewed in (44), spread rapidly over deglaciated areas in Scandinavia, and possibly in Britain also, about 7000 B.C. This species, as noted earlier, is freely nodulated in Europe at the present time, and it is possible to believe that its dominance on the deglaciated terrain was due to its independence of soil nitrogen. In passing it may be noted that in Scandinavia and Britain *Hippophaë* is now largely confined to coastal regions, perhaps because only there can it now find open habitats with loose, easily penetrated soils of relatively high pH (44). Compared with the time sequence of plant succession in Alaska (above), in Europe *Alnus* does not appear to have reached its maximum development until a much longer interval after deglaciation.

Individual species show other adaptations which enable them to grow under conditions which to most plants are unfavourable.

Temperature tolerance. A sufficient immunity to low temperature to allow them to grow within the Arctic Circle is shown by *Alnus incana*, *Myrica gale*, *Hippophaë rhamnoides*, *Shepherdia canadensis* and *Dryas* species, while *Alnus glutinosa* and some other alders approach quite close to that latitude. At the other extreme are casuarinas which in Australia and Indonesia tolerate temperatures of 35° C or more.

Adaptation to extremes of pH. Myrica gale and *Alnus glutinosa* can survive at a soil pH of 4.0. In contrast *Hippophaë* flourishes on sand dunes at pH close to 8.0, while *Shepherdia canadensis* and *Elaeagnus argentea* are stated (9) to tolerate calcareous soils.

Adaptation to badly-drained soils. Myrica gale is again a good example and, as noted on p. 64, may nodulate most freely in wet situation. *Alnus glutinosa* has been said 'to like to have its feet in water', a preference which leads to Loch Lomond and other Scottish lochs being fringed with alders. Some other species of *Alnus*, e.g. *A. rugosa*, have the same preference.

Toleration of drought. Examples are provided by species of *Casuarina* and of *Colletia* which grow in areas of very low rainfall in Australia and South America respectively, while *Purshia* spp. survive in near desert conditions in the U.S.A., though they may not always form nodules (p. 65). Also some species of *Ceanothus*, including *C. cuneatus*, *C. spinosus*, *C. divaricatus* and *C. sorediatus* are typical plants of the chaparral association in the U.S.A.

Tolerance of salt spray. In Australia species of *Casuarina* grow very close to the shore of the Pacific Ocean, while in South Africa *Myrica cordifolia* forms mats close to high-tide mark. As noted already, *Hippophaë rhamnoides* commonly grows on coastal sand dunes in Europe.

Survival of forest fires. It is well-known that the seed of at least some species of *Ceanothus* is markedly heat tolerant, and that in fact there is little germination until the seed has experienced a degree of heat; thus to secure germination under experimental conditions it is customary to immerse the seed in near-boiling water for a period, followed by refrigeration. In consequence the seed germinates freely after a forest fire, and the resulting plants grow strongly in the open exposure left by the destruction of the forest canopy. *Myrica asplenifolia* has also been described as a pioneer plant on burned-over areas in the north-east U.S.A. (9).

Other information concerning habitats is as follows. A survey (45) of coniferous forests of over 70 years of age in the northern Rocky Mountains (Montana and Idaho) revealed that *Shepherdia* was the non-legume genus with a nitrogen-fixing potential most frequently encountered, particularly in forests of *Pinus flexilis* or *Picea*, though *Purshia tridentata* and *Cercocarpus ledifolius* were also of frequent occurrence in forests of *Pinus ponderosa* and *Pseudotsuga menziesii* respectively.

A further report concerning *Myrica asplenifolia* states that it is one of the few natural colonists of nitrogen-deficient waste heaps resulting from coal-mining in Pennsylvania (46); the species spreads over the substrate vegetatively. Of the nine species of *Myrica* that occur in the Republic of South Africa (47), *M. pilulifera* is of tree habit and occurs chiefly at high altitude in the eastern part of the country. Of the remaining, shrubby species *M. serrata* is the commonest and is often found on the banks of streams; the other species mostly occur in dense macchia on steep mountain slopes, though *M. cordifolia* favours coastal situations, as noted earlier in this Section.

The sole European species of *Coriaria*, *C. myrtifolia*, is very abundant in oakwoods in north-east Spain, and often forms a complete ground cover where the tree canopy is not continuous (33).

Reproduction in species bearing Alnus-type nodules

In contemplating the establishment of mixed plantings of a species which forms actinorrhizal nodules and a tree species of normal nitrogen nutrition, the risk that the former will propagate itself by seed or other means at a rate which threatens the welfare of the other partner has to be borne in mind. The information provided below will give some indication of the likelihood of that occurring, and also of measures by which the propagation might be controlled.

In connection with reproduction by seed the disposition of the male and female organs largely decides the possibility of control. This is indicated for most of the genera in Table 4, the explanation of terms used there being as follows:

Hermaphrodite – a flower which has both stamens and ovary.

Monoecious – a species which has unisexual flowers with both sexes being borne on the same plant.

Dioecious – a species which has unisexual flowers, the two sexes being borne on different plants.

It will be noted in the table that the species of certain genera differ from each other in their sexual arrangements, e.g. in the genus *Myrica*. It appears that little could be done to prevent seeding in the upper two categories in the table, though in the remaining categories it might be achieved by propagating vegetatively from plants of known sex. Certain species are, however, rather unpredictable in their behaviour; for example *Myrica gale*, though typically dioecious, occasionally occurs in monoecious form, while unisexual plants have been known to change sex from year to year (48). Some species of *Coriaria* also appear to be variable in their sexual arrangements.

Table 4. Sexual reproduction in plants bearing *Alnus*-type nodules.

Species which have hermaphrodite flowers	Species of the rosaceous genera *Ceanothus* species *Colletia* species *Elaeagnus* species Some *Coriaria* species
Species which are regularly monoecious	*Alnus* species Some species of *Casuarina* Some species of *Myrica*
Species which are regularly dioecious	*Hippophaë* species *Shepherdia* species *Datisca* species Some species of *Casuarina* Some species of *Myrica*
Variable species which can occur in both monoecious and dioecious forms	Some species of *Coriaria* Some species of *Myrica*

Another aspect which has to be considered is the age at which a species commences to bear flowers and produce seed. It is doubtful whether information on this aspect exists anywhere in print for many of these species, though it will doubtless be available locally in respect of species that are raised for horticultural purposes or for amenity planting. From his own experience the author can offer the following information relating to plants grown from seed. In a glasshouse receiving natural light only, *Myrica gale* starts to flower in its second year, but flowering did not occur so early as that in other species of *Myrica*, or in *Coriaria myrtifolia*, *Purshia tridentata*, *Hippophaë rhamnoides* or *Casuarina cunninghamiana*. Plants of *Ceanothus velutinus* and *Shepherdia canadensis* started in the glasshouse and later planted in a garden flowered at an age of 4 years. A plant of *Elaeagnus pungens* in a garden throughout is still not flowering at an age of 6 years. Elsewhere it is reported that *Alnus glutinosa* and *A. rubra* start to form catkins at an age of 3–4 years (49), while in Alaska *Dryas drummondii* flowers at an age of 5 years (31).

Some of the species also spread by vegetative means. In western Scotland *Myrica gale* appears to spread mainly by suckers, for although seed is produced liberally and germinates at a rate of about 10 percent in the glasshouse, the present author has observed singularly few undoubted seedlings in the field, possibly because the seed is taken by birds or other predators. Dr. R.J. Fessenden informed the author that in areas studied by him in Canada and the U.S.A. *M. asplenifolia* also spreads more by suckers than by seed. *Hippophaë rhamnoides* again increases by suckers, especially in loose sand. It is specifically reported (16) that *Chamaebatia foliolosa* spreads sufficiently rapidly by means of rhizomes to threaten adjacent tree seedlings. A young plant of *Dryas drummondii* produces numerous branches which hug the soil surface, forming a circular mat which in Alaska can attain a diameter of 6 m after 40 years' growth (31).

Taxonomy of the nodule endophytes

From the time (about 1865) when the first microscopic studies of non-legume nodules were undertaken, many different names have been proposed for the nodule endophytes, including *Schinzia*, *Plasmodiophora*, *Entorrhiza*, *Tetramyxa*, *Actinomyces*, *Nocardia* and *Streptomyces*. Obviously these reflected the views of the proponents on the identity of the endophytes, a matter on which argument continued until the arrival of the electron microscope. On historical and other grounds it was eventually suggested (50) that a non-committal generic name, *Frankia*, proposed in 1886 (4) in honour of a contemporary European botanist best known for his work on mycorrhiza, was the most valid appellation for the endophytes. At the same time it was proposed (50) that 10 species should be created, chiefly on the basis of symbiotic affinities; thus the endophyte of *Casuarina* nodules would be known as *F. casuarinae*, or that of *Dryas* nodules as

F. dryadis. Also *Frankia* was to be accommodated in a new family, Frankiaceae, placed in the Actinomycetales.

Now that the endophytes of a number of non-legumes have been isolated (Chapter 4) and their actinomycetal identity confirmed, it has been proposed (51) that the use of the generic name *Frankia* should be continued for the present, but that the recognition of any species should be delayed until isolations from further host genera have been made, and their structure, physiology and symbiotic properties examined.

A possible explanation of the taxonomic disparities among plants bearing Alnus-type nodules

It was concluded on p. 56 that despite the taxonomic affinities between certain of the nodulating genera, over-all the disparities are far too great for any attempt to explain the nodulating habit by a common descent to be entertained. As indicated previously (52) the only explanation for this situation that the author can suggest is that these associations between higher plants and actinomycetes arose in a period in the earth's history when for some reason (see below) the level of combined nitrogen in the soil or other substrate available for plant growth had fallen to a level sufficiently low to necessitate that at least a proportion of the then existing angiosperms should acquire an ability to utilise atmospheric nitrogen in their growth. It is suggested that the genera which bear nodules at the present time, or their direct ancestors, formed part of the flora of the postulated period and acquired the nodulating habit then.

Although the origin and early evolution of the angiosperms is wrapped in mystery, a common belief, based partly on such fossil evidence as is available, is that the earliest angiosperms began to appear in the early Cretaceous period, and were woody forms, either trees or shrubs (53); such plants were abundant by the end of that geological period. It is possible that most of them would now be classified in the series Archichlamydeae to which, as noted already, all the nodulating genera belong. No confirmed instances of nodulation have yet been found in Engler's other dicotyledonous series, namely the mainly herbaceous Sympetalae, or among monocotyledonous plants. The fossil record shows little vestige of these plants until the late Tertiary period, i.e. some millions of years later than the mainly woody types. The present author suggests that the period of extreme nitrogen deficiency may have ended before the Sympetalae evolved, and that such a period has not occurred again.

In seeking to explain why a period of nitrogen deficiency should have arisen it may be noted that the evolution of woody angiosperms during the Cretaceous period was accompanied by a marked decline in the distribution of the cycads; provided that they already had the association with cyanobacteria (p. 77) this decline would certainly have resulted in a considerable reduction in the input of newly-fixed nitrogen to the biosphere.

A further possibility arises from recent geochemical evidence (54) which led to the suggestion that some 65 million years ago, i.e. in the late Cretaceous period, an asteroid collided with the earth, with the result that a vast quantity of rock debris and dust was thrown into the atmosphere. The proposers' calculations indicated that this would be sufficient in quantity to obscure the sun and thus turn day into night for a period of several years; all photosynthesis and growth directly dependent on it would cease, with catastrophic effects on the dinosaurs. The present author has considered the possible effects of these events, if they actually occurred, on biosphere nitrogen. Obviously all biological fixation of nitrogen directly dependent on photosynthesis would be halted during the dark period, including fixation in free-living and symbiotic cyanobacteria, together with phyllosphere and rhizosphere fixation. Thus the nitrogen cycle would be out of balance, with the input of newly-fixed nitrogen greatly reduced, though the likely duration of this situation would seem to have been too short for any long-lasting effects to develop. However, serious effects on the availability of nitrogen may have arisen as the debris and dust thrown up by the collision settled. Although the proposers' calculations appear to indicate a mean deposition of only 53 mg of dust per cm^2 of the earth's surface, a subsequent redistribution through the movement of surface water may have resulted in the formation of deep deposits of infertile material, deficient in nitrogen, in lower-lying areas. Perhaps it was to facilitate plant growth in this substrate that the evolution of *Alnus*-type nodules occurred.

NON-LEGUMES THAT BEAR RHIZOBIAL NODULES

Until recently it was generally believed that root nodules inhabited by the bacterum *Rhizobium* were confined to the Leguminosae. This belief had to be abandoned rather hurriedly some ten years ago as a result of the study (55) of nodule which had been detected on the roots of saplings growing as weeds in tea plantations in New Guinea. The saplings were initially identified as a species of *Trema*, a member of the Ulmaceae, but were later re-allocated to another ulmaceous genus, *Parasponia*, the actual species being *P. rugosa* Bl. (56). The nodules, further described in (57), show some resemblance to *Alnus*-type nodules in their external and internal structure, but the micro-organism within the infected cells of the nodules proved to be *Rhizobium*, though its condition differed from that in a legume nodule in that in the majority of infected cells the rhizobia remain enclosed within infection threads. The organism was easily isolated and proved to be able to cause nodulation in some legumes which are normally nodulated by cowpea-type rhizobia. The *Parasponia* nodules were shown to be nitrogen-fixing.

Recently it was stated (58) that *Parasponia andersonii* Planch. growing in the Solomon Islands is also nodulated, and that an old report of the presence of nodules on *P. parviflora* Miq. in Indonesia had been discovered (59); re-inspection of the trees confirmed this, while cytological examination of the nodules (60) showed the endophyte to be again *Rhizobium*. *Parasponia* contains six species distributed in Malaysia and Polynesia.

Actually, a long time before *Parasponia* gained the limelight it had been claimed that members of the angiospermous, dicotyledonous family Zygophyllaceae also bear rhizobial nodules. The family, which falls into the Archichlamydeae, comprises 25 genera and some 240 species occurring chiefly in tropical and sub-tropical regions, and consists mostly of woody, perennial xerophytes and halophytes. The largest genus is *Zygophyllum* with about 100 species inhabiting deserts and steppes from the Mediterranean to central Asia, also southern Africa and Australia. Species of the smaller genus *Tribulus* occur in the southern U.S.A. in such situations as on sand dunes and along railway tracks.

The presence of nodules on the roots of a member of this family appears to have been first noted in 1913 (61) in respect of *Tribulus terrestris* growing in Russia, while the first study of their significance was made at the then Fouad I University, Cairo (62). Nodules had been observed on several species of *Zygophyllum* and also on *Tribulus alatus* and *Fagonia arabica*, all of which grow in the poor sandy soils of Egyptian deserts and also fall within Engler's sub-family Zygophylloïdeae. A photograph of the external appearance of the nodules was provided, and it was stated that the nodule cells contain bodies interpreted as bacteroids resembling those of legume nodules, but no corrobative illustration was included, neither was the general structure of the nodule described. An isolate from the nodules of *Z. coccineum* was obtained on soil-extract mannitol agar, and this again resembled a rhizobium. Pot cultures were set up, using surface-sterilised seed of *Z. coccineum* and sterilised desert soil, half of the pots being inoculated with the isolate. With the latter treatment the plants nodulated and grew healthily, while those in the uninoculated pots did neither of those things, results which suggest that the isolate was indeed the nodule endophyte and that fixation of nitrogen was proceeding. Unexpectedly the same result was obtained when a rich Nile silt was used as the rooting medium, leading the investigator to the conclusion that these plants can grow properly only when they bear nodules, even in a fertile soil.

Five years later a continuation of the above study in the same university though by different workers was reported (63). Further isolations from species of *Zygophyllum*, *Tribulus* and *Fagonia* had now been made, and cross-inoculation trials were set up in pots to find whether the three isolates differed in symbiotic properties. It was found that only when a species was inoculated with what was believed to be its normal endophyte did the plants nodulate and grow well; a photograph of this result was provided, though the nature of the rooting medium was not stated. Next the response of five leguminous species to inoculation with the three isolates was tested. *Arachis hypogaea* nodulated with all three isolates, with the plants growing even better than with their normal endophyte, while *Trifolium alexandrinum* nodulated with two of the isolates and *Phaseolus vulgaris* with one. *Vicia faba* and *Lupinus perennis* failed to respond to any of the isolates.

In a further article (64) work done at the University of Karachi was briefly described. Photographs of the root systems of local material of *Tribulus terrestris*, *Fagonia cretica* and *Zygophyllum simplex* were included and indicate that the nodules are not very conspicuous structures. Isolates were obtained on nutrient agar, yeast-extract mannitol agar and a Congo Red medium from surface-sterilised nodules of the first and last of the above species, and on the basis of cultural characteristics and microscopic examination the isolates were considered to be rhizobia, but no tests of the ability of the isolates to induce nodulation in plants of the *Tribulus* and *Zygophyllum* species were reported. More recently the same authors (65) stated that the general structure of the *Tribulus* nodule is similar to that of *Alnus*, but that the endophyte seen in nodule sections appears to be of rhizobial type. (The present author has received from Mr. Athar a photomicrograph of cells of the *Tribulus* nodule at a magnification of × 800 which, though not of a quality suitable for publication, suggests that the host cytoplasm is packed with structures resembling branched rhizobial bacteroids.) In culture the presumed endophyte of the *Tribulus* nodule resembles a slow-growing strain of *Rh. japonicum*, and trials showed its ability to induce effective nodulation in species of *Vigna*, *Sesbania* and *Cyamopsis*, though the much-needed test of the isolate's ability to induce nodulation in *Tribulus* has been delayed until a method of breaking the dormancy of the seed has been found. The authors are unable to carry out acetylene assays on field nodules through lack of equipment.

Attention will now be turned to a report (66) emanating from the University of Wisconsin by authors whose interest had been aroused by the first report from Cairo (62, cited above). They worked with *Tribulus cistoides*, and starting with seed obtained from Haiti they raised plants in a glasshouse in ordinary potting soil. Nodular structures showing some superficial resemblance to those of legumes formed on the roots, but despite repeated efforts no organism capable of inducing nodulation in test plants of *T. cistoides* could be obtained from surface sterilised nodules, and in most instances the agar plates remained free of any growth. A histological study of the nodules was then made, using sections prepared by a paraffin-wax technique, and this revealed a central vascular strand which sometimes forked once or twice in the apical region of the nodule, and also showed that the swollen nature of the nodule was due to extensive cell division in the pericycle, the deep layer so formed being bounded by an endodermis, a narrow cortex and a periderm. No meristem was present in the nodule.

The cells of the pericycle all contained abundant starch grains, together with other inclusions which at first took the form of minute spheres but in older cells coalesced to form large structureless bodies which were sometimes of spherical shape, 10–15 μ in diameter, at other times irregular in form. The chemical nature of this material could not be established, but because the bodies could not be found in sections of fresh nodules mounted in water, or in aqueous smear preparations, it was concluded that they were artefacts resulting from the precipitation during fixation of some originally dissolved cell constituent. No trace of any endophytic organism was detected in the nodule cells, leading the authors to conclude that no micro-organism is involved in the formation of these nodules or protuberances, that they serve for starch storage and are normal features of the growth of the plants.

The present author offers the following comments on the above studies.

Though their reports were brief and lack important experimental detail, the findings obtained by the workers at Cairo and Karachi, who had access to field material, are mostly mutually consistent and do raise the possibility that zygophyllaceous nodules are of rhizobial type. However, before that can be accepted it is essential that a thorough description, adequately illustrated, of the internal structure and cytology of the nodules should be provided, together with more conclusive evidence of the occurrence of nitrogen fixation in the nodules. The latter could be sought most simply by means of long-term growth trials coupled with nitrogen analyses of the harvested nodulated plants and non-nodulated control plants.

In the Wisconsin study the origin of the soil in which *Tribulus* plants were grown is not clearly stated, but it seems safe to assume that it had not been collected from a field site where nodule-bearing *Tribulus* plants were growing. Thus the investigators' inability to find an endophyte in the nodules may have resulted from its absence in the soil used. Their evidence that nodules can form without any assistance from an endophyte indicates a situation quite different from that obtaining in the development of legume or actinorrhizal nodules, though it may be recalled that podocarp 'nodules' also develop without any stimulus from an endophyte, but are subsequently invaded by a mycorrhizal fungus. Perhaps a similar sequence of events occurs in the Zygophyllaceae, though with a different endophyte, while the arrival of the latter may stimulate the nodule to develop beyond the possibly rudimentary stage studied in Wisconsin. It is also relevant to recall that the particular cells of alder nodules that eventually are invaded by the endophyte are initially filled with starch. The Cairo workers' statements that their uninoculated plants remained free of nodules appears to be in direct conflict with the Wisconsin experience, though it is possible that nodules were present but owing to nitrogen deficiency resulting from the poor sandy soil used they were of barely detectable size. In Wisconsin a fertile soil was almost certainly employed. An alternative explanation of these discrepancies may lie hidden in the statement by the original discoverer (61) of nodulation in *Tribulus* that the nodules are of two distinct sizes.

The important question of the significance of the nodules formed by members of this quite large and widely distributed family will perhaps be answered only through a degree of international co-operation such as has been successful recently in the study of nodules in some other plants.

SYSTEMS THAT INVOLVE NITROGEN-FIXING CYANOBACTERIA

The Gunnera–Nostoc association

The angiospermous, dicotyledonous genus *Gunnera* comprises some 50 species occurring in tropical and southern temperate regions, including Africa, Central America, Tasmania, New Zealand and elsewhere. They are perennial rhizomatous plants which favour damp habitats, while their leaves, which are all radical, measure in some species as much as 3 m across. Some taxonomists place the genus in the family Haloragidaceae along with such genera as *Hippuris* and *Myriophyllum*, while others isolate the genus into a family Gunneraceae. In either case the family falls into the Archichlamydeae.

Some, if not all, species of *Gunnera* display a localised infection of the stem by a cyanobacterium which has been isolated and considered to be *Nostoc punctiforme*; recently it was reported that all 10 species that occur in New Zealand show the infection (67). As will have been indicated in Chapter 2, the micro-organism proliferates intra-cellularly in mucilaginous glandular areas in the outer tissues of the young rhizome, the infected pockets of tissue being linked by vascular traces to the conducting system of the plant.

Excised glandular tissue shows a vigorous reduction of acetylene (67, 68), while young plants of *G. dentata* displayed a marked ability to grow and accumulate nitrogen in a rooting medium free of combined nitrogen. After exposure for 9 hr of whole young plants of the same species to an atmosphere labelled with ^{15}N, assays showed a strong labelling in the glandular tissue and smaller but significant labelling in the leaves, stems and roots (69).

Cycad nodules

The cycads, which are thermophilic plants resembling palms or tree ferns in their growth habit, constitute the class Cycadales which comprises three families, namely the Cycadaceae, Zamiaceae and Stangeriaceae. The first includes the one genus, *Cycas*, of which there are some 20 species; these are found in Madagascar, Indomalaysia, northern Australia and Indonesia, and they include most of the taller-growing cycads. The Zamiaceae is the largest family, comprising 8 genera and 80 species, the genera and their main places of occurrence being as follows; *Zamia* (northern South America), *Macrozamia* (south-east Australia), *Ceratozamia* (Mexico), *Lepidozamia* (east Australia), *Encephalartos* (southern half of the African continent), *Dioon* (Florida), *Bowenia* (north-east Australia) and *Microcycas* (Caribbean islands). The third family includes the one genus *Stangeria*, found in south-east Africa. The cycads were of much wider occurrence in former, warmer eras, particularly about the end of the Triassic and the beinning of the Jurassic, i.e. about 150 million years ago.

Many if not all cycad species are known to form, on the upper parts of their root systems, modified tuberous roots which tend to grow in an upwards direction and branch repeatedly to form coral-like clusters often several cm in diameter and resembling superficially *Alnus*-type nodules. The lobes of these clusters are typically inhabited by a cyanobacterium variously identified in different material as *Nostoc* or *Anabaena*. As will have been indicated in Chapter 2, the micro-organism proliferates in the air spaces between radially elongated cells of what appears to be the middle cortex of the lobe. The route of entry of the organism into the nodule lobe remains uncertain (70).

Earlier reports (71) suggested that the cyanobacterium isolated from cycad nodules fixed nitrogen in culture, while later ^{15}N tests (71, 72, 73) confirmed the occurrence of a vigorous fixation in the nodules, and demonstrated that the fixed nitrogen is translocated to all parts of the cycad plant. There is also evidence suggesting that the fixation represents a significant contribution to the welfare of the relatively few ecosystems of which cycads form part. It can be speculated that in former eras when, as noted, cycads were widespread, the fixation associated with them was of major importance for the biosphere: that, however, presupposes an already-existing association with the cyanobacterium, an aspect on which the examination of fossil material is unlikely to throw light owing to the fleshy, perishable nature of the nodules.

The Azolla–Anabaena association

The genus *Azolla*, formerly assigned to the Salviniaceae but now often separated into its own family, Azollaceae, comprises six species of small aquatic ferns found chiefly in tropical and sub-tropical regions at the present time, though the fossil record shows that in former warmer eras it extended to many other parts of the world. The plants spread extremely rapidly in slowly moving or stagnant water of ditches, shallow pools, swamps and flooded rice-fields. The branched floating stem bears alternate overlapping leaves, each with a stouter upper lobe which stands up from the water, and as will have been described in Chapter 2, during the development of this upper lobe an internal cavity, at first communicating with the exterior through a pore, forms in the tissues near the base of the lobe. This cavity is invaded by a species of *Anabaena*, which also proliferates around the growing point of the stem of the fern, this being the presumed source of leaf infection. The cyanobacterium is additionally present in the megasporocarps of the fern, providing for the infection of the succeeding generation of plants (74).

The evidence for nitrogen fixation in *Azolla* associated with *Anabaena* is provided by the demonstrated ability of the plants to grow and accumulate nitrogen very rapidly in a culture solution free of combined nitrogen (75, 76), and also by the positive results of acetylene assays (74). These findings explain the beneficial effect on yield accruing from the long-standing custom in parts of south-east Asia of introducing *Azolla*, especially *A. pinnata* in its several varieties, into rice fields. The practical details of this operation have been described recently (77).

The Liverwort–Nostoc systems

Of the three liverwort genera to be considered here, *Blasia* and *Cavicularia* are closely related to each other but are remote from the third genus, *Anthoceros*. *Blasia* is usually considered to include only one species, *B. pusilla*, which occurs in Europe, North America, Australia and elsewhere, while *Cavicularia* also is represented by a single species, *C. densa*, occurring in Japan. *Anthoceros* is of world-wide distribution and includes over 50 species, thoug certain of these are allocated by some taxonomists to a separate genus, *Phaeoceros*.

In *Blasia* and *Cavicularia* and at least some of the species of *Anthoceros*, cavities on the lower side of the thallus, possibly formed by invagination, normally contain colonies of *Nostoc* immersed in mucilage (Chapter 2). In all three genera the *Nostoc* is considered to be *N. sphaericum*, and is readily isolated from the thalli and cultured. Trials (78) have shown that other genera of cyanobacteria fail to establish in the cavities of initially uninfected *Blasia*, and that as regards *Nostoc* a rather confused situation exists, since not only did isolates from *Blasia* itself and from *Anthoceros* colonise the cavities, but also an isolate of *N. punctiforme* from *Gunnera*, while cultured *Nostoc* from a cycad nodule and from a lichen did not.

Tests with ^{15}N have shown that nitrogen fixation occurs in thalli of *Blasia* and of *Anthoceros punctatus* containing *Nostoc* (78, 79). *Blasia* infected by *Nostoc* is able to grow in a nitrogen-free medium, and further evidence of transfer of fixed nitrogen from the *Nostoc* to the liverwort was obtained by exposing *Anthoceros* thalli bearing sporophytes to ^{15}N; although the sporophytes are *Nostoc*-free, assay showed them to be clearly labelled with ^{15}N (78).

There is evidence that associations with cyanobacteria also occur in the other section of the Bryophyta phylum, i.e. the mosses. The frequent occurrence of nitrogen-fixing cyanobacteria within the hyaline cells of the leaves of *Sphagnum* has been noted (78), leading to the suggestion that this association is a significant source of fixed nitrogen in bog ecosystems.

Lichen associations

The lichens present an association between a heterotrophic organism – namely a fungus, usually of ascomycete type – and a photosynthetic organism in the shape of a green alga or, less commonly, a cyanobacterium, while in a few lichens both of the latter are present. As will have been explained in Chapter 2, within the lichen thallus these component organisms are arranged in different fashion in different lichens, with the fungus mainly deciding the morphology of the thallus. Tremendous variety is shown in the latter respect, so that some 500 genera and about 18000 species of lichens have been distinguished. Lichens are almost ubiquitous in the land areas of the world, but are especially prominent in the tundras of the polar regions and as colonisers of newly-exposed rock surfaces. The propagation of lichens is chiefly by means of vegetative structures such as soredia which contain both of the constituent organisms.

Lichens with cyanobacteria as constituents account for about 20 percent of lichen genera, though only for some 7 percent on the species basis; those analyses were made with much help from Dr. G.D. Scott on the basis of Zahlbruckner's world list of lichens. The cyanobacterium most commonly involved is *Nostoc*, and it is present, for example, in species of the lichen genera *Collema*, *Lobaria*, *Peltigera*, *Leptogium* and *Stereocaulon*. Other cyanobacteria concerned in lichens include *Calothrix*, *Scytonema* and *Stigonema*, and since these organism fix nitrogen in the free-living state (p. 82) it is an obvious possibility that they will be similarly active in the lichen thallus. That evidence for this is obtainable by a simple technique was shown by the finding (80) that discs of *Peltigera* thallus placed on filter paper moistened with nitrogen-free culture solution doubled their original dry weight and nitrogen content over a period of 14 weeks. By means of the quicker ^{15}N and acetylene methods the occurrence of fixation has been demonstrated in a large number of lichens containing cyanobacteria, while there is also evidence that a large part of the fixed nitrogen is appropriated by the fungal partner (81).

Of particular interest to forestry is a recent study (82) of lichens growing as epiphytes on the branches of Douglas firs in forests in the Cascade Mountains, Oregon. Some 10 lichens containing cyanobacteria occur on the trees, including species of *Lobaria*, *Peltigera*, *Sticta* and *Pseudocyphellaria*, the most abundant being *Lobaria oregana*, of which 10–15 kg dry weight was harvested per tree. From acetylene assay and other data it was calculated that fixation by this lichen amounted to 3–4 kg N/ha/yr, the transfer of this nitrogen to the forest floor being effected by the eventual fall of lichens, which occurs especially during storms, and perhaps also by leaching from living and dead lichens still in position on the trees. Expressed on the same basis the fixation achieved by the lichen *Stereocaulan paschale* growing on the floor of a forest in northern Sweden was, expectedly, much smaller since the lichen covered only 14 percent of the floor (83).

Fixation by the lichens in the sub-arctic tundra of northern Europe is believed to be a significant source of nitrogen to the ecosystem (84). Information – additional to that presented in Chapter 2 – on the physiology of these systems involving cyanobacteria is to be found in references 67, 78, 82 and 85.

THE PHYLLOSPHERE SYSTEM

The concept that the leaf surface could be the site of nitrogen fixation was first proposed about 25 years ago by an investigator who has continued her advocacy of it and whose periodic reviews (86–90) provide a source of information on the topic. The discovery that species of *Beijerinckia* and *Azotobacter* were present in abundance on the usually wet surface of the leaves of a wide range of trees and shrubs in the wet tropics of Indonesia, and the knowledge that the dew and run-off from the leaves contained not only appreciable amounts of sugars, organic acids, minerals – resulting from leaching of the leaves – but also nitrogenous substances, provided the initial basis for the concept. The term 'phyllosphere' was adopted for the environment provided by the wet leaf surface, and it was suggested that the vegetation would benefit by foliar uptake of fixed nitrogen and by root uptake of fixation products carried down in leaf run-off.

Fuller examination of the microflora of the leaf surface has shown that other nitrogen-fixing bacteria are often present in addition to those named above, and that the total population of such bacteria may attain several million per sq. cm. of leaf surface. Also commonly present in Indonesia and under comparable conditions elsewhere are non-fixing organisms including other bacteria, yeasts and other fungi, algae and even lichens, mosses and liverworts, so that the leaf is visibly encrusted. In dorsiventral leaves a microflora is present on both surfaces, but is denser on the adaxial surface. It is believed that phyllosphere fixation is not confined to the tropics but can occur in wet situations elsewhere; it can also occur within the leaf sheath space in grasses and on the surface of mosses and liverworts. Actual evidence for the fixation includes ^{15}N and acetylene data obtained with detached leaves, shoot portions and small whole plants, though in some instances the fixation reported has not been proved to have the leaf surface as its site.

Despite the thick cuticle of most coniferous leaves, tests for phyllosphere fixation in plantations of Douglas fir in England have given positive results (91), though none could be found in Douglas fir in the western U.S.A. (83).

Phyllosphere fixation has obvious parallels with that in the rhizosphere (see next Section), though there is the difference that the former occurs on the surface of an organ which in daytime is steadily evolving oxygen.

Although it is generally accepted that in many soils the activity of free-living nitrogen-fixing bacteria is restricted by low levels of organic matter, the idea has long been entertained that in the immediate vicinity of plant roots the decay of the outer tissues and a possible leakage of carbohydrates from those root surfaces not enclosed by periderm or suberised exodermis, might provide a favourable milieu for such bacteria, leading perhaps to a gain of fixed nitrogen by the plant. Instances of a maintenance of yield which could not be explained otherwise, e.g. in areas used repeatedly for cereal cultivation without nitrogenous fertilisation, were often mentioned in this connection.

Over the last 15 years or so, interest in the possibility of rhizosphere fixation has received a fresh impetus as the result of reports emanating initially from Brazil, based chiefly on the use of the acetylene assay in the examination of root systems for nitrogenase activity. In the earlier of these (92, 93) *Azotobacter* and *Beijerinckia* were described as abundantly associated respectively with the roots of tropical pasture grasses such as *Paspalum notatum* and those of sugar cane. Rapid acetylene reduction was shown by *Paspalum* roots brought in from the field and pre-incubated for some hours under low oxygen tension, and calculation from the data so obtained, and from other observations, even indicated that rhizosphere fixation could satisfy most of the plant's requirement for nitrogen. The rhizosphere bacteria appeared to be firmly attached to the root surface, since vigorous washing did not affect activity. It was suggested that favourable soil temperatures, together with freer exudation of carbohydrates consequent on the high light intensity and the fact that many tropical members of the grass family have the C_4 photosynthetic pathway, contributed to these findings.

In later reports from Brazil (94, 95) it was indicated that in the tropics the bacterium *Spirillum lipoferum* is of still greater importance in rhizosphere associations than the bacteria mentioned above. It is of widespread occurrence in soils, and particularly abundant in the immediate vicinity of the roots of several forage grasses and crops, including sorghum and *Zea mays*. Infected roots again showed marked acetylene reduction, but this bacterium appeared to be located chiefly in the air-spaces of the root cortex. It was suggested that by seed inoculation and the use of crop varieties known to be able to establish effective rhizosphere associations, tropical agriculture could be benefited.

In the light of the above observations many other workers have turned their attention to rhizosphere studies. Inoculation with *Spirillum* in glasshouse and field trials has improved yield in some instances (96) but not in others (97), and while some investigators have detected fixation in natural rhizosphere associations, the levels have been smaller than those reported from Brazil. Wet soil conditions appear to favour activity (98, 99) apparently because of the degree of anaerobiosis which results. Other investigations have demonstrated active rhizosphere fixation in marine angiosperms including *Thalassia*, *Syringodium* and *Zostera* (100) and also in those of aquatic type such as *Hydrocotyle*, *Typha*, *Cyperus* and *Eichornia* (101); in many of these the roots are exposed to relatively anaerobic conditions. Several investigators (97, 99, 102) have warned that pre-incubation of root samples intended for assay can lead to artificially high activity. Also, since practically all workers have used the acetylene assay, the extent to which the plant itself gains nitrogen fixed in its rhizosphere is often uncertain.

In the particular case of forests the opportunities for rhizosphere fixation appear to be restricted, since the surfaces of the younger parts of the root system are commonly pre-empted by the fungal partners of ectotrophic mycorrhizas.

FREE-LIVING NITROGEN-FIXING MICRO-ORGANISMS

By free-living will be understood a micro-organism which is not living in close physical association with another organism and thereby possibly or certainly benefitting metabolically.

The identity and taxonomic position of the micro-organisms to be considered here, namely certain bacteria and cyanobacteria, have been indicated already in Chapter 2. Here it will suffice to note, firstly, a further study (103) of the association of fixation by a species of *Enterobacter* with wet rotting wood in the forest ecosystem, and secondly recent evidence (104) obtained by acetylene and ^{15}N assays that the leafy litter of forests is the site of significant fixation, both in kauri pine forests in New Zealand and Douglas fir forests in north-west U.S.A. A bacterium, at present unidentified, is presumably responsible for the fixation, and it seems possible that the long-appreciated value of leaf-mould in horticulture is partly due to the occurrence of fixation of nitrogen during decay. Rotting wood and leaf litter usually show high C/N ratios and are thus substrates particularly favourable to nitrogen-fixing organisms as compared with those that are dependent on combined nitrogen. Lastly it may be noted that well-preserved fossil cyanobacteria have been found abundantly in Middle Precambrian strata (105) leading to speculation that they then contributed importantly to the plant cover over the earth's land surface.

REFERENCES

1. Hiltner L: Über die Bedeutung der Wurzelknöllichen von *Alnus glutinosa* für die Stickstoffernährung dieser Pflanze, *Landw Versuchst* 46: 153–161, 1896.
2. Hiltner L: Die Wurzelknöllchen der Erlen und Elaeagnaceen, *Forst Naturw Z* 7: 415–423, 1898.

3. Nobbe F, Hiltner L: Ueber das Stickstoffsammlungsvermögen der Erlen und Elaeagnaceen, *Naturw Z Land Forstw* 2: 366–369, 1904.
4. Brunchorst J: Über einige Wurzelanschwellungen, besonders diejenigen von *Alnus* und den Elaeagnaceen, *Unters Bot Inst Tübingen* 2: 150–177, 1886–8.
5. Shibata K: Cytologische Studien über die endotrophen Mykorrhizen. *J wiss Bot* 37: 643–684, 1902.
6. Shibata K, Tahara M: Studien über die Wurzelknöllchen, *Bot Mag Tokyo* 31: 157–182, 1917.
7. Fred EB, Baldwin IL, McCoy E: *Root nodule bacteria and leguminous plants*, Univ. Wisconsin, Madison, U.S.A. 1932.
8. Schaede R: *Die Pflanzlichen Symbiosen*, 3rd ed. Stuttgart, Fischer, 1962.
9. Van Dersal WR: *Native woody plants of the United States*, U.S. Government Printing Office, Washington, 1938.
10. Lawrence DB: Development of vegetation and soil in south-eastern Alaska with special reference to the accumulation of nitrogen. Final Report, Off Nav Res Project NR 160–183, 1953.
11. Bond G: The results of the IBP survey of root-nodule formation in non-leguminous angiosperms. In: *Symbiotic nitrogen fixation in plants*, Nutman PS (ed), International Biological Programme 7, Cambridge, Cambridge University Press, 1976, 443–474.
12. Bond G, Wheeler CT: Non-legume nodule systems. In: *Methods for evaluating biological nitrogen fixation*, Bergersen FJ (ed), Chichester, John Wiley, 1980, 185–211.
13. Melchior H (ed): *A. Engler's Syllabus der Pflanzenfamilien*, Berlin, Gebrüder Borntraeger, 12th ed, 1964.
14. Willis JC: *Dictionary of the flowering plants and ferns*, revised by Airy Shaw HK, London, Cambridge University Press, 8th ed, 1973.
15. Good R D'O: The geography of the genus *Coriaria*, *New Phytol*, 29: 170–190, 1930.
16. Heisey RM, Delwiche CC, Virginia RA, Wrona AF, Bryan BA: A new nitrogen-fixing non-legume: *Chamaebatia foliolosa* (Rosaceae), *Am J Bot* 67: 429–431, 1980.
17. Righetti TL, Munns DN: Nodulation and nitrogen fixation in cliffrose (*Cowania mexicana* var. *stansburiana* (Torr.) Jeps.), *Plant Physiol* 65: 411–412, 1980.
18. Zandee M, Kalkman C: The genus *Rubus* (Rosaceae) in Malesia. 1. Subgenera *Chamaebatus* and *Idaeobatus*, *Blumea* 27: 75–113, 1981.
19. Medan D, Tortosa RD: Nodulos actinomicorricicos en especies argentinas de los generos *Kentrothamnus*, *Trevoa* (Rhamnaceae) y *Coriaria* (Coriariaceae), *Bol Soc Argentina Bot* 20: 71–81, 1981.
20. Focke WO: Species Ruborem, Parts I and II, *Bibliotheca Botan* 17: 1–223, 1910. Part III, *Ibid* 19: 1–274, 1913.
21. Becking JH: Nitrogen fixation by *Rubus ellipticus* J.E. Smith, *Plant Soil* 53: 541–545, 1979.
22. Bond G: Root-nodule symbioses with actinomycete-like organisms. In: *The biology of nitrogen fixation*, Quispel A (ed), Amsterdam, North-Holland Pub. Co, 1974, 342–378.
23. Bond G: Observations on the root nodules of *Purshia tridentata*, *Proc Roy Soc Lond* B 193: 127–135, 1976.
24. Bond G, Becking JH: Root nodules in the genus *Colletia*, *New Phytol* 90: 57–63, 1982.
25. Bond G: Nitrogen fixation in some non-legume root nodules, *Phyton (Buenos Aires)* 24: 1967, 57–66.
26. Mian S, Bond G, Rodríguez-Barrueco C: Effective and ineffective root nodules in *Myrica faya*, *Proc Roy Soc Lond* B 194, 285–293, 1976.
27. Dalton DA, Zobel DB: Ecological aspects of nitrogen fixation by *Purshia tridentata*, *Plant Soil* 48: 57–80, 1977.
28. Pommer E-H: Beitrage zur Anatomie und Biologie der Wurzelknöllchen von *Alnus glutinosa* Gaertn., *Flora* 143: 603–634, 1956.
29. Mowry H: Symbiotic nitrogen fixation in the genus *Casuarina*, *Soil Sci* 36: 409–425, 1933.
30. Raghavan, MS: *Casuarina* plantation technique in the Madras Province, *Indian Forester* 73: 241–260, 1947.
31. Lawrence DB, Schoenike RE, Quispel A, Bond G: The role of *Dryas drummondii* in vegetation development following ice recession at Glacier Bay, Alaska, with special reference to its nitrogen fixation by root nodules, *J Ecol* 55: 793–813, 1967.

32. Wagle RF, Vlamis J: Nutrient deficiencies in two Bitterbrush soils, *Ecology* 42: 745–752, 1961.
33. Bond G, Montserrat P: Root nodules of *Coriaria*, *Nature*, *London* 182: 474–475, 1958.
34. Morrison TM, Harris GP: Root nodules in *Discaria toumatou* Raoul Choix, *Nature*, *London* 182: 1746–1747, 1958.
35. Medan D, Tortosa RD: Nodulos radicales en *Discaria* y *Colletia* (Ramnaceas), *Bol Soc Argentina Bot* 17: 323–336, 1976.
36. Youngberg CT, Wollum AG: Nitrogen accretion in developing *Ceanothus velutinus* stands, *J Soil Sci Soc America* 40: 109–112, 1976.
37. Zavitkovski J, Newton M: Ecological importance of snowbrush, *Ceanothus velutinus*, in the Oregon Cascades, *Ecology* 49: 1134–1145, 1968.
38. Moore, AW: Note on non-leguminous nitrogen-fixing plants in Alberta, *Canad J Bot* 42: 952–955, 1964.
39. Chaudhary AH: Nitrogen-fixing root nodules in *Datisca cannabina* L., *Plant Soil* 51: 163, 1979.
40. Severini G: Sui tubercoli radicali di *Datisca cannabina*, Ann Bot (Rome) 15: 29–51, 1922.
41. Winship LJ, Chaudhary AH: Nitrogen fixation by *Datisca glomerata*: a new addition to the list of actinorhizal diazotrophic plants. In: *Symbiotic nitrogen fixation in the management of temperate forests*, Gordon JC, Wheeler CT, Perry DA (eds), Corvallis, Oregon, Forest Research Laboratory, Oregon State University, 1979, 485.
42. 'Cryptos' (Grimmett RER): The much-maligned Tutu, *Forest & Bird*, August 1944, 4–5.
43. Crocker RL, Major J: Soil development in relation to vegetation and surface age at Glacier Bay, Alaska, *J Ecol* 43: 427–448, 1955.
44. Godwin H: *The history of the British flora*, London, Cambridge University Press, 1956.
45. Jurgensen MF, Arno SF, Harvey AE, Larsen MJ, Pfister RD: Symbiotic and nonsymbiotic nitrogen fixation in Northern Rocky Mountain forest ecosystems. In: *Symbiotic nitrogen fixation in the management of temperate forests*, Gordon JC, Wheeler CT, Perry DA (eds), Corvallis, Oregon, Forest Research Laboratory, Oregon State University, 1979, 294–308.
46. Schramm JR: Plant colonization studies on black wastes from anthracite mining in Pennsylvania, *Trans American Phil Soc* NS 56: 1–194, 1966.
47. Grobbelaar N, Strauss JM, Groenewald EG: Non-leguminous seed plants in Southern Africa which fix nitrogen symbiotically, *Plant Soil* Special Volume: 325–334, 1971.
48. Clapham AR, Tutin TG, Warburg EF: *Flora of the British Isles*. Cambridge, Cambridge University Press, 1952.
49. Hall RB, Maynard CA: Considerations in the genetic improvement of alder. In: *Symbiotic nitrogen fixation in the management of temperate forests*, Gordon JC, Wheeler CT, Perry DA (eds), Corvallis, Oregon, Forest Research Laboratory, Oregon State University, 1979, 322–344.
50. Becking JH: Frankiaceae fam. nov. (Actinomycetales) with one new combination and six new species of the genus *Frankia* Brunchorst, *Internat. J Syst Bact* 20: 201–220, 1970.
51. Lechevalier MP, Lechevalier HA: The taxonomic position of the actinomycetic endophytes. In: *Symbiotic nitrogen fixation in the management of temperate forests*, Gordon JC, Wheeler CT, Perry DA (eds), Corvallis, Oregon, Forest Research Laboratory, Oregon State University, 1979, 111–122.
52. Bond G: The root nodules of non-leguminous angiosperms. In: *Symbiotic associations*, Symposia Soc Gen Microbiol 13, Nutman PS, Mosse B (eds), London, Cambridge University Press, 1963, 72–91.
53. Good R: *The geography of the flowering plants*, London, Longman, 4th ed. 1974.
54. Alvares LW, Alvares W, Asaro F, Michel HV: Extraterrestrial cause for the Cretaceous–Tertiary extinction, *Science* 208: 1095–1108, 1980.
55. Trinick MJ: Symbiosis between *Rhizobium* and the non-legume, *Trema aspera*, *Nature*, *London* 244: 459–460, 1973.
56. Akkermans ADL, Abdulkadir S, Trinick MJ: Nitrogen-fixing nodules in Ulmaceae, *Nature*, *London* 274: 190, 1978.
57. Trinick MJ, Galbraith J: Structure of root nodules formed by *Rhizobium* on the non-legume *Trema cannabina* var. *scabra*, *Arch Microbiol* 108: 159–166, 1976.

58. Trinick MJ: Growth of *Parasponia* in agar tube culture and symbiotic effectiveness of isolates from *Parasponia* spp., *New Phytol* 85: 37–45, 1980.
59. Akkermans ADL, Abdulkadir S, Trinick MJ: N_2-fixing root nodules in Ulmaceae: *Parasponia* or (and) Trema spp.? *Plant Soil* 49: 711–715, 1978.
60. Becking JH: Root-nodule symbiosis betwen *Rhizobium* and *Parasponia* (Ulmaceae), *Plant Soil* 51: 289–296, 1979.
61. Issatschenko BL: Über die Wurzelknöllchen bei *Tribulus terrestris* L., *Bull Jard bot St Petersburgh* 13: 23–31, 1913.
62. Sabet YS: Bacterial nodules in the Zygophyllaceae, *Nature, London* 157: 656–657, 1946.
63. Mostafa MA, Mahmoud MZ: Bacterial isolates from root nodules of Zygophyllaceae, *Nature, London* 167: 446–7, 1951.
64. Athar M, Mahmood A: Root nodules in some members of Zygophyllaceae growing at Karachi university campus, *Pak J Bot* 4: 209–210, 1972.
65. Athar M, Mamood A: Extension of *Rhizobium* host range to Zygophyllaceae. In: *Current perspectives in nitrogen fixation*, Gibson AH, Newton WE (eds), Australian Academy of Science, Canberra, 1981, p. 481.
66. Allen EK, Allen ON: The anatomy of the nodular growths on the roots of *Tribulus cistoides* L., *Soil Sci Soc Amer Proc* 14: 179–183, 1950.
67. Silvester WB: Endophyte adaptation in *Gunnera-Nostoc* symbiosis. In: *Symbiotic nitrogen fixation in plants*, Nutman PS (ed), International Biological Programme 7, Cambridge, Cambridge University Press, 1976, 521–538.
68. Becking JH: Nitrogen fixation in some natural ecosystems in Indonesia. In: *Symbiotic nitrogen fixation in plants*, Nutman PS (ed), International Biological Programme 7, Cambridge, Cambridge University Press, 1976, 539–550.
69. Silvester WB, Smith DR: Nitrogen fixation by *Gunnera-Nostoc* symbiosis, *Nature, London* 224: 1231, 1969.
70. Nathanielsz CP, Staff IA: A mode of entry of blue-green algae into the apogeotropic roots of *Macrozamia communis*, *Amer J Bot* 62: 232–235, 1975.
71. Bond G: Fixation of nitrogen by higher plants other than legumes, *Ann Rev Pl Physiol* 18: 107–126, 1967.
72. Bergersen FJ, Kennedy GS, Wittmann W: Nitrogen fixation in the coralloid roots of *Macrozamia communis* L. Johnston, *Aust J Biol Sci* 18: 1135–1142, 1965.
73. Halliday J, Pate JS: Symbiotic nitrogen fixation by coralloid roots of the cycad *Macrozamia riedlei*, *Aust J Pl Physol* 3: 349–358, 1976.
74. Becking JH: Ecology and physiological adaptations of *Anabaena* in the *Azolla-Anabaena* symbiosis. In: *Environmental role of nitrogen-fixing blue-green algae and asymbiotic bacteria*, Granhall U (ed), *Ecol Bull* 26, Swedish Nat Sci Res Co, Stockholm, 1978, 266–281.
75. Bortels H: Über die Bedeutung des Molybdäns für die Stickstoffbindenden Nostocaceen, *Archiv Microbiol* 11: 155–186, 1940.
76. Saubert GGP: Provisonal communication on the fixation of elementary nitrogen by a floating fern, *Ann Bot Gdns Buitenz* 51: 177–197, 1949.
77. Stewart WDP: Systems involving blue-green algae (Cyanobacteria). In: *Methods for evaluating biological nitrogen fixation*, Bergersen FJ (ed), Chichester, John Wiley, 1980, 583–635.
78. Stewart WDP, Rodgers GA: Studies on the symbiotic blue-green algae of *Anthoceros*, *Blasia* and *Peltigera*. In: *Environmental role of nitrogen-fixing blue-green algae and asymbiotic bacteria*, Granhall U (ed), *Ecol Bull* 26, Swedish Nat Sci Res Co, Stockholm, 1978, 247–259.
79. Bond G, Scott GD: An examination of some symbiotic systems for nitrogen fixation, *Ann Bot London* 19: 67–77, 1955.
80. Scott GD: Further investigation of some lichens for fixation of nitrogen, *New Phytol* 55: 111–116, 1956.
81. Millbank JW: Associations with blue-green algae. In: *The biology of nitrogen fixation*, Quispel A (ed), Amsterdam, North-Holland Pub. Co, 1974, 238–264.
82. Denison WC: *Lobaria oregana*, a nitrogen-fixing lichen in old-growth Douglas fir forests. In: *Symbiotic nitrogen fixation in the management of temperate forests*, Gordon JC, Wheeler CT, Perry DA (eds), Corvallis, Oregon, Forest Research Laboratory, Oregon State University, 1979, 266–275.

83. Huss-Danell K: Nitrogen fixation by *Stereocaulon paschale* under field conditions. In: *Environmental role of nitrogen-fixing blue-green algae and asymbiotic bacteria*, Granhall U (ed), *Ecol Bull* 26, Swedish Nat Sci Res Co, Stockholm, 1978, 216.
84. Kallio S: On the effect of forest fertilisers on nitrogenase activity in two subarctic lichens. In: *Environmental role of nitrogen-fixing blue-green algae and asymbiotic bacteria*, Granhall U (ed), *Ecol Bull* 26, Swedish Nat Sci Res Co, Stockholm, 1978, 217–224.
85. Stewart WDP, Rowell P, Apte SK: Cellular physiology and the ecology of N_2-fixing blue-green algae. In: *Recent developments in nitrogen fixation*, Newton W, Postgate JR, Rodriguez-Barrueco C (eds), London, Academic Press, 1977, 287–307.
86. Ruinen J: Occurrence of *Beijerinckia* species in the 'phyllosphere', *Nature, London* 177: 220–221, 1956.
87. Ruinen J: The phyllosphere. I. An ecologically neglected milieu, *Plant Soil* 15: 81–109, 1961.
88. Ruinen J: The phyllosphere. III. Nitrogen fixation in the phyllosphere, *Plant Soil* 22: 375–394, 1965.
89. Ruinen J: Nitrogen fixation in the phyllosphere. In: *The biology of nitrogen fixation*, Quispel A (ed), Amsterdam, North-Holland Pub. Co, 1974, 121–167.
90. Ruinen J: Nitrogen fixation in the phyllosphere. In: *Nitrogen fixation by free-living micro-organisms*, Stewart WDP (ed), International Biological Programme 6, Cambridge, Cambridge University Press, 1975, 85–100.
91. Jones K, King E, Eastlick M: Nitrogen fixation by free-living bacteria in the soil and in the canopy of Douglas fir, *Ann Bot (London)* 38: 765–772, 1974.
92. Döbereiner J: Nitrogen-fixing bacteria in the rhizosphere. In: *The biology of nitrogen fixation*, Quispel A (ed), Amsterdam, North-Holland Pub. Co, 1974, 86–120.
93. Döbereiner J, Day JM: Nitrogen fixation in the rhizosphere of tropical grasses. In: *Nitrogen fixation by free-living micro-organisms*, Stewart WDP (ed), International Biological Programme 6, Cambridge, Cambridge University Press, 1975, 85–100.
94. Döbereiner J: Physiological aspects of N_2 fixation in grass-bacteria associations. In: *Recent developments in nitrogen fixation*, Newton W, Postgate JR, Rodriguez-Barrueco C (eds), London, Academic Press, 1977, 513–522.
95. Döbereiner J: Influence of environmental factors on the occurrence of *Spirillum lipoferum* in soils and roots. In: *Environmental role of nitrogen-fixing blue-green algae and asymbiotic bacteria*, Granhall U (ed), *Ecol Bull* 26, Swedish Nat Sci Res Co, Stockholm, 1978, 343–352.
96. Smith RL, Schank SC, Bouton JH, Quesenberry KH: Yield increases in tropical grasses after inoculation with *Spirillum lipoferum*. In: *Environmental role of nitrogen-fixing blue-green algae and asymbiotic bacteria*, Granhall U (ed), *Ecol Bull* 26, Swedish Nat Sci Res Co, Stockholm, 1978, 380–385.
97. Burris RH, Okon Y, Albrecht SL: Properties and reactions of *Spirillum lipoferum*. In: *Environmental role of nitrogen-fixing blue-green algae and asymbiotic bacteria*, Granhall U (ed), *Ecol Bull* 26, Swedish Nat Sci Res Co, Stockholm, 1978, 353–363.
98. Day JM, Harris D, Dart PJ, Van Berkum P: The Broadbalk experiment. An investigation of nitrogen gains from non-symbiotic nitrogen fixation. In: *Nitrogen fixation by free-living micro-organisms*, Stewart WDP (ed), International Biological Programme 6, Cambridge, Cambridge University Press 1975, 71–84.
99. Barber LE, Tjepkama JD, Evans HJ: Acetylene reduction in the root environment of some grasses and other plants in Oregon. In: *Environmental role of nitrogen-fixing blue-green algae and asymbiotic bacteria*, Granhall U (ed), *Ecol Bull* 26, Swedish Nat Sci Res Co, Stockholm, 1978, 366–372.
100. Patriquin D, Knowles R: Nitrogen fixation in the rhizosphere of marine angiosperms. *Marine Biol* 16: 49–58, 1972.
101. Silver WS, Jump A: Nitrogen fixation associated with vascular aquatic macrophytes. In: *Nitrogen fixation by free-living micro-organisme*, Stewart WDP (ed), International Biological Programme 6, Cambridge, Cambridge University Press, 1975, 121–125.
102. Sloger C: Associative nitrogen fixation involving cereals and grasses. In: *Current perspectives in nitrogen fixation;* Gibson AH, Newton WE (eds), Australian Academy of Science, Canberra, 1981, 321.

103. Buckley BM, Triska FJ: Non-symbiotic nitrogen fixation in small wood decomposition. In: *Symbiotic nitrogen fixation in the management of temperate forests*, Gordon JC, Wheeler CT, Perry DA (eds), Corvallis, Oregon, Forest Research Laboratory, Oregon State University, 1979, 473–474.
104. Silvester W: Asymbiotic nitrogen fixation and molybdenum deficiency in forests. In: *Current perspectives in nitrogen fixation*, Gibson AH, Newton WE (eds), Australian Academy of Science, Canberra, 1981, 508.
105. Fogg GE, Stewart WDP, Fay P, Walsby AE: *The blue-green algae*. Academic Press, London, 1973.

4. Isolation and culture of nitrogen-fixing organisms

CAROLYN V. CARPENTER and LINDA R. ROBERTSON
Weyerhaeuser Technology Center, Tacoma, WA 98477, USA

INTRODUCTION

Dinitrogen-fixing microorganisms, living in symbiosis with host plants, play a major role in restoring nitrogen to agricultural and forest soils. While legumes are widely used in agriculture, actinorhizal or non-leguminous plants are the dominant nitrogen-fixing species in forest ecosystems. These woody plants include a variety of trees and shrubs which harbor the nitrogen-fixing endophyte in root nodules (1). Many actinorhizal plants fix atmospheric nitrogen at rates comparable to the legumes and consequently play an important role in replenishing nitrogen in forest soils (2, 3, 4, 5).

While the Rhizobium bacteria from leguminous plants have been studied extensively (6), information on the nitrogen-fixing symbionts from actinorhizal plants is limited due to the failure of researchers to isolate an endophyte that fulfilled Koch's Postulates (7, 8, 9, 10). The recent isolation of an actinomycetous, nitrogen-fixing endophyte from *Comptonia peregrina* (11) has led to the successful isolation and *in vitro* cultivation of microsymbionts from several actinorhizal systems (12, 13).

The objective of this chapter is to present the methodologies currently available for the isolation, cultivation, and maintenance of the nitrogen-fixing symbionts commonly found in forest ecosystems. Due to the abundant literature already published on the Rhizobia from legumes, the chapter concentrates chiefly on techniques for the isolation and culture of the actinorhizal endophytes from non-leguminous systems. While the isolation procedure described draws heavily on the authors' experience with endophytes from *Alnus* species, it is hoped that the same techniques are applicable to other actinorhizal nodules. Throughout the text, the nitrogen-fixing endophyte from *Alnus* is referred to as a *Frankia* spp. This is in accordance with Lechevalier and Lechevalier's assignment of the actinorhizal endophytes to the genus *Frankia* of the family *Frankiaceae* (14).

REVIEW OF ASEPTIC TECHNIQUES

The successful isolation and cultivation of endophytes from actinorhizal root

Gordon, J.C. and Wheeler, C.T. (eds.). Biological nitrogen fixation in forest ecosystems: foundations and applications

ISBN 90-247-2849-5.
Printed in The Netherlands

nodules depends largely on the practice of good aseptic technique. Standard microbiological procedures for excluding contamination from the surrounding environment must be employed at all times during the initial isolation. In accordance with these basic principles, the following guidelines should be observed.

— Prior to the isolation, sterilize all glassware, pipettes, mortars, pestles, and centrifuge tubes by autoclaving at 121° C for 30 min. Small sized glassware and centrifuge tubes can be conveniently sterilized and stored in disposable, self-seal syringe envelopes. Sterilize contaminated utensils after use.

— Sterilize all water and liquid media by autoclaving at 121° C for 20 min. Increase the sterilization time for volumes over 1 L.

— When a container of sterile water or liquid media is first opened, flame the mouth of the flask before removing the contents. This action prevents the introduction of contaminating microorganisms from the outside of the container.

— To reduce air-borne contamination, perform the isolation in a 'clean' area. Executing the entire operation within a laminar flow hood is an excellent way to prevent unwanted air-borne contamination.

ISOLATION OF FRANKIA

Preparation of sucrose gradients

The technique described in this section is a modification of the sucrose density fractionation used by Baker, Torrey and Kidd for the isolation of the nitrogen-fixing endophyte from *Alnus viridis* spp. *crispa* (12). It employs the basic biochemical technique of using a liquid density gradient for the separation of particles or macromolecules which differ in size, density, and permeability to the gradient solute. An excellent account of the theory and potential uses of such gradients may be found in the chapter on Zonal Centrifugation by Cline and Ryel in *Methods of Enzymology*, Vol. XXII (15).

In the *Frankia* isolation, a three-layer discontinuous sucrose density gradient is used to separate the desired endophyte from contaminating bacteria, fungi, and nodule debris. The materials required for the gradient preparation include:

Sterile sucrose solutions:

30% (w/v)
45% (w/v
60% (w/v)

Sterile centrifuge tubes (polyallomer or cellulose nitrate) with a capacity of approximately 15 ml

Sterile 10 ml pipettes

Small test tube rack

Sterile 5 ml syringes and cannula (needle) 14 ga × 10.2 cm

Note: Polyallomer centrifuge tubes are autoclavable; the cellulose nitrate are not.

Procedure

1. Place desired number of centrifuge tubes into the test tube rack.
2. Pipette 6 ml of 30% sucrose into each tube. (The lightest or top layer of the gradient is placed in the tube first and will be displaced by the heavier layers of greater density.)
3. Fill the syringe with 3 ml of the 45% sucrose solution. Carefully lower the needle along the side of the tube until the tip of the needle is at the bottom of the tube (Figure 1). Slowly inject the 45% sucrose. Carefully withdraw the syringe from the tube; avoid rapid movements, which create air bubbles or turbulence within the gradients.
4. In like manner, fill the second syringe with 3 ml of the 60% sucrose solution, lower the needle to the bottom of the tube and slowly inject the 60% sucrose. This is the most dense layer and will displace the two lighter layers. The completed gradient should now contain three layers of sucrose with the heaviest (60% sucrose) at the bottom of the tube.
5. Store the gradients in the test tube racks in a 'clean' area until the samples are ready to be applied to the top of the gradients.

Note: Baker uses a gradient of higher densities composed of three layers of sucrose of 33, 55, and 87 g/100 ml, respectively (personal communication). In this gradient, the actinomycete endophytes band predominantly in a clearly visible band at the interface between the lower two layers of the gradient. This band also contains, however, a large number of unwanted fungal contaminants. In our laboratory, we have had good success using the gradient of lighter densities described above. With our gradient, a high percent of the fungi – as well as actinomycetes – sediment into a pellet of cell debris at the bottom of the tube while the smaller, fragmented pieces of the endophyte remain suspended in the upper part of the gradient. When these fractions are plated in *Frankia* medium, many small, well-defined colonies of the endophyte are obtained with little or no fungal contamination.

Selection of root nodules for endophyte isolation

In alder the most active preparations generally result from young nodules found on the periphery of the root system. While the older nodule clusters near the root collar may be very large, the younger nodules that are visibly 'plump' and well hydrated generally yield the best preparations of viable endophyte.

The optimum seasons for isolating viable endophyte from the actinorhizal plants is still under investigation. A study in the authors' laboratory suggests that early spring and fall are the best times to isolate the endophyte from *Alnus rubra*. These data contrast with the period of greatest nitrogenase activity which peaks during late spring and summer (16). The presence of viable endophyte in spring and fall may coincide with the incidence of increased root growth and new nodule synthesis.

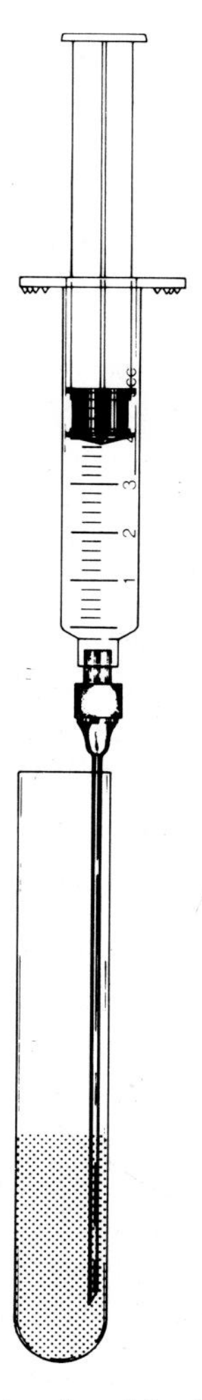

Fig. 1. Proper position of needle and syringe for adding bottom layers of a discontinuous sucrose gradient.

Isolation of endophyte from root nodules

The procedure outlined below basically results from the sucrose gradient technique of Baker *et al.* (12) and has been used successfully to isolate active, nitrogen-fixing endophyte from several alder species. Hopefully, the same techniques can also be used to isolate endophyte from other actinorhizal species. Materials required for this isolation include:

Fresh alder nodules, 1–2 gm
Sterile water
30% (v/v) chlorine bleach (5.25% NaOCl)
Tween 20 (polyoxyethylene-20 sorbitan mono-laurate) brand, surfactant
Sterile 125 ml Erlenmeyer flask
Sterile mortar and pestle
Sterile Pasteur pipettes ($\sim$2 ml)
Sterile glass funnel containing Miracloth brand, all-purpose cloth and glass wool (packed loosely in upper portion of funnel stem)
Sterile 10 ml graduated cylinder.
Sucrose density gradients containing three layers of 30%, 45%, and 60% sucrose
Refrigerated centrifuge with swinging bucket rotor

Procedure

1. Wash loose dirt and soil from freshly excised nodules under running tap water. Blot nodules dry with paper towels.
2. Suspend approximately 0.8 g (fresh weight) washed nodules in 50 ml sterile water + 1 drop of Tween 20 in a 125 ml Erlenmeyer flask. Shake vigorously for 15 min.
3. Remove water from the flask and add 50 mL of 30% chlorine bleach (v/v) containing one drop of Tween 20. Shake nodules in the bleach solution for an additional 15 min.
4. Remove the chlorine solution from the nodules and carefully wash nodules three times with sterile water.
5. Place washed nodules in the sterile mortar. Add approximately 1–2 mL sterile water to the nodules. Grind the nodules thoroughly until the preparation appears to be a homogeneous paste.
6. Add approximately 3–5 mL sterile water to the ground nodules. Filter the resulting slurry serially through the Miracloth filter and glass wool.
7. Dilute the filtrate to 10 mL with sterile water.
8. Layer 2.0 mL of this filtrate on the top of a sucrose density gradient consisting of three layers of sucrose: 3 mL of 60% (w/v), 3 mL of 45% (w/v), and 6 mL of 30% (w/v). (See section on preparation of the gradients.)
9. Cover each tube with caps, previously soaked in 70% ethyl alcohol.

10. Place capped tubes securely into the cups of a swinging bucket rotor. Before starting the centrifuge, weigh the loaded cups to be certain the rotor is evenly balanced.

11. Centrifuge at 5° C for 4 hr at 100,000 xg or for 16 hr at 5000 xg. If a high speed centrifuge is not available, the 16-hr centrifugation at 5000 xg gives a very acceptable separation.

12. After centrifugation carefully remove tubes from the rotor. Secure the first tube in a pinch clamp that is fastened to a ring stand.

13. Swab the bottom of each tube with cotton or gauze saturated with 70% ethyl alcohol.

14. Carefully puncture the bottom of the tube with a sterile needle. Collect 1 mL fractions (approximately 20 drops/mL) into labelled petri dishes. Generally, the first eight fractions contain the endophyte; whether additional fractions are collected is optional.

15. After all the desired fractions have been collected from each tube, add approximately 25 mL of *Frankia* medium (cooled to 48° C) to each plate. Swirl plates carefully to disperse the 1 mL fraction uniformly throughout the media.

16. Seal plates with Parafilm brand, laboratory film and incubate inverted in the dark at 28° C for 4–6 weeks. After six weeks, examine plates for the appearance of small (<1 mm in diameter) white *Frankia* colonies growing within the agar (see Section 4.5 on Morphology of *Frankia* on selected media).

Note: Any colonies that appear within the first two weeks and exhibit fluffy aerial mycelium are probably contaminants. All new isolates should be considered potential human pathogens and handled with proper aseptic technique to prevent contamination of personnel.

Alternative methods for isolation of Frankia

For over 70 years scientists have attempted to isolate viable endophyte from the root nodules of actinorhizal plants (8, 9, 10, 17, 18). While many actinomycete and *Streptomyces* spp. had been isolated in these studies (8, 9, 10, 18, 19), Callaham and his coworkers (11) were the first to isolate a viable endophyte that fulfills all the demands of Koch's Postulates. A review of all these studies is quite beyond the scope of this chapter; therefore, this discussion is limited to a brief summary of three techniques which offer practical alternatives to the sucrose density procedure described above.

Serial dilution

Serial dilutions are used frequently in microbiology to obtain single isolates from samples containing mixed populations of organisms. The procedure consists of making successive dilutions of the original sample which has been weighed and suspended in a known amount of sterile water (20). For soil samples or crushed nodule suspensions, it is helpful to use 0.15% water agar as the diluent. Dilutions required for isolating organisms from the rhizosphere frequently range from 10^{-4} to 10^{-7}. A very comprehensive study on a crushed nodule suspension from *Alnus crispa* v. *mollis* Fernald was performed by Lalonde *et al.*, but the isolate failed to infect host plant seedlings (10).

Microdissection

This technique was used successfully by Berry and Torrey at Harvard Forest to isolate the actinomycetous endophyte from *Alnus rubra Bong* (13). After surface sterilization, nodule lobes were sliced into 1–3 mm thick transverse sections under a dissecting microscope. Endophyte filaments or clusters were then removed from individual host cells with dissecting needles. Cell debris was removed by filtration through nylon mesh and low speed centrifugation, and the washed endophyte clusters were grown in yeast-extract-dextrose broth. A more detailed account of this procedure is found in the original presentation by Berry and Torrey in 'Symbiotic Nitrogen Fixation in the Management of Temperate Forests' (13).

Sucrose centrifugation without a gradient

Burggraaf and his co-workers have recently reported the successful isolation of several *Frankia* strains from *Alnus glutinosa* and *Myrica gale* using a short centrifugation with a single concentration of sucrose (21). A root-nodule homogenate was centrifuged in 60% sucrose (w/v) at 25,000 xg for 30 minutes. The *Frankia* endophyte was then isolated from the sediment, which was resuspended on the top layer of a double agar layer plating system (21).

CULTURE ON SELECTED MEDIA

Frankia are microaerophilic organisms that grow well either in broth or in solid agar. When grown in broth, the culture requires no mechanical agitation; swirling the culture once a week by hand provides adequate aeration for normal growth. If *Frankia* is grown in solid medium, it is best to use pour plates so that the endophyte can grow within the agar.

For routine subculture of *Frankia*, several modified yeast-extract media may be used. The *Frankia* medium of Baker and Torrey (Table 1) is excellent for growing *Frankia* cultures in either broth or in agar plates (22). In broth the medium remains clear so that gross morphology of the culture may be readily observed. The QMOD medium, modified from Quispel (23, 24), is also good for the subculture of new isolates (Table 2). Lalonde and Calvert reported very rapid growth of the *Comptonia* isolate in QMOD medium. These authors detected new

Table 1. Composition of the medium used for isolation of *Frankia* species (From Baker and Torrey, 1979).

Component	*Concentration*
Yeast extract (Difco)	0.5% (w/v)
Dextrose (Difco)	1.0% (w/v)
Casamino acids (Difco)	0.5% (w/v)
Vitamin B_{12}	1.6 mg/l
Micronutrient salts:	
Component	Final *Concentration*
H_3BO_3	1.5 mg/l
$ZnSO_4 \cdot 7H_2O$	1.5 mg/l
$MnSO_4 \cdot H_2O$	4.5 mg/l
$NaMoO_4 \cdot 2H_2O$	0.25 mg/l
$CuSO_4 \cdot 5H_2O$	0.04 mg/l
Agar (optional)	0.8% (w/v)
Final pH is adjusted to 6.4	

Table 2. Composition of an effective growth medium for the cultivation *in vitro* of Frankia isolates (From Lalonde and Calvert, 1979).

QMOD medium (modified from Quispel, 1960)

Component	Concentration (per liter)
K_2HPO_4	300 mg
NaH_2PO_4	200 mg
$MgSO_4 \cdot 7H_2O$	200 mg
KCl	200 mg
Yeast Extract (BBL)	500 mg
Bacto-Peptone (Difco)	5 g
Glucose	10 g
Ferric Citrate (Citric Acid and Ferric Citrate, 1% solution)	1 mL
Minor Salts*	1 mL
H_2O, Deionized to:	1 L
Adjust pH to 6.8–7-0 with NaOH or HCl, then add $CaCO_3$	100 mg
Lipid Supplement**	0.5–50 mg
Agar, if used	15 g
Mix thoroughly and pour 15 mL per tube, autoclave for 20 min.	

* Minor Salts (g/liter): H_3BO_3, 1.5; $MnSO_4 \cdot 7H_2O$, 0.8; $ZnSO_4 \cdot 7H_2O$; 0.6; $CuSO_4 \cdot 7H_2O$, 0.1; $(NH_4)_6Mo_7O_{24} \cdot 4H_2O$, 0.2; $CoSO_4 \cdot 7H_2O$, 0.01.
** Lipid Supplement: Dissolve 500 mg of L-α-lecithin (commercial grade from soybeans, 22% phosphatidyl choline, P-5638 from Sigma Chemical Company, St. Louis, MO) in 50 ml of absolute ethanol, and add 50 ml of distilled water.

Frankia growth in less than 4 days following inoculation of the broth (24). Recent work on the physiology of the *Frankia* isolate from *Alnus viridis* spp. *crispa* indicates that (a) the complex lipid supplement in QMOD what can be replaced by Tween 80, and (b) the endophyte is able to grow with Tween 80 as the sole carbon source (25). Additional studies now indicate that when grown *in vitro*, the endophyte does not utilize glucose as a carbon source (26).

When maximum sporulation is desired for microscopy work, the endophyte should be grown in Ormerod's Defined Succinate Medium (27).

MORPHOLOGY OF FRANKIA ON SELECTED MEDIA

Gross morphology on agar pour plates

Preliminary examination of agar isolation plates should begin 5–6 weeks after inoculation. Plates with *Penicillium* contamination and those with excessive numbers of eubacteria are discarded. The remaining plates are first examined visually either in indirect sunlight or under florescent lights. The *Frankia* colonies isolated from nodules of *Alnus rubra* and *Alnus sinuata* show many of the characteristics listed in Table 3. At this time, the colonies are very small; a typical colony from *Alnus rubra* is often less than 1 mm in diameter. While the first examination occurs at six weeks, the plates should be kept for 12 weeks to allow sufficient time for slow growing *Frankia* colonies to become visible. An excellent summary of the morphological properties of several *Frankia* spp. is given in the recent paper by Lechevalier, Horriere, and Lechevalier (28).

Gross morphology in broth

Individual *Frankia* colonies may be picked out of agar plates using a sterile Pasteur pipette that is fitted with a rubber bulb. The agar plug is transferred aseptically to a tissue grinder and homogenized with 1–2 mL of sterile *Frankia* broth. The resulting suspension is transferred to a flask containing approximately 250 mL of *Frankia* broth. After three weeks of incubation at 28° C, fine filamentous growth can be seen at the bottom of the flask. The broth above this blanket of cells should remain clear, and no growth should be evident at the air-liquid interface. Cultures which exhibit growth at this interface probably

Table 3. Morphological characteristics of *Frankia* isolates from *Alnus* on agar plates.

Morphology indicating colony is a *Frankia spp.*	Morphology indicating colony is *not* a *Frankia spp.*
Colony size: 0.5–3 mm	>5 mm at six weeks
Growth*: filamentous	colonies with defined edges
: colonies submerged in agar	presence of aerial mycelia on top of agar
Cell size: 0.3–1.2 μm in diameter	
Cell structures: presence of vesicles and/or sporangia submerged in the agar	aerial sporangia formed

* The endophyte isolated from *Elaeagnus umbellata* will grow on the surface of agar, but does not form true aerial mycelia (12).

contain a contaminating actinomycete. The typical filamentous growth of *Frankia* isolates growing in broth cultures is shown in Figure 2.

Light microscopy

A microscopic examination is done on the original colonies in agar to determine cell diameter, the presence of filamentous growth, and the presence or absence of sporangia. At 5–6 weeks, a young colony will generally have 5–60 sporangia below the surface of the agar. In isolates from *Alnus rubra*, the sporangia have an irregular shape similar to that of a peanut; they may be either terminal or intercalary and range in size from 10–30 μm in length.

For direct examination of agar plates, a long-working distance condensor and long-working distance objective of 20 × is recommended. Phase contrast enhances visualization of the cell structures. In microscopes without phase capability, it is possible to see unstained specimens by closing down the aperture diaphragm.

If numerous colonies have been obtained during an isolation, it may be desirable to sacrifice one for more intensive study with the light microscope. The following method is a simple way to examine the sporangia in greater detail:

1. Remove a *Frankia* colony from the agar using a Pasteur pipette fitted with a rubber bulb.

Fig. 2. Typical filamentous growth of the *Alnus rubra* endophyte growing in *Frankia* broth. Picture was taken 4 weeks after inoculation.

2. Expel the agar plug onto a glass slide. Add one drop of water to the slide.
3. Heat slide very gently to soften the agar.
4. Place a coverslip over the specimen. Press the coverslip gently with a blunt object to flatten the colony.

The resulting slide is examined with a 20× objective and is suitable for photomicrography. The sporangia depicted in Figure 3 were prepared in the above manner.

MAINTENANCE OF CULTURES

Broth

Frankia cultures are quite stable in broth and do not require frequent transfers to retain viability. Cultures maintained in *Frankia* media (see section on Selected media) for 6–12 months grow readily after transfer to fresh media (Robertson and Carpenter, unpublished observations). While this method provides ready access to the endophyte, the sheer bulk of many culture flasks may present a storage problem. Due to the relatively recent isolation of *Frankia* spp (11, 12, 13), the effect of long-term storage in broth on the genetic stability of these strains is

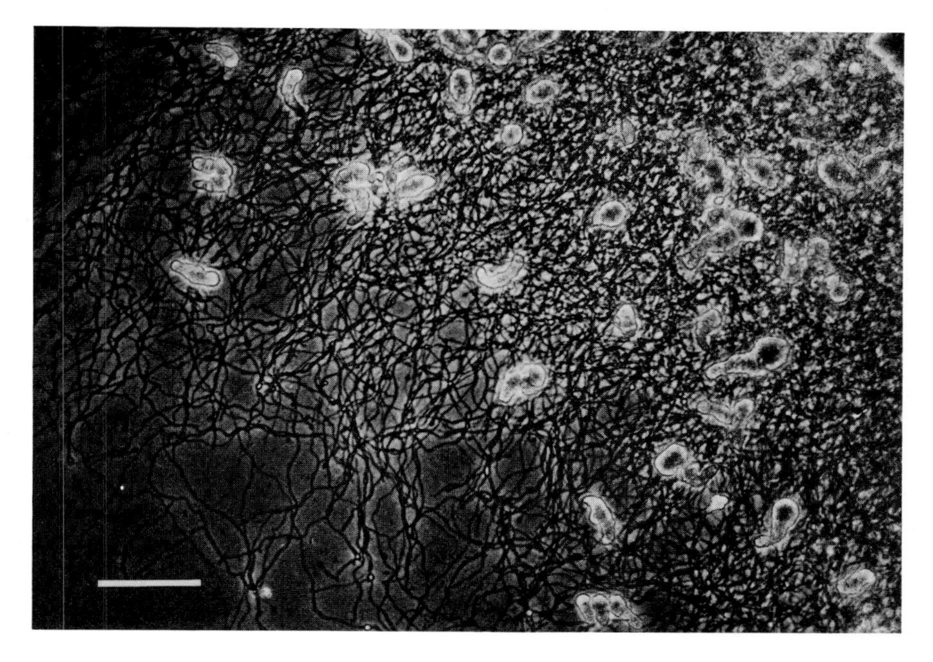

Fig. 3. Peanut-shaped sporangia from a six-week-old colony of the *Alnus rubra* endophyte growing on the initial agar isolation plate. Bar = 60 μm.

still unknown. Therefore the processes of lyophilization and liquid nitrogen storage are probably more appropriate for long-term preservation.

Cryogenic storage

Storage in liquid nitrogen provides a simple method for preserving cultures as well as a rapid means for reactivation. Cultures are placed in sterile cryotube vials that fit into aluminum canes. The canes are then immersed in the liquid nitrogen. The obvious disadvantage of this method is the necessity of keeping the container filled with liquid nitrogen. A large-capacity refrigerator, however, such as the 29L Union Carbide model XR24, needs to be filled only once every three months after initial cool down.

The procedure for freezing a sample is quite simple:

1. Fill sterile, self-standing cryotubes (2.0 mL capacity) with 1 mL *Frankia* cells + 20% glycerol in broth. Colonies growing in solid agar can be transferred directly in an agar plug and covered with a small amount of sterile broth.
2. Cover the loaded vials and place into the aluminum canes.
3. Carefully lower the canes into the liquid nitrogen. (Caution: Extreme care must be taken at this point to avoid splashing liquid nitrogen on the operator. Eye protection is required.) Replaced cover on the canister.

To reactivate the frozen cultures:

1. Remove vials from the liquid nitrogen and place in lukewarm water to thaw. Do not let water cover the caps or contamination of the culture may result.
2. Remove cap from vial, flame the opening *lightly*, and transfer contents of the vial aseptically to *Frankia* broth. If the original material was an agar plug, the specimen should be ground in a tissue grinder prior to transfer to broth.
3. Typical growth should be seen after three weeks incubation.

Lyophilization (freeze drying)

Lyophilization is the most common technique used to maintain the genetic stability of a bacterial culture for long periods of time. The procedure involves the use of a high vacuum with a desiccant to (a) provide quick freezing conditions, and (b) dehydrate the sample over a period of 1–8 hours (29). Major equipment required includes a lyophilization apparatus, vacuum pump and a natural gas-oxygen torch for cutting and sealing the glass specimen tubes. A detailed account of the methodology involved in lyophilization can be found in the references by Pridham and Lyons (30) and Vincent (29). A new publication on lyophilization techniques from the American Type Culture Collection also provides practical suggestions (31).

This technique requires a great deal of skill and patience on the part of the operator; therefore it is recommended that the first attempt be done under supervision on a sample whose loss would not be devastating to the laboratory. In any case, multiple samples are always made so the culture collection can be tested periodically for viability and purity.

Soil culture

Dried sterilized soil is sometimes used to maintain cultures of bacteria including those of the nitrogen-fixing Rhizobia (32). The soil used is generally a sieved sandy loam which has undergone four 30-minute sterilizations at 121°C on alternate days. Approximately two grams of the sterilized soil in a test tube are then inoculated with 2 ml of a broth culture. The tubes are plugged and allowed to air dry for 2–3 weeks (30).

While many laboratories do store bacteria in soil culture, the authors hesitate to recommend this method as the only way to store *Frankia* isolates. This reservation stems in part from the report by Pridham and Lyons who do not recommend soil culture for long-term preservation of Actinomycetales (30).

INOCULATION AND NODULATION OF SEEDLINGS

Preparation of vegetative clones

After the endophyte has been isolated and cultivated in pure culture, additional characterization is required to determine infectivity and host specificity. The immediate goal of these experiments is to determine if the isolate will infect seedlings or cuttings of the host plant and induce the formation of active, nitrogen-fixing nodules. While either seedlings or vegetative cuttings may be used for this purpose, vegetative clones in a cross-fertilizing species like alder provide a more uniform genotype in the host. The cloning procedure described below has been used with several species of alder and is basically the technique described by Monaco, Ching, and Ching for cloning red alder genotypes (33). The recommended rooting medium is sterile horticultural grade perlite:vermiculite, medium grade (1:1 v/v).

Note: If seedlings are used for inoculation experiments, the seeds should first be surface-sterilized with hydrogen peroxide (33).

Procedure

1. Collect green stem cuttings with two leaves from the desired host plant.
2. Dip cuttings into 8000 ppm indole-3-butyric acid (IBA). For Sitka alder, a dip in Rootone 10 brand, rooting compound induces better cooting (Robertson, unpublished observations).
3. Dust the stem with 10% Benomyl, systemic fungicide.

4. Place cuttings into the perlite-vermiculite rooting medium. The cuttings should be kept under intermittent mist throughout the rooting period. Generally, 10 weeks is required for root formation.

Inoculation of rooted cuttings with cultured endophyte

The following equipment is required for inoculation of rooted cuttings:

Paper towels
Sterile Pasteur pipettes
Sterile 100 mL beaker
Sterile 10 mL pipettes
Sterile tissue grinder
Sterile water
Table-top clinical centrifuge
Sterile graduated conical centrifuge tubes
Sterile rooting media (perlite:vermiculite, 1:1 v/v)
Three-month old *Frankia* culture in liquid broth

Procedure

1. Remove rooted cuttings from rooting medium. Wash root system under running water until roots are free of perlite and vermiculite. Blot gently with paper towels.
2. With a 10 mL pipette, transfer approximately 10 mL of the *Frankia* suspension from the culture flask into the tissue grinder.
3. Homogenize the suspension until it appears to be homogeneous.
4. Carefully transfer *Frankia* suspension into a sterile 10 mL centrifuge tube. Centrifuge at high speed for 10 minutes.
5. After centrifugation, remove supernatant with a sterile Pasteur pipette.
6. Resuspend the pellet in approximately 10 mL of sterile water. Transfer suspension back to the tissue grinder and homogenize a second time until the suspension again appears to be homogeneous.
7. Transfer suspension to a 100 mL beaker. Add approximately 40 mL sterile water (see section on quantifying the inoculum).
8. To inoculate cuttings, dip roots of each cutting into the endophyte suspension for 10 seconds. Immediately replant inoculated cuttings into sterile growth media. *Note:* Although the growth media used here is initially sterile, no attempt is made to maintain absolute sterility after the cuttings are inoculated and replanted. The cuttings are maintained in a clean greenhouse environment, but are open to the air and are watered with tap water.
9. For uninoculated controls used additional cuttings which are dipped in sterile water in place of the endophyte solution.

Quantification of the inoculum

When inoculating cuttings with a suspension of the endophyte, it is necessary to quantify the amount of *Frankia* in the suspension. While the filamentous growth of *Frankia* cultures prohibits the use of a haemacytometer for cell counts, the following parameters are useful for estimating total cell mass:

1. Dry weight
2. Packed cell volume following centrifugation
3. Turbidity or optical density measurements

This is a very convenient and rapid method for estimating cell mass and can be readily calibrated with dry weight or viable cell counts. For example, a suspension of the red alder endophyte which gives an optical density reading of 0.239 at 550 nm yields a dry weight of 0.12 mg/ml and a viable plate count of 9.0×10^5 colonies/mL (Carpenter and Robertson, unpublished observations).

4. Plate count

This requires the plating of serial dilutions of the original endophyte suspension (20). While this is the most time-consuming method for quantifying the inoculum, the colony counts obtained are a direct measure of viable endophyte.

5. TOC (Total Organic Carbon)

Blom, Roelofsen, and Akkermans use TOC as a means for determining growth yields of *in vitro* grown *Frankia* cultures (25). Cell suspensions are homogenized by sonication before TOC analysis.

KOCH'S POSTULATES

While careful examination of the morphology of a root nodule isolate may strongly suggest the presence of a *Frankia* spp., positive identification of the isolate as the true agent of infection requires a more rigorous exercise. In microbiology the final test for distinguishing the agent of infection from a secondary invader or other adventitious microbe involves meeting the criteria set forth in Koch's postulates (35). Although originally formulated for identifying human and animal pathogens, the postulates apply equally well to organisms isolated from plant lesions or nitrogen-fixing root nodules.

If Koch's postulates are applied to a *Frankia* isolate, the following criteria must be fulfilled:

1. The endophyte is regularly found in the root nodules of a given plant species.
2. The endophyte can be isolated in pure culture on artificial media. The one assumption implicit in Koch's postulates is that the infecting organism may be grown in pure culture on artificial media.
3. A pure culture of the endophyte will induce the formation of active,

nitrogen-fixing nodules on greenhouse seedlings or cuttings of the host plant.

4. The endophyte can be reisolated from the root nodules of the inoculated plants. By fulfilling Koch's postulates with the endophyte from *Comptonia peregrina* (11), Callaham *et al.* opened the way for the successful isolation of microsymbionts from several actinorhizal systems.

ISOLATION AND CULTIVATION OF RHIZOBIUM

Since the *Rhizobium* bacteria were first isolated from root nodules by Beijerinck in 1888 (6), numerous isolates from other legume species have been cultivated and studied *in vitro*. A review of the extensive literature in this field is clearly beyond the scope of this chapter; the authors therefore recommend that Vincent's book entitled 'A Manual for the Study of Root-Nodule Bacteria' (29) be used to supplement the procedure described below. This is an excellent book which gives detailed instructions on all phases of the *in vitro* cultivation of Rhizobia. The references edited by Bergersen (36) and Quispel (37) also give comprehensive accounts of the legume systems.

The procedure for isolating Rhizobia is quite simple and depends primarily on the use of good aseptic technique. The isolation requires the following materials:

Fresh nodules from a leguminous plant
Sterile water
30% (v/v) chlorine bleach (5.25% NaOCl)
Tween 20 (polyoxyethylene – 20 sorbitan mono-laurate) brand, surfactant
Sterile 125 mL Erlenmeyer flask
Sterile micro glass tissue grinder (~0.2 mL capacity) or sterile mortar and pestle
Sterile Pasteur pipettes (~2 mL)
Petri plates containing 20 mL of yeast extract mannitol agar containing 0.002% actidione (29)
Wire inoculating loop
Bunsen burner

Procedure

1. Wash loose dirt and soil from freshly excised nodules under running tap water. Blot nodules dry with paper towels.
2. Suspend the washed nodules in 50 mL sterile water + 1 drop Tween 20 in a 125 mL Erlenmeyer flask. Shake vigorously for 15 min.
3. Remove water from the flask and add 50 mL of 30% chlorine bleach containing one drop of Tween 20. Shake nodules in the solution for an additional 15 min.
4. Remove the bleach solution from the nodules and carefully wash nodules 3 times with sterile water.

5. Place 1 nodule lobe in the tissue grinder or sterile mortar. Grind thoroughly. Add approximately 0.2 mL sterile water to the nodules.
6. With a sterile Pasteur pipette, transfer 1 drop of the ground nodule suspension to the surface of a petri plate containing yeast extract mannitol agar.
7. Sterilize the wire inoculating loop by holding it in the Bunsen burner flame until the loop and wire are red hot. Allow the wire loop to cool slightly.
8. Spread the crushed nodule suspension over the surface of the agar with the inoculating loop.
9. Incubate plates at 26° C. Depending on the *Rhizobium* strain, colonies should appear within 3–10 days.

Note: Vincent recommends the use of 0.1% acidifed mercuric chloride ($HgCl_2$, 1 g; Conc. HCl, 5mL; water 1 L) for surface sterilization of the nodules (29). Treatment with 30% chlorine bleach has worked equally well in the authors' laboratory.

REFERENCES

1. Bond G: The results of the IBP survey of root-nodule formation in non-leguminous angiosperms. In: *Symbiotic nitrogen fixation in plants*, Nutman PS (ed), IBP7, Cambridge, Cambridge University Press, 1976, p 443–474.
2. Berg A, Doerksen A: Natural fertilization of a heavily thinned Douglas-fir stand by understory red alder. Corvallis, Res Note No 56, Forest Research Laboratory, Oregon State University, 1975, p 3.
3. Miller RE, Murray MD: The effects of red alder on growth of Douglas-fir. In: *Utilization and management of alder*, Briggs DG, DeBell DS, Atkinson WA (eds), Portland, Oregon. USDA Forest Service Gen Tech Report PNW-70, 1978, p 283–306.
4. Newton M, El Hassen BA, Zavitkovski J: Role of red alder in western Oregon forest succession. In: *Biology of alder*, Trappe JM, Franklin JF, Tarrant RF and Hansen GM (eds), Portland, Oregon. Pacific Northwest Forest and Range Exp Stn, 1968, p 73–84.
5. Tarrant RF, Miller RE: Accumulation of organic matter and soil nitrogen beneath a plantation of red alder and Douglas-fir. *Soil Sci Soc Amer Proc* 27: 231–234, 1963.
6. Burns RC, Hardy RWF: *Nitrogen fixation in bacteria and higher plants*. New York, Springer-Verlag, 1975.
7. Borm L: Die Wurzelknöllchen von *Hippophaë rhamnoides* und *Alnus glutinosa*. *Bot Arch* 31: 441–488, 1931.
8. Krebber O: Untersuchungen über die Wurzelknöllchen der Erle. *Arch Mikrobiol* 3: 588–608, 1932.
9. Youngberg CT, Hu L: Root nodules on mountain mahogany. *Forest Sci* 18: 211–212, 1972.
10. Lalonde M, Knowles R, Fortin JA: Demonstration of the isolation of a noninfective *Alnus crispa* v. *mollis* Fern. nodule endophyte by morphological, immunolabelling and whole cell composition studies. *Can J Microbiol* 21: 1901–1920, 1975.
11. Callaham D, Del Tredici P, Torrey JG: Isolation and cultivation *in vitro* of the actinomycete causing root nodulation in *Comptonia*. *Science* 199: 899–902, 1978.
12. Baker D, Torrey JG, Kidd GH: Isolation by sucrose-density fractionation and cultivation *in vitro* of actinomycetes from nitrogen-fixing root nodules. *Nature* 281: 76–78, 1979.
13. Berry A, Torrey JG: Isolation and characterization *in vivo* and *in vitro* of an actinomycetous endophyte from *Alnus rubra* Bong. In: *Symbiotic nitrogen fixation in the management of temperate forests*, Gordon JC, Wheeler CT, Perry DA (eds), Corvallis, Forest Research Laboratory, Oregon State university, 1979, p 69–83.

14. Lechevalier M, Lechevalier H: The taxonomic position of the actinomycetic endophytes. In: *Symbiotic nitrogen fixation in the management of temperate forests*, Gordon JC, Wheeler CT, Perry DA (eds), Corvallis, Forest Research Laboratory, Oregon State University, 1979, p 111–123.
15. Cline GB, Ryel RB: Zonal Centrifugation. In: *Methods in enzymology* Vol XXII *Enzyme purification and related techniques*, Jakoby WB (ed), New York, Academic Press, 1971, p 168–204.
16. Wheeler CT, Lawrie AC: Nitrogen fixation in root nodules of alder and pea in relation to the rupply of photosynthetic assimilates. In: *Symbiotic nitrogen fixation in plants*, Nutman PS (ed), Cambridge, Cambridge University Press, 1976, p 497–510.
17. Bottomley WB: The root nodules of *Ceanothus americanus*. *Ann Bot* 29: 605–610, 1915.
18. Pommer E: Über die isolierung des endophyten aus den wurzelknöllchen von *Alnus glutinosa* Gaertn. und über erfolgreiche reinfektionsversuche. *Deutsch Bot Gesell* 72: 138–150, 1959.
19. Wollum AG II, Youngberg GT, Gilmour CM: Characterization of a *Streptomyces* sp. isolated from root nodules of *Ceanothus velutinus* Dougl. *Soil Sci Soc Am Proc* 30: 463–467, 1966.
20. Johnson LF, Curl EA: *Methods for research on the ecology of soilborne plant pathogens*. Minneapolis, Burgess Publishing Company, 1972.
21. Burggraaf AJP, Quispel A, Tak T, Valstar J: Methods of isolation and cultivation of Frankia species from actinorhizas. *Plant and Soil* 61: 157–168, 1981.
22. Baker D, Torrey JG: The isolation and cultivation of actinomycetous root nodule endophytes. In: *Symbiotic nitrogen fixation in the management of temperate forests*, Gordon JC, Wheeler CT, Perry DA (eds), Corvallis, Forest Research Laboratory, Oregon State University 1979, p 38–56.
23. Quispel A: Symbiotic nitrogen fixation in non-leguminous plants. V. The growth requirements of the endophyte of *Alnus glutinosa*. *Acta Bot Neerl* 9: 380–396, 1960.
24. Lalonde M, Calvert HE: Production of *Frankia* hyphae and spores as an infective inoculant for *Alnus* species. In: *Symbiotic nitrogen fixation in the management of temperate forests*, Gordon J, Wheeler CT, Perry DA (eds), Corvallis, Forest Research Laboratory, Oregon State University 1979, p 95–110.
25. Blom J, Roelofsen W, Akkermans ADL: Growth of *Frankia* AvcI1 on media containing Tween 80 as C-source *FEMS Microbiol Lett* 9: 131–135, 1980.
26. Blom J, Harkink R: Metabolic pathways for gluconeogenesis and energy generation in *Frankia* AvcI1. *FEMS Microbiol Lett* 11: 221–224, 1981.
27. Tjepkema JD, Ormerod W, Torrey JG: Vesicle formation and acetylene reduction activity in *Frankia* sp. CPI1 cultured in defined nutrient media. *Nature* 287: 633–635, 1980.
28. Lechevalier MP, Horriere F, Lechevalier H: The biology of *Frankia* and related organisms. In: *Developments in Industrial Microbiology* 23, Underkofler LA (ed), Arlington Va. 22209. Society for Industrial Microbiology 1982, p 51–60.
29. Vincent JM: *A manual for the practical study of root-nodule bacteria* IBP Handbook No. 15, London, Blackwell Scientific Publications, 1970.
30. Pridham TG, Lyons AJ: Methodologies for *Actinomycetales* with special reference to Streptomycetes and Streptoverticillia. In: *Actinomycete Taxonomy*. SIM Special Publication No. 6, Dietz A, Thayer DW (eds) Arlington VA 22209. Society for Industrial Microbiology 1980, p 152–224.
31. Hatt H (ed): *Laboratory manual on preservation freezing and freezedrying as applied to algae, bacteria, fungi and protozoa*. Rockville, Maryland, American Type Culture Collection, 1980.
32. Jensen HL: The viability of lucerne rhizobia in soil culture. *Nature* 192: 682, 1961.
33. Monaco PA, Ching TM, Ching KK: Rooting of *Alnus rubra* cuttings. *Tree Planters Notes* 31(3): 22–24, 1980.
34. Trappe JM: Strong hydrogen peroxide for sterilizing coats of tree seed and stimulating germination. *J For* 59: 828–829, 1961.
35. Davis BD, Dulbecco R, Eisen HN, Ginsberg HS: *Microbiology* 3rd Ed. Hagerstown, Harper and Row 1980, p 7.
36. Bergersen FJ (ed): *Methods for evaluating biological nitrogen fixation*. Chichester, John Wiley and Sons, 1980.
37. Quispel A (ed): *The biology of nitrogen fixation. Frontiers of biology*. Vol 33 Amsterdam, North-Holland Publishing Company, 1974.

5. Biochemical, physiological and environmental aspects of symbiotic nitrogen fixation

ROBERT O.D. DIXON* and CHRISTOPHER T. WHEELER**

* *Department of Botany, The Kings Buildings, University of Edinburgh, Mayfield Road, Edinburgh EH9 3JH*
** *Department of Botany, The University, Glasgow. G12 8QQ*

Over the last eighty years, an extensive body of data has built up, describing symbiotic nitrogen fixation in biochemical and physiological terms. This information can now be drawn on for many agricultural legumes to suggest with some confidence combinations of host plant variety and microsymbiont strain which, together with the appropriate plant management practices, will give good returns of biologically fixed nitrogen under particular growth conditions. Our knowledge of similar procedures which may enhance symbiotic nitrogen fixation in non-agricultural systems is much less complete, however. Indeed, in systems involving actinorhizal species, the basic experimental data on which rational programmes for improvement of the symbiotic association may be founded are still quite fragmentary. This chapter presents an overall view of those aspects of the biochemistry and physiology of symbiotic nitrogen fixation which are basic to the development of programmes designed to improve the symbiotic association and to the interpretation of the results obtained from such programmes. The experimental evidence discussed has been selected to indicate areas of research which should be of fundamental concern to forest physiologists interested in the profitable introduction of symbiotic nitrogen fixation into the forest environment.

BIOCHEMISTRY OF NITROGENASE

It is not relevant, here, to discuss the biochemistry of nitrogenase in any detail, such a discussion is given by Yates (1) but it is necessary to point out those features which are pertinent to an understanding of nitrogen fixation within the root nodule. While the principal concern is with the requirement for an anaerobic site for the nitrogenase proteins and with the high energy demand of the fixation process it can be seen that a number of other properties are also of importance.

The enzyme complex consists of two proteins which are easily separable and which have different prosthetic groups and properties. One protein is a non-haem iron protein with iron-sulphur clusters similar in many respects to the ferredoxins while the other, as well as having iron-sulphur clusters, has an iron molybdenum cofactor. These proteins have no agreed nomenclature and conse-

Gordon, J.C. and Wheeler, C.T. (eds.). Biological nitrogen fixation in forest ecosystems: foundations and applications

ISBN 90-247-2849-5.
Printed in The Netherlands

quently have a number of names. The molybdenum iron protein has been called component 1 or protein 1 in view of the fact that it is first eluted from the column when the proteins are separated. This is unfortunate as it is the second protein sequentially in the action of the protein complex. Other names are molybdo-ferredoxin, to indicate its non-haem iron and molybdenum content, azofermo, and nitrogen reductase, to indicate that nitrogen is reduced by this protein. The iron protein has been called component 2, protein 2, azoferrodoxin, azofer and nitrogenase reductase, to indicate that its mode of action is to reduce the iron molybdenum protein. In the subsequent discussion they will be referred to as MoFe and Fe proteins respectively.

The mode of action of the enzyme

The function of the two proteins is to accept electrons from an electron donor and then reduce molecular nitrogen to ammonia. The electrons for the reduction must come from a high energy, low reducing potential electron donor such as ferredoxin or flavodoxin with an E_o' of about -400 mv. ATP and magnesium are also required. The sequence of events is as follows: The Fe protein is first reduced and it then binds ATP and Mg to form a MgATP Fe protein complex. This complex is then able to reduce the MoFe protein and the bound ATP is hydrolysed in the process. The reduced MoFe protein then donates electrons to the substrate that is bound to it.

The Fe protein has an E_o' of -294 mv and on binding MgATP this midpoint potential is reduced to approximately -400 mv (2) or -490 mv (3). The change in midpoint potential denotes an increase in energy of the electrons. This is accomplished by the change in environment of the iron-sulphur clusters due to a change in protein conformation when MgATP is bound. Evidence for such a conformational change is that the protein is more sensitive to oxygen inhibition and more reactive to compounds that can bind iron such as α-dipyridyl and also to sulphydryl reagents such as *DTNB* showing that the iron-sulphur clusters are more exposed. Two ATP molecules bind to the Fe protein and two ATPs are hydrolysed for every electron transferred. It can thus be considered that the energy associated with the terminal phosphate bond of ATP is transferred to the electrons giving them greater reducing power. However, there are indications that ATP may be hydrolysed also during substrate reduction as, under certain circumstances, the ATP:2e ratio can differ with the substrate (1). The substrates that nitrogenase can reduce are given in Table 1. The natural substrates are, of course, confined to protons and nitrogen.

There are a number of inhibitors of nitrogenase. One would imagine that alternative substrates would inhibit the reduction of nitrogen in a competitive manner but this is not the case. Nitrogen inhibits the reduction of acetylene competitively but the reverse, acetylene inhibition of nitrogen reduction, is non-

Table 1. Substrates for nitrogenase.

Substrate	Products
N_2	$2NH_3$
$2H^+$	H_2
N_3^-	N_2, NH_3
N_2O	C_2H_4
HCN	CH_4, NH_3, CH_3NH_2
CH_3CN	C_2H_6, NH_3
CH_3NC	CH_4, C_2H_6, C_2H_4, CH_3NH_2
CH_2CHCN	NH_3, C_3H_6, C_3H_8

competitive. Non-substrate inhibitors, such as hydrogen and carbon monoxide inhibit the reduction of different substrates differently. Carbon monoxide inhibits nitrogen, azide, acetylene and cyanide reductions while proton reduction remains unaffected. Hydrogen competitively inhibits nitrogen reduction, does not compete with azide, acetylene and proton reductions and enhances the reduction of HCN. This evidence, together with similar evidence using other inhibitors has led to the conclusion that different substrates may be reduced at different sites on the enzyme. Another piece of evidence given to support this is the work of Smith *et al.* (4) who combined the MoFe protein of *Klebsiella* and the Fe protein of *Clostridium* and found that there was a lag before the reduction of acetylene, while proton reduction was linear from the start. The rate of acetylene reduction did not reach the same rate as that of proton reduction until 25–30 minutes had elapsed. The rate of nitrogen reduction to ammonia was lower still, being about 10% of the rate of proton or acetylene reduction. This heterologous system, however, reduced substrate at only 12% of the rate of the homologous pair of *Klebsiella* proteins so that perhaps too much should not be deduced from it. With possibly two molybdenum iron-sulphur clusters in the protein, there may be room for more than one active site. Burris (5) points out the possibility that the substrates themselves may modify the sites, which would alter the interaction with inhibitors.

Hydrogen evolution

In the absence of inhibitors and in the presence of nitrogen both protons and nitrogen are reduced. Protons are always reduced, thus evolving hydrogen; under atmospheric nitrogen a minimum of 2 protons are reduced for each nitrogen molecule reduced. This minimum evolution stays constant for both *in vivo* and *in vitro* assays and has thus been seen as an essential part of the mechanism by which nitrogen is reduced. If the nitrogenase is supplied with optimal supplies of ATP and reductant the electron flow through the enzyme system will be at a maximum. This electron flow is not reduced if the supply of nitrogen is cut. In these circumstances less nitrogen is reduced and more hydrogen evolved as a result of proton reduction. This property can be used to assay nitrogenase activity as the total electron flow can be estimated by depriving nitrogenase of substrates other than protons by placing it under argon or helium and measuring the amount of hydrogen produced.

The output of hydrogen from nitrogenase can have important physiological consequences as hydrogen inhibits the reduction of nitrogen but not that of protons. Should hydrogen accumulate, nitrogen reduction will be inhibited but nitrogenase will still produce hydrogen and consume reducing power and ATP and thus be a drain on the organism's resources. The production of hydrogen may be affected by a number of factors:

1. As has been indicated above, hydrogen inhibition will cause a greater amount of hydrogen to be produced.
2. The ratio of the component proteins, Fe:FeMo protein, will affect hydrogen production. This ratio will normally be optimal *in vivo* but can be affected by water stress (6) and also by high oxygen levels which may inactivate the Fe protein, which is more oxygen sensitive, and thus give an unbalanced ratio. In nodules supraoptimal oxygen levels are only likely to be obtained under experimental conditions in which extra oxygen is supplied. It has been shown for *Azotobacter* (7) that very low oxygen concentrations may also affect the amount of hydrogen produced.
3. ATP:ADP ratios can affect hydrogen output. The lower the ratio the greater the amount of hydrogen evolution. The higher hydrogen output at lower oxygen concentrations in *Azotobacter* (7) can perhaps be attributed to this effect as under partial anaerobiosis less ATP will be formed.
4. pH can affect proton reduction. This has only been shown *in vitro* and in these cases the variation in pH necessary to achieve substantial changes in hydrogen output would not be expected under *in vivo* conditions (8).
5. Nitrogen concentration. As the previous discussion shows, lack of nitrogen will lead to more protons being reduced. The low fluxes of nitrogen and the high concentration of nitrogen in air make it unlikely that diffusion will be unable to maintain a high concentration of nitrogen at the enzyme site in natural situations.

These factors are considered when hydrogen evolution is discussed again later.

Acetylene reduction

Acetylene reduction provides a simple, cheap and convenient assay for nitrogenase activity. At a concentration of 10% in the gas phase acetylene inhibits both nitrogen reduction and proton reduction so that the ethylene produced is a measure of the total electron flux through the enzyme. As the flame ionisation detector on a gas chromatograph is so much more sensitive than the katharometer detector necessary for hydrogen, acetylene reduction provides a more sensitive assay for total electron flux than does the measurement of hydrogen evolution under argon.

The hydrogen production by nitrogenase must be taken into account when converting acetylene reduction assays to estimates of nitrogen fixation as such. The conversion ratio is not three, as it would be if the number of electrons required for the reduction of each substrate is considered, but four, as for each nitrogen reduced 8 electrons are transferred through the nitrogenase, two of them to protons.

$$N_2 + 8e + 8H^+ \rightarrow 2NH_3 + H_2 \qquad (1)$$

Experimentally a number of different conversion figures have been obtained. This is because in root nodules more hydrogen may be produced than is shown in equation 1 and also because, in some systems, the nitrogenase may not be fully saturated with acetylene. Because of these difficulties it is essential to check acetylene reduction figures with actual nitrogen fixations assays.

Electron transport to nitrogenase

If the Fe protein has a midpoint potential of −294 mv (2) then NAD(P)H, midpoint potential −320 mv, should be capable of reducing it. This is not so. An electron donor of lower reducing potential is necessary. In *Azotobacter* the electron donor to nitrogenase is a flavodoxin. In order that electrons can be donated to nitrogenase from this, the ratio of fully reduced to semireduced (hydroquinone:semiquinone) has to be high thus demonstrating the need for a low potential electron donor (9). It was estimated that a reducing potential lower than −460 mv was required.

The nitrogenases of anaerobes such as *Clostridium* have electrons transferred to them by way of ferredoxins. While a ferredoxin has been identified in *Azotobacter* it has not proved capable of transferring electrons to nitrogenase and must have some other function. Flavodoxin has been identified also in bacteroid but more recently (10) a ferredoxin has been found in *Rhizobium japonicum* bacteroids which is closely similar to clostridial ferredoxin and acts in the transfer of electrons to nitrogenase.

What is perhaps of more interest than the identity of the actual electron donor to nitrogenase is the way in which these low potential electron donors are reduced in aerobic systems. Clearly they cannot be reduced by the normal biochemical reactions in which NAD or NADP furnish the electron acceptors. Veeger et al (11) have shown for both *Azotobacter* and *Rhizobium* bacteroids that energised membranes are necessary for nitrogenase activity and that membrane potentials are important. Haaker and Veeger (12) have produced a scheme which proposes that the midpoint potentials are altered by changes in pH over the pH gradient across an energised membrane. This scheme does not seem very satisfactory as it involves changes in pH over 4 pH units which is unlikely to be achieved in practice. Chemosynthetic bacteria have the same problem. *Nitrobacter* for instance takes electrons from the redox couple NO_2^-/NO_3^-, E_o' 421 mv, which is far too high a potential to reduce the pyridine nucleotides, E_o' − 320 mv, which are necessary for carbon dioxide fixation and other cell reduction reactions. The reducing power of the electrons is raised by reversed electron flow whereby the electrons are transferred through the electron transport chain in the reverse direction and ATP is hydrolysed to provide the energy for this process (13). One of the main pieces of evidence given for the role of membrane energisation in the provision of reductant for nitrogenase, is that nitrogenase activity in inhibited by an uncoupler at much lower concentrations that that at which the level of ATP concentration is affected in the cell (12). Veeger and his colleagues have however failed to test the effect of the uncoupler on nitrogenase itself, also it is conceivable that reversed electron flow may be more sensitive to uncoupling than flow in the normal direction. The dependence of nitrogen fixation on membrane energisation makes the investigation of this system difficult because of the need for the maintenance of intact membranes.

CARBON METABOLISM

The supply of ATP, which in aerobes will be mainly the result of oxidative phosphorylation, and the supply of electrons depend upon the metabolism of carbon compounds transported into the nodule. The metabolism of carbon compounds is not known in detail for either legume or non-legume root nodules. Such information that we have is mainly derived from the metabolism of isolated bacteroids or vesicles or from free living organisms. The carbon substrates for nitrogen fixation in legume nodules have been thought, until recently, to be the citric cycle and glycolysis intermediates, succinate and pyruvate, as these supported nitrogenase activity in isolated bacteroids whereas glucose and sucrose did not. A recent report (14) shows that both glucose and sucrose support acetylene reduction in isolated bacteroids at low oxygen tensions and that succinate requires a higher oxygen concentration. It has been shown that free living rhizobia have the citric acid cycle.

The isolation of the endophyte, *Frankia*, has enabled progress to be made with non-legume nodule physiology. Blom and Harkink (15) found that *Frankia*, isolated from *Alnus viridis*, could not take up and utilise glucose although it grew on free fatty acids and the fatty acid residues of some detergents as the sole carbon sources. Growth with Tween and added glucose still did not induce the organism to use glucose and it was found that the organism did not possess the glycolytic enzymes, hexokinase, pyruvate kinase and pyruvate dehydrogenase. It did possess the enzymes concerned with the glyoxylate cycle, isocitrate lyase and malate synthetase and it also possessed all the citric cycle enzymes looked for. However, Baker *et al.* (16) cultured the same strain of *Frankia*, Avcl1, on a medium containing yeast extract, glucose and amino acids. Tjepkema *et al.* (17) were able to culture an isolate from *Comptonia* on a similar medium and by changing the carbon source to succinate and omitting any nitrogen source were able to obtain vesicles and nitrogen fixation in culture. Thus although progress has been made, the nutrition of the endophyte in the host plant remains unclear.

The phloem in alder contains sucrose (18) which is thus the main form in which carbon is delivered to the nodule. It is unlikely to be fats. It is possible then that the plant partially metabolises sugars and that compounds such as succinate are utilised by the endophyte in the nodule. This idea is strengthened by the fact that no glyoxylate cycle enzymes were found in the isolated endophyte from nodules.

OXYGEN SUPPLY AND OXYGEN PROTECTION OF NITROGENASE

Aerobic nitrogen fixers have the problem that oxygen needs to be supplied at a high rate in order to meet the large demands of the nitrogenase for ATP and at a low concentration to keep the nitrogenase enzyme in an anaerobic environment. Two factors are mainly concerned with this in legume root nodules, nodule structure and leghaemoglobin.

Tjepkema and Yocum (19) first demonstrated that the structure of the nodule was important when they found that the rate of respiration of the central tissue of the nodule is almost entirely determined by the rate of oxygen diffusion to the respiratory enzymes therein. They based this conclusion on three lines of evidence. Firstly the oxygenation of the leghaemoglobin was low which indicated a low oxygen concentration in the central core; secondly, respiration was dependent upon oxygen concentration: at 23° C there was a linear increase in respiration rate as the oxygen was raised to 0.9 atmospheres and at 13° C there was a rise in respiration rate up to 0.5 atmospheres oxygen. This showed that at both temperatures respiration was limited by oxygen at 0.2 atmospheres. Thirdly they found that there is a band of cells round the cortex which lacked intercellular space, thus forcing diffusion through the liquid phase. As diffusion is 10,000 times slower in the liquid phase than in the gas phase this layer of aqueous diffusion provides a large resistance to oxygen flow into the central core. These findings were further confimed (20) by the demonstration that the intercellular spaces in the cortex were not contiguous with the intercellular spaces of the central tissue. They infiltrated nodules with indian ink and found that the ink was confined to the outer cortical tissue. When the nodules were cut in half and infiltrated with indian ink they found that two thirds of the central tissue was infiltrated through the intercellular air spaces. By using an oxygen microelectrode they confirmed that the concentration of oxygen was low in the central tissue. In half nodules where oxygen could readily penetrate through the interior air spaces they found that oxygen remained high and that infiltrating these air spaces with water reduced the oxygen concentration in the tissue. The effects of water infiltration were, however, less in smaller nodules. More recently Sinclair and Goudriaan (21) have presented a theoretical analysis of gaseous diffusion. They concluded that gas spaces must exist in nodules for adequate oxygen distribution but that these spaces should not be in contact with the atmosphere. They therefore put forward the hypothesis that there should be a layer of cells without intercellular spaces in the cortex to form a diffusion barrier to oxygen. They estimated that this layer should be about 45 μ. They also calculated that, as the flux of nitrogen into the nodule was much lower than that of oxygen and the concentration of nitrogen much higher on the outside, nitrogen fixation would not be limited by the diffusion of nitrogen. The presence of this barrier has now been shown in pea and lupin root nodules (22). It was shown that these nodules have much less intercellular space than soya bean root nodules, 1% and 5% respectively (23). It was found also that in pea and lupin root nodules the intercellular spaces in the central tissue were not continuous as found by Tjepkema and Yocum for soya bean but were separate and discrete and that indian ink infiltration did not occur even though the nodules were halved. It is possible that the intercellular spaces in smaller soya bean nodules were not continuous also since water infiltration had less effect on oxygen concentration than in the large ones.

This diffusion resistance, although it enables bacteroid respiration to reduce the oxygen concentration to low levels in the central core, is not enough by itself fully to control oxygen supply. Oxygen has to diffuse through to the centre of the nodule so that fixation can occur through the whole body of the nodule. It is for this that haemoglobin is necessary.

Leghaemoglobin has a number of points of interest which will be discussed before its function in the nodule is considered. It is a monomeric haemoglobin with a molecular weight of about 16,000. It is thus like one of the subunits of mammalian haemoglobin and is similar in this respect to myoglobin. It is the only known haemoglobin in the plant kingdom and thus it is intriguing to speculate as to how it comes to be found in plants. Has it arisen by convergent evolution as a result of selection pressures of the symbiosis or has it been acquired at some time as a result of gene transfer by natural genetic engineering? There can be no direct evidence to answer these questions. There is a reasonable amount of homology of amino acid sequence with mammalian haemoglobin (24) but this may be discounted as evidence for common origin since the same sequences may be necessary for the same function. More convincing evidence that the leghaemoglobin gene has been acquired rather than evolved within the symbiosis itself comes from the study of the structure of the leghaemoglobin gene. It has been shown that the gene is not one continuous sequence of bases but has three intervening sequences of bases that are not transcribed. The leghaemoglobin gene has been cloned on plasmids and hybrilised with the messenger RNA of the gene. The bases pair where appropriate and the intervening sequences, which cannot pair with the messenger, appear as loops which have been seen under the electron microscope (25). Intervening sequences are found in all genes for the globin component of haemoglobins which have so far been examined. The others have two intervening sequences and not three as found in the leghaemoglobin gene. However, two of the three intervening sequences occur in exactly the same place as the intervening sequences of the other globin genes. This suggests that the leghaemoglobin gene is of common genetic origin with the animal genes and thus leghaemoglobin was obtained by genetic transfer rather than arising by convergent evolution.

Leghaemoglobin is synthesized by cooperation of both symbionts. The plant synthesises the protein part of the molecule whereas the bacterium synthesises the haem group. Leghaemoglobins are typical of the plant rather than of the bacteria (26). Different plants may have a different number of leghaemoglobin molecules. *Lupinus luteus* has four and *Ornithopus sativus* has two or three (26). The different types of leghaemoglobin in each plant differ from each other in a few amino acids only. The leghaemoglobins from soya bean have been most intensively studied. This has leghaemoglobins which have been labelled a, c_1 and c_2 respectively. Leghaemoglobin a differs from the two leghaemoglobins c in 6 amino positions and c_1 and c_2 differ from each other in one amino acid. These minor variants have probably arisen by gene duplication and subsequent mutation (27).

The amount of haem in nodules is large. The concentration of leghaemoglobin, averaged over the central core of the nodule, is about 0.7 mM (28) which is considerably more than any other haem protein. As both plants and bacteria synthesize haem, for haem proteins other than leghaemoglobin, the evidence for the site of the synthesis of the haem component of leghaemoglobin must be quantitative.

Two enzymes that initiate the synthesis of the tetrapyrrole ring are δ-aminolevulinic acid synthase (ALAS):

$$\text{Succinyl CoA} + \text{glycine} \rightarrow \delta\text{-aminolevulinic acid} + CO_2 + H_2O \qquad (3)$$

and δ-aminolevulinic acid dehydrase (ALAD):

$$2\ \delta\text{-aminolevulinic acid} \rightarrow \text{porphobilinogen} + 2H_2O \qquad (4)$$

Nadler and Avissar investigated the activity of these two enzymes in root nodules in both the bacteroid and plant fractions (29). They found that ALAS could only be detected in the bacteroid fraction and that its activity rose during nodule development with the same proportional increase to haem concentration. ALAD is found in both plant and bacteroid fractions with a larger part in the plant. However, the activity of the plant enzyme declines whereas the activity of the bacteroid enzyme stays constant during nodule development. A more direct pointer to the role of bacteroids is given by the fact that whereas ALAD ıs found in the plant fraction of both effective and ineffective nodules (29). The enzyme is only found in the bacteroids of effective nodules. This would seem reasonable evidence that the bacteroids synthesise the haem.

Because of this dual synthesis there is some doubt as to the location of the leghaemoglobin in the nodule cell. One suggestion is that it lies in the space between the bacteroid and the membrane envelope (28). Another is that it is confined to the cytoplasm of the host cell (30). The evidence in each case is produced by different methods and in each case looks equally convincing. It remains possible also that it is in both locations. While it is feasible for the leghaemoglobin to occupy only the space between the membrane envelope and the bacteroids in nodules such as soya bean, where this space occupies about 38% of the volume of the cell (31), it is difficult to envisage such a confined location in the nodules of legumes such as pea and clover where the membrane envelope lies quite close to the bacteroid cells. Such space would be only a very small fraction of the cell volume. Consideration of the role of leghaemoglobin would suggest that it must occur over a larger volume.

The role of leghaemoglobin is concerned with the requirement for keeping the oxygen concentration in the nodule low, in order that nitrogenase is not inhibited, while the bacteroids need a high rate of respiratory activity for nitrogen fixation and maintenance. Thus oxygen has to be distributed at a high rate and at a low concentration within the nodule. The maintenance of the low oxygen concentration is achieved by the structure of the nodule and the high rate of respiration which utilises the oxygen as it diffuses in. Inside the central core there are air spaces which will distribute the oxygen to the nodule cells, although in smaller nodules the amount of space is limited. The cells in the nodule are large and the intercellular spaces are only in contact with a small proportion of the cell surface so that diffusion of oxygen into and within the nodule cells is a problem. This suggests that the role of leghaemoglobin is one of facilitated diffusion.

Fick's law states that

$$dw = -D\frac{dc}{dx}dt \tag{5}$$

where w = flux, the quantity diffusing across 1 cm^2 in time dt when the concentration gradient is $\frac{dc}{dx}$. Thus we can see that the amount that diffuses is dependent upon the concentration gradient. Because of the very high binding power of leghaemoglobin (it is half saturated at an oxygen concentration of 7×10^{-8} M), and its high concentration within the nodule, (about 7×10^{-4} M) even though it has, on average, only about 20% bound oxygen the bound oxygen concentration

$(2 \times 10^{-4}$ M) will be very much higher than the free oxygen $(1 \times 10^{-8}$ M). Thus even though the diffusion constant for free oxygen (1.8×10^{-5}) is much larger than the diffusion constant for leghaemoglobin (about 1×10^{-6}) it can readily be calculated that oxygenated leghaemoglobin will diffuse up to 1,000 times as fast as free oxygen.

Because of the differences in the physiology of the endophytes the oxygen relations of non-leguminous root nodules differ from those of legume root nodules. Non-leguminous root nodules have no haemoglobin although they have a high concentration of haem. The extra haem in this case may be associated with enzymes such as peroxidase and *o*-diphenol oxidase which the nodules contain in high amounts. No correlation has been found between the activities of these enzymes and nitrogenase and they may be concerned with lignification of the nodule (32).

Another important difference is that the intercellular spaces round the infected tissue of non-legume nodules such as those of alder is in continuity, through the lenticels, with the air. Tjepkema (33) considers that the gas phase in these spaces will have a similar oxygen content to the air as there is no diffusion barrier as in legume nodules. However, Wheeler *et al.* (32) found that while the outer cortex had a large amount of intercellular space, where the infected cells occurred there were fewer spaces and that spaces were not found between infected cells. Perhaps there might be some resistance to gaseous diffusion offered by the plant structure but not so severe as in legume root nodules. Tjepkema (33) investigated the nodules of *Myrica gale* (bog myrtle) with an oxygen microelectrode and found that there were places in the nodule with a low oxygen concentration. He ascribes these regions to ones where there was infected tissue. Although the root surface is permeable to oxygen, oxygen may be provided from the nodule roots which grow upwards from the nodule. A large amount of intercellular space in these roots is consistent with a role in gas transport. With the nodules in water and the nodule roots in air, cutting off these roots made no difference to nitrogenase at 20% oxygen. However when the oxygen was reduced to 5% there was a 70% reduction in activity after cutting off the roots. The nodule roots could be important for aeration in natural situations under waterlogged conditions.

Physiological experiments have shown that nitrogen fixation in non-legume root nodules is not limited by the diffusion of oxygen whereas in legume nodules such as those of soya bean a respiratory quotient of about 1.0, is evidence that respiration in these nodules is limited by a physical process such as diffusion. The nodules of *Alnus* have nitrogenase activity with a greater temperature dependence suggesting that diffusion is not limiting the activity. The effect of oxygen also differs and is consistent with the fact that diffusion of oxygen is not a limiting factor. Oxygen concentrations above 20% enhance nitrogen fixation in soya bean root nodules and inhibition only becomes apparent when the concentration of oxygen in the gas phase exceeds 50% and the rate of respiration continues to rise at oxygen concentrations higher than this. Non-leguminous root nodules tend to have optimal fixation rates at 20% oxygen and higher levels are inhibitory.

With high levels of oxygen the need for a haemoglobinlike oxygen carrier is lessened as these are only efficient at facilitating diffusion of oxygen when they are partially saturated, which implies a low oxygen concentration. It is questionable whether non-leguminous root nodules possess such a carrier. If absent, then the high oxygen content of the intercellular spaces would result in an oxygen concentration in the infected host cells high enough to inhibit nitrogen fixation. Resistance to oxygen and thus protection of the nitrogenase would then reside in the endophyte itself. The ability of *Frankia* to fix nitrogen in culture in contact with air (17) must reflect the ability of the organism to do the same in symbiosis. It has been suggested that the vesicles act like the heterocysts in blue green algae. The exact nature of this protection has not yet been found but the vesicles, like the heterocysts, have a thick wall which may be relatively impermeable to oxygen. The fact that non-legume root nodules fix nitrogen with the same sort of efficiency, with regard to carbon metabolism, as legume root nodules would argue against a form of respiratory protection as used for *Azotobacter* where the use of carbon for such a purpose lowers its efficiency considerably.

HYDROGENASE AND RELATIVE EFFICIENCY IN ROOT NODULES

Hydrogenases are present in most nitrogen fixing organisms and as this correlation between nitrogen fixation and the possession of hydrogenase became noticed it was naturally looked for in legume root nodules. Hydrogenase was found in pea root nodules first but, for unknown reasons this early report (34) was not able to be confirmed. Later, hydrogenase was again found in pea root nodules because by chance, the same strain of *Rhizobium* ONA 311 was used (35, 36). This hydrogenase was shown to be essentially similar to the hydrogenase present in *Azotobacter* (37) and is unidirectional. In the nodule it only catalyses the uptake of hydrogen. The reaction is as follows:

$$H_2 \rightarrow 2H^+ + 2e \qquad (6)$$

The electrons are passed to electron acceptors and down the oxidative phosphorylation pathway to the terminal electron acceptor oxygen. During this passage ATP is synthesised (37). Thus the hydrogen, produced by the nitrogenase, can be reutilised by the bacteroids and some of the ATP utilised in its production can be recouped.

The question remains as to the principal function of hydrogenase. It has been postulated that there are three possible functions for hydrogenase in the nodule:

1. That the oxidation of hydrogen serves to utilise excess oxygen and thus assists in the maintenence of nitrogenase in an anaerobic environment.
2. As there is a high demand for ATP in the nitrogen fixation process and as ATP is synthesised as a result of the action of hydrogenase this saving will make the process more efficient.
3. As hydrogen is an inhibitor of nitrogenase, hydrogenase could serve to reduce or prevent accumulation of hydrogen at the nitrogen reduction site. It may be assumed that in the legume root nodule the only hydrogen available as substrate for the hydrogenase will be that produced by the nitrogenase. It can thus be calculated that only 6% of the total oxygen consumed can be attributed to hydrogenase. The same figure has been calculated for *Azotobacter* and for *Rhizobium* (7, 37). Thus the contribution to energy saving and to saving of carbohydrate for respiratory protection is fairly small. The postulate that hydrogenase prevented hydrogen inhibition was put aside at an early stage because it was thought that the high diffusivity of hydrogen would in itself be sufficient to maintain hydrogen at a low level within the nodule. Evidence that this may not in fact be so will be discussed below.

Schubert and Evans (38) assayed the hydrogen evolved from root nodules of a number of species of legumes and non-legumes. They found that a large number of legumes evolved a substantial amount of hydrogen compared with the amount of nitrogen fixed. This demonstrated that legume root nodules were not very efficient since ATP and reducing power, that could be utilised for the reduction of nitrogen, were being expended on the production of hydrogen that was of no use to the plant. Some of the nodules examined however, gave off little or no hydrogen which was evidence that they contained hydrogenase. In their paper they developed the concept of relative efficiency (RE):

$$\text{Relative Efficiency} = 1 - \frac{\text{rate of } H_2 \text{ evolution in air}}{\text{rate of } H_2 \text{ evolution in A, } 80\%\ O_2 20\% \text{ or rate of } C_2H_2 \text{ reduction}} \qquad (7)$$

Values of RE lower than 0.75 indicate that energy is being wasted as more than the minimum amount of hydrogen is being evolved. Values higher than 0.75 indicate that some or all of the hydrogen is being recycled by means of hydrogenase.

Of 19 species of legume examined only two species had hydrogenase and they both had the same strain of *Rhizobium*, R 32H1, whereas 4 of 5 species of non-legume nodules evolved barely any hydrogen thus indicating hydrogenase activity. These species were *Alnus rubra*, *Purshia tridentata*, *Eleagnus angustifolia* and *Myrica californica*. The non-legume that had nodules of lower efficiency, possibly without hydrogenase, was *Ceanothus velutinus*. This paper stimulated interest in the efficiency of nodules and hydrogenase and a lot of work has since been done by Evans and his associates and others (39, 40).

The improvement of RE by hydrogenase is self evident as little or no hydrogen will escape when it is present. What is of concern is the low RE of those nodules which do not possess hydrogenase. The average RE of the 17 legume species was only 0.52 (38). Why does this differ so much from 0.75, which is the RE found for nitrogenase *in vitro* under optimum conditions? The factors which are involved in altering the amount of hydrogen evolved by nitrogenase, or altering its RE, have been briefly discussed. They are ATP:ADP ratio, nitrogen concentration, oxygen concentration, pH, Fe:FeMo protein ratio and hydrogen concentration. An examination of how RE changes in legume root nodules can help to decide which of these factors is most likely to be of effect. There is, at the moment, no direct evidence for the involvement of any one of these factors. However a knowledge of the mechanism which affects the RE is necessary in order both to understand the role of hydrogenase and to assess the energy required for fixation and that conserved by hydrogenase action.

Changes in efficiency have been observed during growth and development of nodulated plants and with changes in light intensity (41, 42). From these studies it is evident that efficiency is lowest when nitrogen fixation activity is at its highest. This would preclude the ATP:ADP ratio as a controlling factor of RE since a high nitrogenase activity indicates a high ATP:ADP ratio which in turn would give a high RE not a low one. It has been calculated that nitrogen is not limiting in nodules (21). The oxygen concentration in nodules is low and they have been shown to be oxygen limited, nevertheless lack of oxygen has been shown not to be the cause of low RE. Experiments in which extra oxygen was added, with nitrogen at the normal partial pressure, showed enhancement of nitrogenase activity and decreased RE compared with control nodules in air (43). This is another example of efficiency being inversely proportional to activity.

Pea root nodules lose their activity after harvest and the opportunity was taken to assay changes in RE with changes in activity. Averaging over a number of experiments, the activity declined from 16 umoles ethylene/g fresh weight/hr to 10.6 umoles ethylene/g fresh weight/hr. The RE rose as activity declined from 0.39 to 0.66. As these changes occurred during an experimental period of 30 minutes it would not be expected that the Fe:FeMo protein ratio or the pH would alter by a sufficient amount to effect these changes. Furthermore when bacteroids were isolated from other nodules of the same set of plants and assayed for RE it was found that when dilute they had a RE of 0.75 although more concentrated preparations exhibited a lower RE. Because washed bacteroids were used in these experiments it is unlikely that soluble inhibitors in the nodule extracts were responsible for this increase in RE with dilution.

Having eliminated the other factors which could alter RE we are left with hydrogen concentration. All these experimental facts are consistent with the hypothesis that hydrogen can accumulate and give a low RE by inhibiting the nitrogenase. It would appear that the oxygen protection mechanism that provides a diffusion resistance to gases entering the nodule is obtained at the price of also limiting the diffusion of gases out of the nodule. Under abnormal conditions of course some of the other factors might operate (6) but it would seem that under normal conditions hydrogen inhibition is the main factor.

Hydrogenase is present in non-leguminous nodules (44) but while of high activity earlier in the season, when nitrogenase was most active, the activity declined during the autumn so that then only part of the hydrogen was recycled. In view of the fact that the nodules of non-leguminous host plants do not have the diffusion barrier which is present in legume root nodules one would, at first sight, assume that the role of the hydrogenase differed from that of legumes if the above discussion is correct. When the hydrogenase is at its most active it was shown to be able to take up hydrogen from the external atmosphere. Roelofsen and Akkermans (44) postulate that as well as the synthesis of ATP by recycling hydrogen from hydrogenase the nodules are also able to utilise hydrogen which is present in wet bog soil where these plants grow. If in the soil there is hydrogen to be taken up this would be added to the hydrogen produced by the nitrogenase unless there were a hydrogenase present so that there is still the possibility that the function of the hydrogenase is the same in both groups of plants.

AMMONIA ASSIMILATION AND AMINO ACID SYNTHESIS

The product of nitrogen fixation is ammonia (equation 1). This product has to be rapidly assimilated into organic combination in order to prevent feedback inhibition of nitrogenase synthesis and also because of its toxic properties.

It is well known that fixed nitrogen compounds inhibit the overall process of nitrogen fixation. The first effect is that of preventing the formation of the symbiosis. This process is not well understood and is not pertinent here but what is of interest is that nitrogen fixation can continue in a nitrogen rich environment of its own making. When ammonia is supplied to free living organisms such as *Azotobacter* it rapidly depressed the level of nitrogenase activity. This is the result of two effects: 1) Ammonia inhibits the flow of reducing equivalents and also possibly ATP by lowering the membrane potential across the *Azotobacter* cell membrane. This is a short term reversible effect and activity is recovered when ammonia is removed (45). 2) The second mechanism is common to other fixing systems as well and is due to the fact that ammonia represses the synthesis of nitrogenase so that as enzyme is broken down and inactivated, activity declines because of the lack of synthesis of fresh enzyme. In contrast to *Azotobacter* it has been found that there is no effect of ammonia on bacteroids of *R. leguminosarum*. The reason for this is that the bacteroids do not accumulate ammonium as it is actively excreted from the cells. Under anaerobic conditions, when active excretion could not take place, it proved possible to load bacteroids with ^{14}C methyl ammonium. On return to aerobic conditions this was excreted from the cells. When the bacteroids were loaded with methyl ammonium under anaerobic conditions and then returned to aerobic conditions under which nitrogen fixation was possible, it was found that nitrogen fixation was inhibited by 80% over that of the control. This confirms that lack of inhibition of nitrogenase by ammonia is not due to a different nitrogenase system but that ammonia is actively removed from the cells and thus does not accumulate to inhibit; also, of course, ammonia supplied to the cells cannot enter them (45). Bergersen (46) found that most of the nitrogen fixed as ^{15}N was excreted into the medium as ammonium by soja bean bacteroids.

Other evidence is concerned with the activity of enzymes that incorporate ammonia into organic combination. The enzymes which catalyse the following reactions are:

Glutamine synthetase

$$\text{Glutamate} + NH_3 + \text{ATP} \rightarrow \text{Glutamine} + \text{ADP} + \text{Pi} \qquad (8)$$

Glutamate synthase, glutamine oxoglutarate aminotransferase (GOGAT)

$$\text{2-oxoglutarate} + \text{glutamine} + \text{NAD(P)H} + H^+ \rightarrow 2\ \text{glutamate} + \text{NAD(P)} + H_2O \qquad (9)$$

Glutamate dehydrogenase

$$\text{2-oxoglutarate} + NH_3 + \text{NAD(P)H} + H^+ \rightarrow \text{glutamate} + \text{NAD(P)}^+ + H_2O \qquad (10)$$

In free living rhizobia the mechanism of ammonia assimilation is dependent upon the nitrogen supply. Under conditions of nitrogen limitation, ammonia is assimilated by glutamine synthetase and GOGAT. This enables ammonia to be utilised at low levels as glutamine synthetase has a much lower K_m for ammonia, 0.02 mM (47), than does glutamic dehydrogenase, 10 mM (48). Under carbon-limiting conditions most of the ammonia is assimilated by glutamate dehydrogenase and GOGAT levels are reduced to zero (49). In root nodule bacteroids all three enzymes are present but glutamine synthetase is present at very low levels. Glutamate dehydrogenase is also at a low level in comparison with GOGAT activity. The glutamate synthetase activity in bacteroids is not sufficiently high to cope with the amount of nitrogen fixed. The levels of both of these enzymes does not rise in the bacteroids (50). This evidence, then, is in accord with the idea that the bacteroids excrete ammonia into the plant cytoplasm where it is then incorporated by plant enzymes.

A very similar system operates in actinorhizal nodules of *Alnus*. No detectable activity of glutamine synthetase was found in isolated vesicle clusters but large amounts were found in the whole nodule nomogenate. The situation with regard to GOGAT is less clear as proportionately large amounts were found in the vesicle clusters (51) Akkermans *et al.* thus concluded that this enzyme together with glutamate dehydrogenase, which was also present, were of endophyte origin. It is possible that there is a shuttle of nitrogen compounds between endophyte and host and that ammonia assimilation is cooperative but more evidence will be needed to be sure of this.

As compounds must cross membranes to be loaded or unloaded from the phloem and xylem transport streams, the transport of sugars in the phloem and amino acids in the xylem are energy dependent processes. It makes sense therefore to carry compounds that are rich in nitrogen from the nodule to the shoot as in this way less carbon has to be cycled. The actual compounds that act as nitrogen transporters vary with the plant species. Lupins export asparagine predominantly, 76% of the nitrogen transported being in this compound and 14% in glutamine. Thus 90% of the nitrogen is exported as amides (52). Asparagine is synthesised in the plant by asparagine synthetase. The activity of this enzyme is not detectable in the bacteroids. The nitrogen for amide synthesis can come from either ammonia or glutamine.

Asparagine synthetase

$$\text{Aspartate} + \text{glutamine} + \text{ATP} \rightarrow \text{asparagine} + \text{glutamate} + \text{AMP} + \text{PPi} \quad (11)$$

or

$$\text{Aspartate} + NH_3 + \text{ATP} \rightarrow \text{asparagine} + \text{AMP} + \text{PPi} \quad (12)$$

The fact that the nodule has a high level of glutamine synthetase, with a $K_m NH_3$ of 0.02 mM (25), and that the $K_m NH_3$ for asparagine synthetase is 3–5 mM led to the conclusion that the synthesis of asparagine arises from the donation of the amide group of glutamine rather than directly from ammonium (52). Aspartate can be generated from glutamate using the amino transferase reaction thus regenerating 2-oxoglutarate for reuse in the GOGAT reaction.

Several species of legumes transport a large fraction of their nitrogen as ureides, allantoin and allantoic acid (53) which is reviewed in detail elsewhere (54). The synthesis of purines from which these compounds are derived is given in Figure (1) Evidence that ureides are obtained from purines as shown in Figure (2) rather than from some other pathway is that enzymes that convert inosine to allantoin and allantoic acid have been found in sufficient quantity, in the plant fraction of the nodules, to account for the amounts of ureides synthesised (55). These enzymes are located either in the plant cytosol or organelles (56).

Interest in the ureides of nodules is recent and no work has yet been published on the synthesis of purines in nodules. The only information available about purine synthesis in plants is from wheat germ and pea seedlings. (57–59). One difference between plant purine synthesis and that of animals and micro-orga-

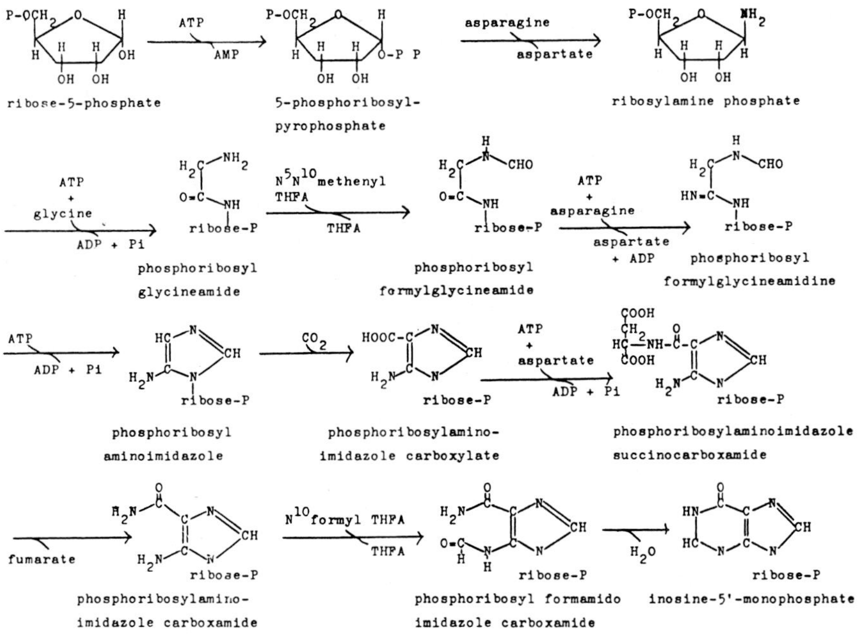

Fig. 1. The pathway of synthesis of inosine monophosphate.

nisms is that asparagine is the nitrogen donor for the first step instead of glutamine. Whether it occurs in the later step, when nitrogen is incoporated from amide, is uncertain as these steps have not been worked out in detail. It is possible that this is so since work from Pate's group (53) has shown for species that do not synthesize ureides significantly, such as peas and lupins, that the nitrogen source N_2, NH_4, NO_3^+ or urea, makes no difference to the types of compunds exported in the xylem sap. However plants which export ureides do so when nodulated but to a considerably less extent when nitrate is the nitrogen source. When nitrate is being used the balance of compounds is similar to that of species which do not transport ureides and there is a high asparagine to glutamine ratio. When fixing nitrogen and exporting ureides the balance of the remaining nitrogen has a low asparagine:glutamine ratio. They suggest that one possible explanation for this is the utilisation of asparagine for ureide synthesis, thus leaving less for export.

Another differences between plants and most other organisms is that hypoxanthine, the free base of inosine, and xanthine are oxidised by a dehydrogenase rather than an oxidase (Equations 13 and 14) and the energy is thus conserved as two moles of NADH + H^+. Thus although much energy must be expended in enzyme synthesis to catalyse the many steps in ureide synthesis the recapture of energy by xanthine dehydrogenase and the high N:C ratio of 1:1 give a favourable energy balance to the nodule and a minimum amount of carbon cycling (55).

Xanthine dehydrogenase

$$\text{Hypoxanthine} + NAD^+ + H_2O \rightarrow \text{xanthine} + NADH + H^+ \quad (13)$$

$$\text{Xanthine} + NAD^+ + H_2O \rightarrow \text{uric acid} + NADH + H^+ \quad (14)$$

Analyses of the amino acids present in a number of non-legume nodules have been reported (60–62) and an analysis of the concentration of nitrogen transport compounds in the xylem sap of both *Alnus glutinosa* and *Myrica gale* is given here in Table 2. (63). It can be seen that citrulline, the largest component in the nodule extract with 64% of the free amino nitrogen, was also responsible for the

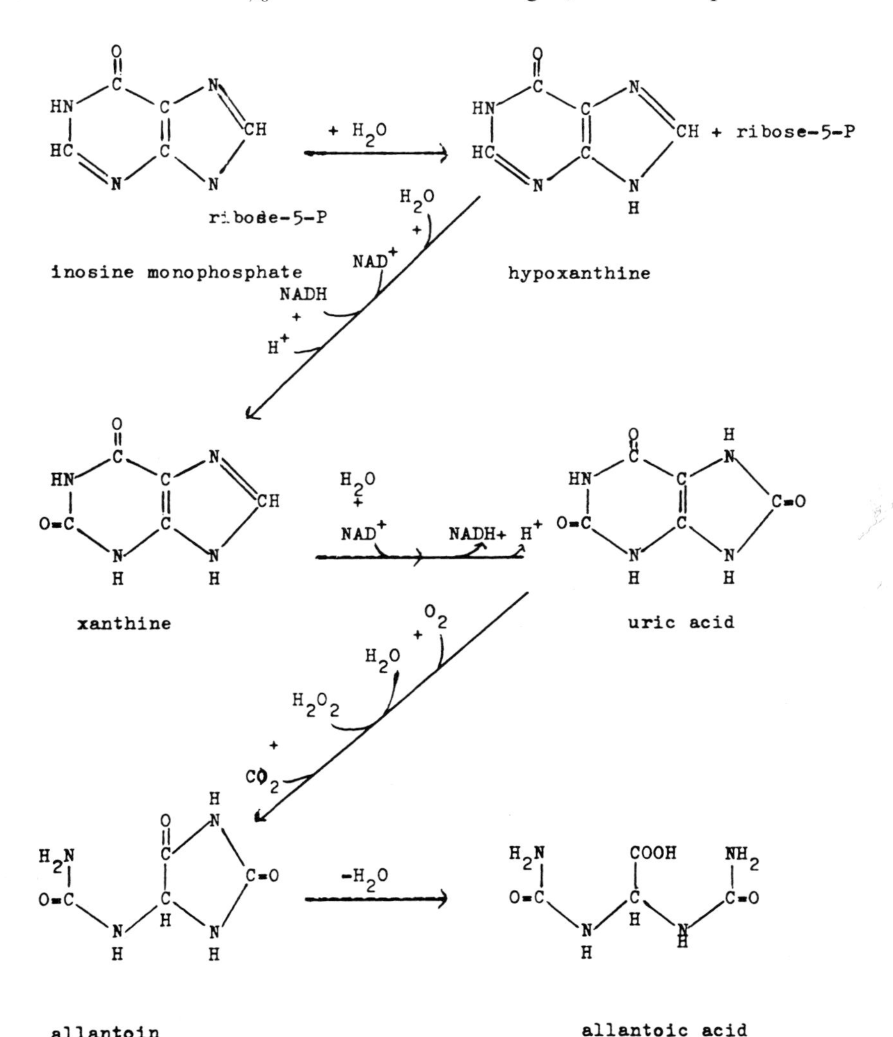

Fig. 2. The synthesis of ureides from inosine monophosphate.

Table 2. Major amino acids of actinorhizal nodules and bleeding sap.

	Nodules μmole g^{-1} fresh wt.	Bleeding sap μmole g^{-1}
Alnus glutinosa		
Aspartic	0.04	0.14
Glutamic	0.02	0.33
Glutamine	0.07	1.50
Asparagine	Not detected	
Citrulline	1.17	4.04
Alanine	0.01	1.34
γ aminobutyric	0.04	0.43
Myrica gale		
Aspartic	0.03	1.17
Glutamic	0.04	1.88
Asparagine + Glutamine	1.65	9.91
Serine	0.56	0.27
γ aminobutyric	0.01	0.03

Nodules were harvested from 9 plants at 13.00 h and bleeding sap from 20 plants between 15.00 and 17.00 h. Analysis essentially as described in reference 61.

greatest amount of nitrogen transport. 88% of the free amino nitrogen in the xylem sap. In the nodules of *Myrica gale* the amides contain 85% of the soluble nitrogen and 75% of the nitrogen in the xylem sap is present in the form of amides. The proportion of nitrogen in the amino compounds in the nodule extracts is then probably a fair reflection of the proportion of nitrogen present in these compounds in the xylem. The predeominant amino compound in most species was asparagine, which comprised 86% and 81% of the total amino acids in *Ceanothus* and *Myrica cordifolia* respectively.

The pathway of citrulline synthesis is as follows:

$$\text{Glutamate} + \text{acetylCoA} \rightarrow \text{N-acetylglutamic acid} + \text{CoA} \qquad (15)$$

$$\text{N-acetylglutamate} + \text{NADH} + H^+ \rightarrow \text{N-acetyl glutamic semialdehyde} + H_2O + NAD^+ \qquad (16)$$

$$\text{N-acetylglutamic semialdehyde} + \text{glutamate} \rightarrow \text{N-acetyl ornithine} + \text{2-oxoglutarate} \qquad (17)$$

$$\text{N-acetyl ornithine} + H_2O \rightarrow \text{ornithine} + \text{acetate} \qquad (18)$$

$$\text{Ornithine} + \text{carbamoyl phosphate citrulline} + \text{Pi} \qquad (19)$$

The final enzyme which catalyses this series of reactions has been assayed in *Alnus* nodules and has been shown, as expected, to be in the plant fraction (51). The enzyme which synthesises carbamoyl phosphate has not been well characterised in plants owing to its instability (64) so that the precise reaction is not known. Bacteria and animal enzymes use ammonia as the nitrogen donor whereas some fungi may use glutamine. O'Neal and Taylor (64) found that the enzyme in peas used glutamine amide nitrogen preferentially to ammonia: Km glutamine = 0.4 mM, Km ammonia = 3 mM. So one might conclude that the following reaction which occurs in *Agaricus* also occurs in plants.

$$HCO_3^- + H_2O + \text{glutamine} + 2ATP \rightarrow \text{carbamoyl phosphate} + + 2ADP + Pi + \text{glutamate} \quad (20)$$

THE ENERGY REQUIREMENTS FOR NITROGEN FIXATION

First of all the theoretical cost of nitrogen fixation will be considered and then studies that have attempted to assess the energy requirements by measuring respiration and nitrogen gain will be discussed.

Energy is used for the fixation of nitrogen in two forms, energy in reducing electrons and energy as ATP. In order to simplify the discussion the energy present in the electrons will be treated as ATP units since, if the reducing power were not used for the reduction of substrate, it could be used to generate ATP by oxidative phosphorylation. For the purposes of this discussion a P/O ratio of 3 will be assumed. (The P/O ratio is the ratio of the number of molecules of inorganic phosphate esterified per atom of oxygen consumed. It is therefore a measure of the efficiency of oxidative phosphorylation). The amount of energy is commonly quoted as gC/gN. Figures for this will be calculated from the theoretical data assuming 36 moles ATP per mole of glucose.

The fixation of nitrogen at its optimum with an RE of 0.75 is given by equation (1) and consumes 4 electron pairs and 4 ATP per electron pair which equals 28 ATP equivalents per N_2 or 2.0 gC/gN. When the RE drops to 0.5 we have the following:

$$N_2 + 12H^+ + 12e \rightarrow 2NH_3 + 3H_2 \quad (21)$$

The cost is now for 6 electron pairs and is 42 ATP equivalents per N_2 or 3.0 gC/gN.

The role of hydrogenase in energy saving

Hydrogenase takes up hydrogen evolved from nitrogenase and can resynthesise some of the ATP used in its production. With an RE of 0.75, 3 ATPs might be saved in this way and the cost will reduce to 25 ATP equivalents or 1.78 gC/gN. With an RE of 0.5, 9 ATPs might be recovered and the cost will be reduced to 33 ATP equivalents or 2.36 gC/gN. However, should the theory that the lower RE is caused by hydrogen inhibition in the latter example RE will revert to 0.75 and the cost will be 25 ATP equivalents or 1.78 gC/gN. Thus if hydrogen inhibition is the cause of the lower RE there will be twice as great a saving as if the cause of the lower RE were other than this. A knowledge of the reason for lowered RE is obviously essential if we are to understand the effect of hydrogenase.

The synthesis of transport compounds

The assimilation of ammonia into organic combination also requires energy. The amount of energy will depend upon which nitrogen compounds are synthesised and exported. Common to all systems is the synthesis of glutamic acid by glutamine synthetase and GOGAT (equations 8 and 9). This will utilise 1 ATP and 1 NADH + H^+ per N and so another 8 ATP equivalents per N_2 or 0.57 gC/gN must be added for nitrogen incorporated into the α position. The amide nitrogen of glutamine requires 1 ATP and that of asparagine 3 ATP as in this reaction (equation 11) ATP is hydrolysed to AMP and pyrophosphate. It is also necessary to add 1 ATP for the incorporation of CO_2 to replenish the citric acid cycle by anaplerotic reactions or for the formation of oxaloacetic acid from pyruvate for aspartic acid synthesis. Assuming an average composition of xylem sap as before with asparagine 76%, glutamine 14% and other amino acids 10% a further 7.8 ATP equivalents per N_2 or 0.56 gC/gN are required. Thus the minimum estimate for the fixation of nitrogen and assimilation into transport compounds will be 33 ATP equivalents or 2.34 gC/gN.

Similar calculations can be done for ureide-producing nodules using the schemes in Figures 1 and 2 and making the assumption that 1 ATP equivalent is used for the synthesis of formyl tetrahydrofolic acid and 4 ATP equivalents are required for the synthesis of methenyl tetrahydrofolic acid. This gives 23 ATP equivalents for the formation of inosine monophosphate. 6 ATP equivalents are recovered by xanthine dehydrogenase giving a total of 8.5 ATP equivalents per N_2. Averaging the figures given by Pate *et al.* (53) then 86% of the nitrogen is transported as ureides and about 4 and 5% as asparagine and glutamine respectively. This requires expenditure of 7.9 ATP equivalents per N_2 or 0.56 gC/gN, which is approximately the same as the value when asparagine and glutamine are the main transport components. The higher nitrogen: carbon ratio will however give savings in transport costs. However, glutamine can be further metabolised in the shoot to produce energy whereas the carbon in the ureides has little energy content. These theoretical energy costs calculated on the assumption that P/O = 3 are summarised in Table 3 and compared here with similar calculations assuming P/O = 2. These latter data are not in agreement with Schubert and Ryle (65) who did not take account of hydrogen evolution in their calculations.

The other compound of quantitative importance is citrulline. Taking the reactions given in equations 15–20 and assuming that acetyl CoA is 1 ATP

Table 3. Theoretical energy costs of nitrogen fixation.

		gC/gN P/O = 3	gC/gN P/O = 2
Nitrogenase	RE = 0.75	2.0	2.57
	RE = 0.5	3.0	3.85
nitrogenase	RE = 0.75	1.78	2.36
+			
hydrogenase	RE = 0.50	2.21	3.2
	RE = 0.5*	1.78	2.36
Assimilation	Amides	0.56	0.72
into transport compounds	Ureides	0.48	0.63

* Energy required if hydrogen inhibition is the cause of a low RE and is removed by the action of hydrogenase.

equivalent it can be calculated that 9.3 ATP equivalents per N_2 are consumed. Clearly citrulline is less energy efficient as far as the nodule is concerned than other means of nitrogen transport.

If the assumptions made are correct, these theoretical estimates will give the minimum amount of energy that is required for nitrogen fixation. Allowance has also to be made for nodule growth and maintenance and for the transport of substances in the phloem and the loading of solutes into the xylem. Overall costs must therefore be established empirically.

With the growing interest in the economy of crop growth and also in the effects of different nitrogen sources on crop growth a number of attempts have been made to assess the total energy required for nitrogen fixation. These have led to wildly different estimated ranging from 0.3 to 14.5 gC/gN (42, 65). One of the reasons for this is that it has not been possible in the past to assay nitrogenase activity, oxygen uptake and carbon dioxide evolution at the same time on the same tissue. Most estimates have been made using carbon dioxide evolution as a measure of carbohydrate consumed, assuming an RQ of 1.0. The use of a new technique throws doubt upon estimates based on this. Direct mass spectroscopy as described by Lespinat *et al.* (66) enabled the estimation of acetylene reduction, carbon dioxide evaluation and oxygen uptake at the same time on samples of soya bean and pea root nodules (67). In all the experiments the RQ was less than 1.0 and this resulted in a higher estimate of the cost of fixation than estimates based on carbon dioxide. This study gave averaged values of 7.0 and 10.5 gC/gN with an RE of 0.75 and 0.5 respectively when carbon dioxide was taken as a measure of carbohydrate consumption and corresponding figures of 8.9 and 13.3 gC/gN when oxygen was used as a measure. Most values based on carbon dioxide vary between 6 and 7 gC/gN (42, 65) and the few measures quoted for oxygen between 10.0 and 12.6 gC/gN. Thus the theoretical cost of nitrogen fixation varies between a third and a fifth of that used by nodules. The overheads are high!

The savings than can be achieved by the possession of hydrogenase have to be considered in the context of the overall cost of nitrogen fixation including these overheads. As there are no separate estimates of the total cost of fixation for nodules with and without hydrogenase this must be a theoretical estimate. The overall respiration has been calculated to be about 5 O_2 per electron pair (66). This interestingly is the same as that for *Azotobacter* when operating at maximum efficiency (7). On this basis the savings by hydrogenase amount to 2.5% and 5% of the energy for RE values of 0.75 and 0.5 respectively. However if hydrogen inhibition is relieved by hydrogenase so that an RE of 0.75 is attained the savings rise to 33%. Savings of 2.5 and 5% would be difficult to demonstrate whereas a saving of 33% should clearly show. Estimates of the potential of hydrogenase for saving energy have been done over long periods by comparing growth of plants with nodules with and without hydrogenase and increases in dry matter and nitrogen have been obtained. As an example Schubert *et al.* (68) compared two strains of R. *japonicum*, USDA 31 without hydrogenase and USDA 110 with hydrogenase, in soya beans. They found that plants inoculated with USDA 110 gave increases in total drymatter of 24%, in percent nitrogen of 7% and of total nitrogen fixed of 31%. A similar study with cow peas which were inoculated with a hydrogenase positive strain or a hydrogenase negative strain gave similar but not so dramatic results with an 11.5% increase in dry matter and 15% increase in overall nitrogen content. Bearing in mind that the relative efficiencies of the nodules in these experiments were much higher over most of the growth period than 0.5 and that there may be synergistic effects these are the sort of differences that one would anticipate from the considerations given above. Schubert and Ryle (65) estimated the savings by hydrogenase to be between 10 and 15% but they did not take the overhead costs into account.

A comparison with nitrate reduction

The theoretical cost of nitrate reduction in plants is a little less than that of nitrogen fixation. The reduction to ammonia requires 8 electrons,

$$NO_3^- + 8e + 10H^+ = NH_4^+ + 3H_2O \qquad (28)$$

which with a P/O ratio of 3 is equivalent to 12 ATP per N or 24 ATP per N_2. This compares with 28 ATP per N_2 for nitrogen fixation or 25 ATP if hydrogenase is involved. If the same overhead costs of ammonia assimilation and amino acid transport are involved one would expect to find little or no difference in total energy cost and no difference in plant growth due to the difference in nitrogen source. There is however the added complication that an unknown amount of energy is required for NO_3^- uptake by the roots and that a proportion of the nitrate may be reduced in the leaves. As the leaves are closer to the sinks for nitrogen transport costs will be reduced. Also as nitrite may be directly reduced as a result of photosynthesis under conditions of carbon limitation at least, one might expect that some nitrite and nitrate reduction will be performed at no cost to carbon. The question then really, if there is to be a major difference between the two sources of nitrogen, is whether the plants are limited for carbon dioxide for any appreciable time. In temperate climates carbon dioxide limitation is only likely to arise under conditions of water stress when the stomata close. It is not certain how water stress may also affect nitrate reduction. One must also consider that under these conditions there may be a shortage of nitrogen acceptors if carbon fixation is not occurring and this may also have an effect on nitrate reduction. There is obviously uncertainty on all these points so that it is not possible to do calculations concerning the effect of photosynthesis. If nitrate is reduced at the expense of carbon fixation the cost to the plant will be the same as if that carbon was respired to achieve the same end.

Work carried out in a number of laboratories to find whether or not there is any difference in the cost of these two forms of nitrogen to the plant has been summarised by Schubert and Ryle (65). What comes out clearly is that the methods of assessment affect the conclusions that are reached. Experiments in which carbon dioxide fluxes are measured all suggest that nodulated plants have higher rates of respiration than plants dependent upon nitrate and this indicates an extra expenditure of energy for nitrogen fixation. Two papers using growth analysis with the two forms of nitrogen come to the conclusion that there is little or no difference in total cost to the plant between the two forms of nitrogen. The possible reasons for the differences in costings, arrived at by different experimental approaches, will not be considered here but the growth experiments would seem to answer the practical question as to whether it is better to inoculate legume crops with an appropriate strain of *Rhizobium* or apply nitrogen fertiliser. The IBP experiments described by Nutman (69) put this to test in an agricultural situation. Application of nitrogen fertiliser to uninoculated plots without other fertiliser addition gave an increase of 76% in yield whereas a similar addition of nitrogen fertiliser to plots which had been inoculated and supplied with lime, phosphorous and potassium gave an increase in yield of only 24%, the yield in this instance being expressed as the amount of nitrogen in the crop. These experiments show that fertiliser added to inoculated legumes increased the yield but that the added fertiliser was much less cost effective than when added to the uninoculated crop.

INTERACTION WITH THE ENVIRONMENT

An understanding of the limitations imposed on biological systems capable of symbiotic nitrogen fixation by the physical and chemical components of the environment is essential for the design of genetic improvement programmes and efficient forest management practices which will maximise the input of symbiotically fixed nitrogen into the forest ecosystem. The environment interacts with nitrogen fixing systems at three levels. Firstly, by effects on the free-living microsymbiont in the soil; secondly, by interaction with the microsymbiont during infection and subsequently during its multiplication and activity in the nodule; thirdly, through effects on the physiology and metabolism of the host plant which may influence any or all of the stages which produce and support the activity of the functional nodule.

If the microsymbiont of interest is *Rhizobium*, then environmental effects on the free-living micro-organism may be monitored fairly readily by application of standard culture technique for assay of the nature and number of soil-borne cells (70, 71 and Chapter 4). Isolation and culture techniques for *Frankia* which will allow the distribution of these organisms in the soils to be studied with equal facility have yet to be devised. Particular problems for experimental design are posed after the microorganism enters into symbiotic association with its host plant due to the difficulty of separating direct effects of the environment on the nitrogen fixing process from those which act through other aspects of host plant physiology and metabolism which support the root nodule function, such as photosynthesis, translocation, etc. The usual experimental approach to this question involves comparison of nodulated plants with non-nodulated plants, provided with combined nitrogen and grown under the same environmental conditions. It is then presumed that some of the differences between the two sets of plants are the result of effects on nodule development or function *per se*. This assumption is, of course, not always valid for apart from the technical difficulties involved in supplying non-nodulated plants combined nitrogen at the same rate as the nodulated plants fix nitrogen, the two groups of plants always develop rather obvious differences of form, particularly in their root development. For example, in a series of experiments in which nodulated and non-nodulated *Alnus glutinosa* were compared the roots of non-nodulated plant always comprised 10–20% more of the plant total fresh weight, even though shoot growth of the two sets of plants was maintained at similar rates by controlling the levels of mineral nitrogen supplied to the non-nodulated plants (72). Obviously, it could be argued that differences in the response of two such groups of plants to a particular environmental stress might result from differences in root form rather than from effects special to the nodules. Satisfactory procedures to overcome this problem have not been devised. Comparison of nodulated plants with non-nodulated plants supplied with a range of concentrations of nitrogenous salts, thus producing plants with different root to shoot ratios, might assist the interpretation of comparative experiments of this type.

Much of our knowledge of the relationships between the environment and biological nitrogen fixation in nodulated plants stems from the study of the agriculturally important legumes, many of which because of their convenient size and rapid growth, form good experimental subjects. Perennial nitrogen fixing species, which are likely to be of most use in forestry because of their ability to provide an input of biologically fixed nitrogen into the ecosystem over many years, have been made less well studied. These may be herbaceous perennial legume species such as lupins (73), some woody legumes such as *Acacia*, *Robinia* (74), or the woody non-legumes, with their nitrogen fixing association with *Frankia* (75). This last diverse group of plants has particular potential for the input of nitrogen and perhaps also as timber species in their own right in many forest situations.

Any factor in the environment which effects the growth of the host plant is likely to affect also the development and function of the root nodules. The factors which generally have most effect on the nitrogen fixing process are considered individually below, although the precise effect of each factor requires individual investigation for each association. Because of the difficulties associated with isolation of *Frankia*, virtually all the information concerning adaptation to the environment by the free living microsymbiont relates to *Rhizobium*, although there is now good reason to suggest that such variability exists in *Frankia* also.

Temperature

Strains of *Rhizobium*, adapted to the particular temperature regime of their natural environment, have been described both for cold and for warm climates (e.g. nodule isolates from clover in sub-arctic Scandinavia; isolates from cowpea in the tropics (76, 77). The geographical origin of the host plant is also important in determining temperature limits for nodulation. In species relatively tolerant of cold conditions, such as *Vicia faba* or *Lotus corniculatus* it is 4–6° C whereas temperatures above 18° C are necessary for nodulation of tropical and subtropical species. (78, 79).

Detailed studies of temperature effects on the development of nitrogen fixing nodule are confined to relatively few legumes. The processes of infection and nodule differentiation are particularly affected by low temperature. The result being a reduction in infection thread growth in *Trifolium glomeratum* and an increase in the period required for nodule initiation in this species, in *Trifolium subterraneum* and in *Lotus corniculatus* (78, 79, 82). Temperatures slightly above those necessary for infection and growth have also been shown to be necessary for nodule initiation in the tropical pea variety 'Iran' (81).

Nodule number is not usually greatly affected by temperature except at temperatures close to the lower limit for nodulation where both the numbers and amount of nodule tissue produced by the plant are reduced (78, 81, 83). A moderate reduction in temperature can actually increase the proportion of nodules formed on the plant although this may then function at reduced activity

(84, 85, 86). For example, in *Trifolium subterraneum* a greater volume of active bacteroid tissue was formed at root temperatures of 11° C compared with 15° C or 19° C but with a lower rate of nitrogen fixation per unit of bacteroid tissue (83). This somewhat surprising effect of temperature on the development of nitrogen fixing tissues is probably due to differential temperature effects on the activity of the nodule meristem and on differentiation of the bacteroid containing tissues. Retardation of the rate of bacteroid differentiation and senescence at lower temperatures has been demonstrated in *Trifolium subterraneum* (87). A larger overall size of the nodules of *Vicia faba* grown at low temperatures is due to a change in the balance between nodule meristematic activity and the rate of differentiation of the nitrogen fixing tissues, so that nodules are larger at each stage of their development (88). This production by the plant of a larger volume of nodule tissue can help to compensate for the lower specific activity of nitrogen fixation in nodules at these lower temperatures (79).

High temperature inhibition of nodulation and nitrogen fixation also affects different stages of nodule development as well as affecting directly the expression of nitrogen activity. Thus, it has been suggested that in peas and beans (81) inhibition of root hair function may be a cause of reduced nodulation at 30° C, while in *Trifolium subterraneum* temperatures of 30° C reduced the period when the nodule was active in nitrogen fixation by accelerating bacteroid degeneration (89). Upper temperature limits for nodulation are higher in tropical than temperature species, although as with low temperatures, the precise effects vary with species of host plant and with *Rhizobium* strain (79).

Apart from a description of delay of nodulation of *Ceanothus velutinus* at low (10–15° C) and inhibition at high (greater than 31° C) temperatures (90), the effect of temperature on the development of actinorhizal nodules has been little studied. Several studies of the responses of nitrogenase activity to temperature have been made, mostly by means of the acetylene reduction assay, which have shown significant differences in temperature optima, particularly between temperate and tropical species. As discussed earlier nitrogenase activity in actinorhizal species tends to be more sensitive to temperature change than in legumes, which show rather broader temperature optima over the range 5° C to 25–30° C with maximum activity between 20–30° C (91). Temperate legumes tend to show optima at the lower end of this range and tropical/sub-tropical species at the upper end. In temperate actinorhizal species such as *Alnus glutinosa*, *Myrica gale* and *Hippophae rhamnoides*, optima around 20–25° C have been demonstrated (91–93) while optima of 35° C or higher have been shown for sub-tropical *Casuarina* (94). Environmental adaptation has been clearly demonstrated for some legumes e.g. a low temperature optimum of 12–15° C for *Astragalus alpinus* in Northern Scandinavia (95) but the only evidence favouring similar adaptation for actinorhizal species is the slightly higher temperature optimum for acetylene reduction of glasshouse grown *Alnus glutinosa* (92) compared with field grown plants (96) and a higher temperature optimum for *Alnus rubra* in the warmer soils (in summer) of plants of the same species in Scotland. (97, 98).

By contrast to the short-term studied described above, which have examined immediate effects of temperature on nitrogenase activity, information concerning temperature effects on nodulated plants over longer periods is much more limited. In the latter type of experiment, a range of whole plant activities which support nitrogen fixation, would be affected by the particular temperature regime. Short term measurements of nitrogenase activity e.g. by acetylene reduction assay, might be expected to underestimate the effects of temperature on nitrogen fixation compared with estimates derived from measurements of nitrogen accumulation over a long growth period. However, comparison of data obtained with *Trifolium subterraneum* does not show large differences in response to temperature between measurements of nitrogenase activity (by acetylene reduction) when plants previously grown at favourable temperature were incubated at the required temperatures, or by nitrogen accumulation when plants were grown in combined nitrogen-free media for 20 days at different root temperatures (Fig. 3). Thus the rate of acetylene reduction at 24° C was a little less than twice that at 8° C, while N accumulation at 24° C was just over twice that at 8° C. It would seem, therefore, that contrary to what might be expected, short term measurements of nitrogenase activity may provide a reasonable guide to the relative amounts of nitrogen which may be fixed by legumes grown at different root temperatures. It remains to be determined whether the greater sensitivity to temperature of nitrogenase activity in nodules of actinorhizal species is shown also when temperature acts on the whole nodulated plant in long term growth experiments.

Light

Nodule development and function is influenced by light intensity, quality and duration. Changes in the rate of photosynthesis following changes in light intensity affects the supply of photosynthates to the nodules, which as discussed above, affects the availability of reductant, ATP and carbon skeletons required for the assimilation of nitrogen (100). Light wavelength discrimination though a phytochrome mediated mechanism is also suggested by the demonstration of far red light inhibition of pea nodulation and its reversibility by red light (101). This latter observation has important implications for root nodulated plants introduced into the forest understorey as a cover crop and to increase soil fertility. In such plants, nitrogen fixation may be inhibited not only by a reduction in photosynthesis due to shading (102) but also by inhibition of nodulation due to the increased proportion of far red light in the sub-canopy light spectrum following preferential absorption of red light by the canopy (103, 104).

While further work is required to confirm the importance of phytochrome-

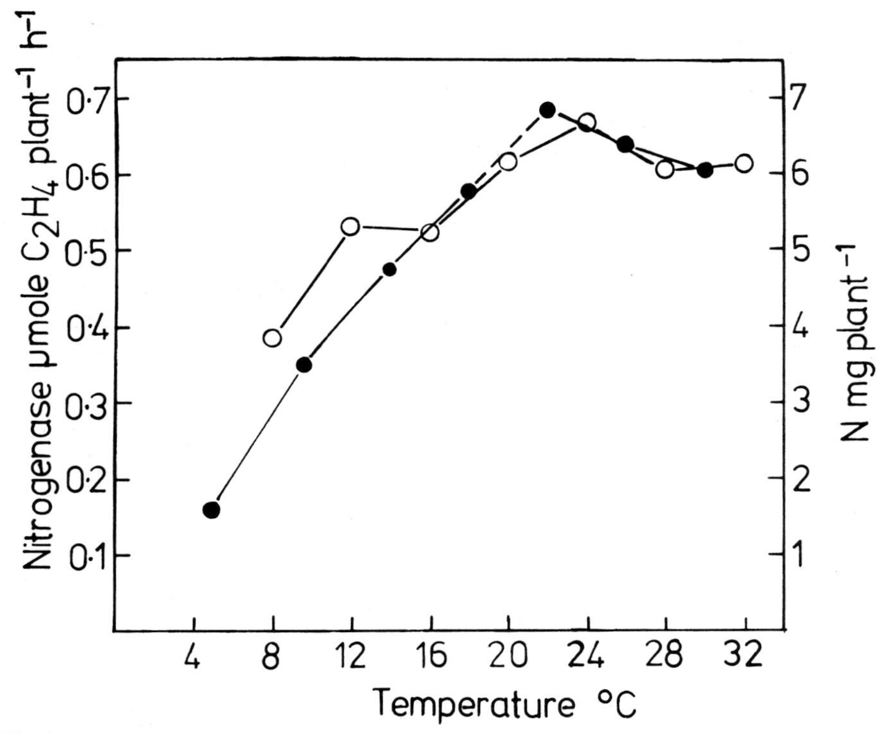

Fig. 3. Effects of temperature on nitrogen fixation. *Trifolium subterraneum* c.v. Tallarook, inoculated with *Rhizobium trifolii* strain TA1 and grown for 28 days at different root temperatures before analysis for total N ●—●, or grown at favourable temperatures before assaying for acetylene reduction at the appropriate temperature ○—○. Redrawn from refs. 99 and 79. The long-term growth data are from two separate experiments, the results from which are separated on the graph by the broken line.

mediated responses for nodulation in natural situations, the general relationships between photosynthesis and nitrogen fixation is well established. Interruption in the supply of photoassimilates to the nodules will reduce or virtually eliminate nitrogenase activity, sometimes within 24 h (105–107). Similarly, treatments designed to stimulate photosynthesis may enhance nitrogen fixation e.g. supplemental light (108, 109) or enrichment with CO_2 of the air supply to the host plant (110, 111). An increase of up to 5 fold in the nodule mass and nitrogen fixation of field grown Soybeans has been achieved by enrichment with CO_2 of the air supplied to plants growing in special open top enclosures (112). Two sets of experiments will be described which serve to illustrate how differences in host plant photosynthesis can affect nitrogen fixation during plant growth and, conversely, how differences in effectivity in nitrogen fixation of the micro-symbiont can influence host plant photosynthesis. In the first experiment (113) reciprocal intervarietal grafts of roots and shoots of eight soybean genotypes, each inoculated with the same strain of *Rhizobium japonicum*, gave large differences in nodule weight per plant between shoot genotypes, which in turn was positively correlated with total nitrogen fixation per plant (Fig. 4a). Differences in the photosynthetic rate for each shoot genotype resulted primarily from differences in leaf area since the photosynthetic rate per unit leaf area was quite similar between genotypes. Enhanced nodule development and hence higher rates of nitrogen fixation by these plants thus seemed to be related directly to the ability of the shoot to provide photoassimilates for nodule function.

In the second experiment (114), the increasing amounts of nitrogen fixed in peas inoculated with strains of *Rhizobium leguminosarum* differing in symbiotic

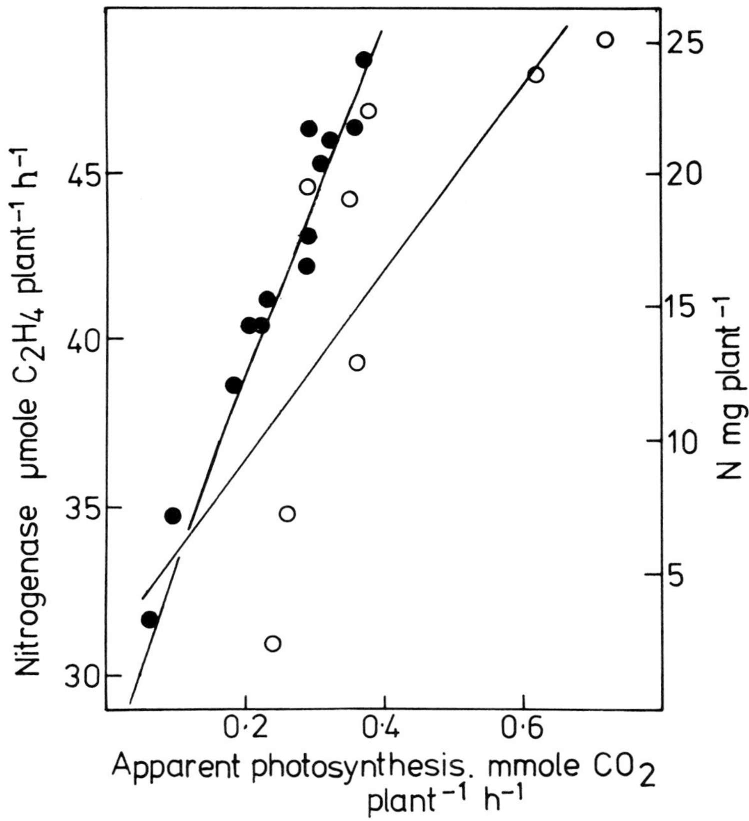

Fig. 4. Relationships between photosynthesis and nitrogen fixation. a) Shoot genotype effects (photosynthesis) on nodule activity (acetylene reduction) of eight soybean varieties subjected to reciprocal intervarietal grafts of root and shoot. Plants were inoculated with the same strain of *Rhizobium japonicum* ○—○. b) Photosynthesis and nitrogen fixed over a 25 day growth period of *Pisum sativum* cv. Alaska inoculated individually with different strains of *Rhizobium leguminosarum* ●—●. Redrawn from refs. 113 and 114.

effectivity, was accompanied by a curvilinear increase in whole plant apparent photosynthesis and in plant dry weight at harvest (Fig. 4b). A decrease in the relative amount of nitrogen allocated to the root and an increase in the relative nitrogen content of the leaves was a feature of plants inoculated with the most effective *Rhizobium* strains. Good correlations between leaf nitrogen content and rate of photosynthesis have been found for several species e.g. soybean, black poplar, spruce (115). Although the reasons for the stimulation of photosynthesis by nitrogen is not clear, the mechanisms may involve changes in leaf chlorophyll content or in ribulose bisphosphate carboxylase (115).

The precise response of nitrogen fixation to changes in light depends on complex interactions with other environmental factors such as temperature and moisture, and host plant factors such as age, height and content of storage compounds. The complexity of these interactions is shown clearly by the varied changes in nitrogenase activity which have been reported in the literature to follow diurnal fluctuations in light intensity. Changes in temperature, rather than in light intensity, often appear to be the major factor controlling daily changes in nitrogenase activity (116–119). In some experiments, however, diurnal fluctuations have been demonstrated which cannot be ascribed wholly to temperature difference. For example, a three fold diurnal variation in the acetylene reduction rate of peas grown at constant temperatures, which was ascribed to a decline in available carbohydrate in the nodules in the dark period (120); a 2.5 fold increase in acetylene reduction by field nodules of *Alnus incana* at 15.00 h compared with 0.700 h which was accompanied by little fluctuation in soil temperature (121); a 4–6 fold increase in acetylene reduction by field nodules of young *Alnus rubra* at midday compared with midnight; and a doubling of the nitrogenase activity of the nodules of young *Alnus glutinosa* plants over the midday period when temperature variation was held to within 2° C. (123). This last variation was not detected in older plants, growing under the same conditions (124) nor were significant diurnal variations found in field nodules of *Alnus glutinosa* (96). Moisture stress has been suggested as an important factor influencing diurnal changes in acetylene reduction in four year old *Alnus glutinosa*, with positive responses to fluctuations in solar radiation or temperature being most marked under conditions of ample soil moisture (123). Using the ^{15}N method for measuring nitrogen fixation and from separate measurements of the effect of temperature on nitrogen fixation a major portion of the higher day time fixation rates of *Casuarina* nodules was suggested to be due to the effects of light (94). This variability in response to environmental change between different species and even for the same species under different conditions, suggests that it is not possible to predict with confidence the pattern of daily fluctuations in nitrogen fixation in nodulated species except under clearly defined and maintained conditions of both plant and environment. Under the relatively constant conditions of light temperature and humidity obtained by growing plants in controlled environment cabinets, diurnal rates of nitrogen fixation frequently remain relatively constant.

In the nodulated plant, the root nodules are continually competing with other organs for a share of the available photosynthates to support their growth and metabolism. The factors which determine the distribution of assimilates between different organs are not understood, but obviously competition between 'sinks' will change during growth and with the development of different organs as the plant changes from the vegetative to reproductive state. In some legumes, under some growth conditions, nitrogen fixation declines during fruit development, probably as a result of competition for photosynthate between nodules and fruits (108, 126, 127). On other occasions, a decline in nitrogenase activity has not been demonstrated until the plant senesces following seed maturation (119, 128). Although there may well be intrinsic differences between species, or varieties of the same species, in their ability to supply photoassimilates in sufficient quantities to satisfy the demands of both the nodules and of other competing 'sinks', environmental and cultural factors also appear to play an important part in mediating the competition for photoassimilates between different organs. Thus, a maximum in nitrogenase activity observed shortly after flowering in *Vicia faba* became less pronounced as plant density increased and factors such as soil moisture content and shading wielded a major influence on activity (129, 130). Although shading reduced activity, the nodules senesced later due to delayed senescence of the leaves (130).

Annual changes in nitrogen fixation have been described for several woody, actinorhizal species e.g. *Alnus* (96, 98, 131), *Myrica gale* (132) *Purshia tridentata* (133), *Hippophae rhamnoides* (134). The precise pattern of change in nitrogenase activity obviously will vary considerably depending on the local environment of the study site.

Data for two perennial shrubby species are compared in Fig. 5. The annual cycle of nitrogenase activity in nodulated deciduous perennials depends primarily on interactions between photoperiod, photosynthetic leaf area and soil temperature, which determine the time of commencement and cessation of nitrogen fixation.

Myrica gale was growing on poorly drained hill sites in Scotland and acetylene reducing activity was first detected in June when soil temperatures rose to about 10° C. Maximum nitrogenase activity was attained in July/August and then declined to zero in November, when soil temperature had fallen to about 6° C. In *Purshia tridentata* on elevated sites in N.W. America, nodule activity was detected in May or June, when the leaves were fully emerged, and reached a maximum in June of July. Activity was severely curtailed by drought in July/ August. In evergreen perennials nitrogen fixation may be detected throughout the winter provided that soil temperatures do not fall too low (97). Interactions of light and temperature stimulate maximum nitrogenase activity in mid-

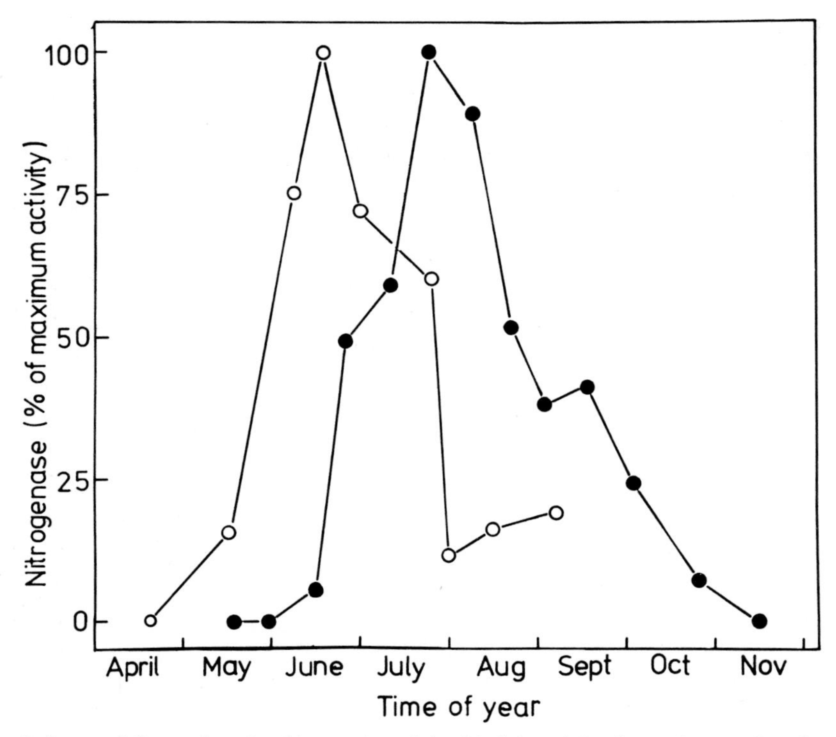

Fig. 5. Seasonal fluctuations in nitrogenase activity. Nodules of *Purshia tridentata* ○—○ *were from central Oregon, U.S.A. and of Myrica gale* ●—● from a hill site in Perthshire, Scotland. Redrawn from refs. 133 and 132.

summer, but here droughting may severely inhibit nodule activity and limit the period over which high rates of nitrogenase activity are maintained. Nodules activity in the period up to mid-summer is probably important in supplying nitrogen to support new growth but after this may be used to replenish stores of nitrogen in storage rhizomes (132) or in shoots (98) for overwintering for support of new spring growth.

Little is known of the fluctuations in nodulation and nitrogen fixation which occur during the whole life cycle of the perennial nodulated plant. For example, it is not known how the development or reproductive growth or plant to plant interactions influences nitrogen fixation, or indeed how the build-up of soil nitrogen over a period of years under a stand of perennial nitrogen fixing plants affects nodulation. These remain topics for future investigation.

pH

Rhizobium varies widely in its tolerance of extremes of pH, with the slower growing species tending to show most tolerance of acidic conditions to about pH 3.5. Rhizobia of the 'cowpea' group generally can best withstand acidic soils with the tolerance shown by different species decreasing through *R. lupini*, *R. japonicum*, *R. leguminosarum* and *R. trifolli* to *R. meliloti* (135). Raising the pH of acidic soils by liming stimulates multiplication of rhizobia in the soil and in the rhizosphere and hence improves nodulation (136).

The critical pH for nodulation of most legumes lies above pH 4.5 to 5.5 (133) and is usually higher than the critical pH for growth of the host plant species (134). Root hair infection and nodule function are inhibited by low pH at levels which depend on the particular host – *Rhizobium* combination (139–141). Adverse effects of low pH on nodulation can often be moderated by liming (142). As discussed later, many of the adverse effects of unfavourable pH are due to effects on the availability of mineral ions for host plant nutrition.

Frankia has not yet been isolated directly from soils but a bioassay of different soils has suggested pH as one factor affecting variability within this group of microorganisms (98). In these experiments, *Alnus glutinosa* seedlings were grown in soils from existing alder stands for a period sufficient to allow nodulation. The seedlings were then transferred to nitrogen-free water culture for further growth so that soil nitrogen had minimal effect on total plant nitrogen. Total nitrogen content of nodulated plants after further growth for several weeks suggested that nodules of highest specific nitrogen fixation were formed on seedlings initially infected in soils of pH 4.5 to 6.5, while in soils lower than pH 4.5 the nodules were of lower specific activity, suggesting less effective forms of the endophyte. Exceptions to these general observations showed that other factors additional to pH are also important determinants of the effectivity of soil-borne *Frankia*. A major advantage of actinorhizal plants over leguminous species lies in the diversity of genera in the former group, within which are plants suited to a wide variety of ecological niches. Limits of pH for nodulation and nitrogen fixation range from pH 3.3 for *Myrica gale* to pH 9 for *Coriaria myrtifolia* (Table 4) but as with legumes, nodulation is more sensitive to pH than host plant growth (143, 144).

Extrusion of protons from the roots of nodulated legumes (150, 151) and actinorhizal species such as alder (152) can generate considerable acidity in the rooting medium when nitrogen fixation provides a major part of the nitrogen for growth. A mechanism based on the operation of an electrogenic pump, by which the uptake by plant roots of different forms of nitrogen may influence rhizosphere pH has been reviewed recently (153). Briefly, the conditions for ion uptake are created by proton extrusion from the roots through the action of a reversible ATPase, orientated asymmetrically in the root cell plasmalemma so that H^+ are actively translocated from the cell. This process is summarised by Equation 29; the precise numbers of protons extruded from the cell for each ATP hydrolysed is not known.

Table 4. Effects of pH upon nodulation, nitrogen fixation and growth of some actinorhizal nodulated plants.

Species	Nodulation	N fixation and plant growth	Plant growth with combined N	References
Myrica gale	Poor at 3.3 Good at 5.4–6.3	Best at 5.4	Best at 3.3 with NO_3^-	143
Alnus glutinosa	Zero at 3.3	Best at 4.2–5.4	Best at 4.2–6.3 with NO_3^- and 4.2–7.0 with NH_4^+	145 146
Alnus rubra	Poor at 3.5 Good at 4.5–6.5	Best at 4.5–6.5	Good 5.5–7.5 with NH_4NO_3*	98
Casuarina cunninghamiana	0 – poor at 4–5 Good at 6–7	Best at 6–7	No pH damage at 4.0 (seedlings with NH_4^+)	147
Hippophae rhamnoides	0 – poor at 4.2 Best 5.4–7.0	Best at 5.4–7.0	Best at 5.4–7.0 with NH_4^+	145 148
Shepherdia canadensis (*Hippophae* inoculum)	0 at 5.0 Best at 6.0	—	Successful growth at 5.0 with NH_4^+	149
Coriaria myrtifolia	0 – poor at 5.0 Best at 6–9	Best at 7–9	Good growth at 5–9 with NO_3^-	144

* Unpublished data of Wheeler & McLaughlin.

$$ATP^{4-} + H_2O + x[H^+]_{inside} \rightarrow ADP^{3-} + HPO_4^{2-} + H^+ + x[H^+]_{outside} \quad (29)$$

The electrochemical potential difference of protons thus established across the plasmalemma serves to support entry of cations into the cell. Internal OH^- generated by the H^+ extrusion process, can then be counter transported from the cell in exchange for anions from the root environment.

NO_3^- is readily transported across the plasmalemma relative to other anions and when NO_3 is in good supply there is a balance of cation and anion uptake by the plant roots. Under conditions of low NO_3^- availability cation uptake by the plant may exceed anion uptake. Consequently, the rate of proton extrusion will exceed the counter transport of OH^- from the cell so that acidification of the rhizosphere will ensue. Such a mechanism, operating in a nodulated plant relying on nitrogen fixation for its nitrogen, would result in imbalance of inorganic cation over anion uptake. This has been demonstrated for nodulated soybeans, where inorganic cation uptake was more than fivehold anion uptake in plants relying solely on nitrogen fixation, while analysis of the xylem sap of nodulated plants showed that the cation excess was reduced by feeding plants NO_3^- (153). In order to maintain cytoplasmic pH within physiological levels, cytoplasmic OH^- not exchanged for anions from the environment may be combined with CO_2 to form HCO_3^- which may be utilised in the cell in the synthesis of organic anions.

These can be readily deposited in the xylem to act as counterions to inorganic cations in the xylem sap.

Acidification during nitrogen fixation in soils low in nitrate can be considerable. As an example it has been calculated that a yield of alfalfa of 10 tonnes ha^{-1} dry weight which fixed 270 kg/ha N and contained 120 meq/100 g excess base would produce acidity equivalent to 600 kg $CaCO_3$ ha^{-1} (151). Rhizosphere acidification of this magnitude could present problems for nodulation and even nodule function, particularly in soils low in calcium (155). In addition to and, for perennial deciduous species, perhaps even more important than physiological acidification by the root system which accompanies nitrogen fixation is the acidification of soil by bacteria during nitrification of ammonia, released by ammonification of organic nitrogen during decay of shed plant parts (equ. 30).

$$2NH_4^+ + 3O_2 \rightarrow 2NO_2^- + H_2O + 4H^+ \overset{+O_2}{\rightarrow} 2NO_3^- \quad (30)$$

The acidifying effect of *Alnus crispa* on the uppermost horizons of a glacial till in Alaska is illustrated in Fig. 6. The pH was reduced from 8 to 6 within 20 years and

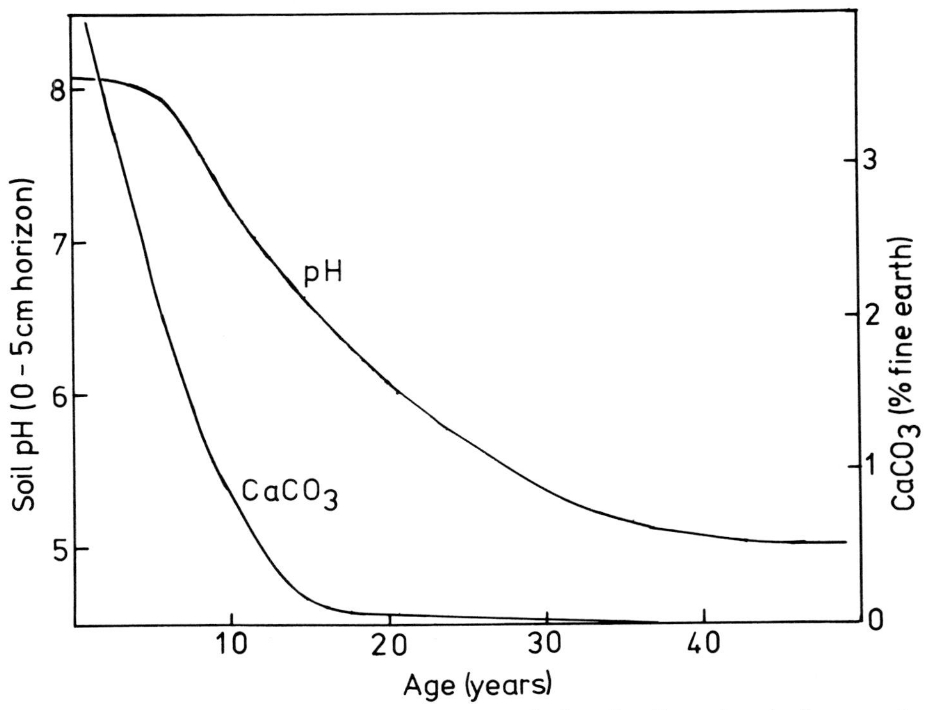

Fig. 6. Changes in pH and calcium carbonate content of soils under *Alnus crispa*. Analyses are of the 0 to 5 cm horizon under stands of different ages on post-glacial morainic debris in Glacier Bay Alaska. Redrawn from ref. 154.

was accompanied by rapid leaching of calcium carbonate so that the concentration in fine earth fell from 4% to less than 1% over the same period. Changes in pH and calcium carbonate were much less under another nodulated actinorhizal plant (*Dryas*), as well as under stands of *Populus* and *Salix* (155).

Mineral nutrition

The availability of mineral elements for satisfactory nodulation must be at least sufficient to support plant growth when assimilating combined nitrogen, with the obvious exception of the mineral nitrogen itself. Robson (157) lists four main experimental approaches which can be used to assess whether symbiotic nitrogen fixation has a special requirement for a particular element, over and above that required for host plant growth. In summary, these involve:

1. Alleviation of the effect of a particular mineral nutrient deficiency when either combined nitrogen or the nutrient under investigation is supplied to a nodulated plant. If growth is increased by both treatments then it is likely that symbiotic nitrogen fixation has a greater requirement for the nutrient than the host plant.
2. An increase in nitrogen concentration of the nodulated plant when the particular nutrient deficiency is corrected. Careful comparison with plants fed combined nitrogen is necessary to discriminate between effects of nutrient deficiency on nitrogen metabolism other than nitrogen fixation.
3. A change in nodule distribution, weight or number following correction of nutrient deficiency.
4. Demonstration of increased nitrogenase activity, prior to increased growth, following alleviation of nutrient deficiency.

Experimental evidence obtained by various investigators suggests a unique requirement for symbiotic nitrogen fixation for only one element, cobalt, which as discussed below is required for certain aspects of the metabolism of the microsymbiont.

Numerous formulations of mineral salts for the growth of nodulated plants in nitrogen-free conditions have been described in the literature. The composition of two of these is shown in Table 5. One has been used in the culture of legumes and the other has been used extensively by Bond and other workers for culture of actinorhizal species. Control of pH of the nutrient medium is of importance for the successful culture of nodulated plants for the study of mineral nutrition. As noted in the previous section, pH not only directly affects nodulation but when markedly above or below neutrality has considerable influence on the availability or absorption of certain ions by the roots. In acid media the availability of phosphorus and molybdenum is reduced and there is increased leaching from soils of several major and trace elements such as calcium, magnesium, potassium, boron and copper. Also, the increased availability of certain minerals such as

Table 5.

Nutrient solution 1	Concentration		*Nutrient solution 2*	Concentration	
	$g \cdot l^{-1}$	mmole l^{-1}		$g \cdot l^{-1}$	mmole l^{-1}
CaSO4	0.8	5.8	KCl	0.75	10
K_2HPO_4	0.4	2.3	$CaSO_4 \cdot 2H_2O$	0.5	2.9
$MgSO_4 \cdot 7H_2O$	0.2	0.8	$MgSO_4,7H_2O$	0.5	2.0
			$Ca_3(PO4)_2$	0.25	0.8
			$Fe_3(PO4)_2 \cdot 8H_2O$	0.25	0.5
	mg l^{-1}	μmole l^{-1}		mg l^{-1}	μmole l^{-}
Fe (as chelate)	5	89.0	$MnCl_2 \cdot 4H_2O$	0.40	2.02
$MnCl_2$	1.81	14.4	H_3BO_3	0.62	10
H_3BO_3	2.86	46.1	$Na_2O \cdot 4SiO_2$	0.43	1.4
$ZnCl_2$	0.11	0.8	KMnO4	0.40	2.5
$CuCl_2 \cdot 2H_2O$	0.05	0.29	$CuSO_4 \cdot 5H_2O$	0.055	0.2
$Na_2MoO_4 \cdot 2H_2O$	0.025	0.10	$ZnSO_4 \cdot 7H_2O$	0.055	0.2
			$Al_2(SO_4)_3$ $16H_2O$	0.055	0.1
			$NiSO_4 \cdot 6H_2O$	0.055	0.25
			$CoCl_2$ $6H_2O$	0.055	0.2
			$TiCl_4$	0.055	0.3
			$Li_2SO_4 \cdot H_2O$	0.035	0.37
			$SnCl_2 \cdot 2H_2O$	0.035	0.15
			KI	0.035	0.2
			KBr	0.035	0.39
			$Na_2MoO4 \cdot 2H_2O$	0.030	0.1
Originally used for growth of *Medicago tribuloides* in sand culture (157).			Based on Crone's formula, used originally in a slightly different form for soybeans in water culture (150) and used extensively by Bond and co-workers for growth of actinorhizal plants.		

manganese, aluminium and heavy metals can lead to the uptake of quantities toxic to plant growth (158, 159). Poor mycorrhizal development under very acid conditions may further limit mineral uptake. Lime-induced iron deficiency, leading to tissue chlorosis, is the most obvious condition to develop in alkaline pH.

Calcium interacts strongly with pH in its effects on infection and nodulation so that requirements for calcium are greater at low pH, and vice versa. Addition of calcium has been shown to increase nodulation of several legumes in moderately acid conditions (161) with the earliest stages of injection, prior to infection thread elongation, being particularly affected (140, 161, 162). Responses to calcium differ greatly between different strains of *Rhizobium* and different plant species and part of the effect of increased calcium supply may be to increase numbers of *Rhizobium* in the rhizosphere (154, 158). The precise way in which calcium ions affect the initiation of infection is not known but could be by some modification of the host cell wall which aids the attachment of *Rhizobium* (131). It has been suggested that H^+ extruded from the root during ion uptake may compete with Ca^{++} for interaction at the infection sites so that nitrogen fixing plants which show greatest net H^+ extrusion, and hence have a larger excess of cations over anions, may have the largest calcium requirements for nodulation (153).

Among actinorhizal plants, as mentioned above, alders show rapid acidification of their growth medium and ability to reduce soil base content, in particular calcium (Fig. 6) and magnesium (154, 163). This may stem in part from increased leaching of minerals from soils due to the presence of acid by-products of the decomposition of the nitrogen-rich alder leaf litter but could also be a response to high rates of H^+ extrusion from the roots consequent upon nodule activity. A study of the cation/anion balance of nodulated and non-nodulated alders under different conditions of mineral nutrition would be of particular interest. Improved nodulation of another actinorhizal species, *Ceanothus velutinus*, has also been noted following dressings with lime (164).

Of the other major elements required for normal growth, nodulated species probably also have a higher requirement for phosphorus compared with plants fed mineral nitrogen especially in acid soils (165). The common presence in legumes of endomycorrhizae has been shown to aid phosphorus nutrition (166–168) and to increase nodulation and nitrogen fixation of the host species. Mycorrhizae of actinorhizal species have been less well studied, but ectomycorrhizae and/or endomycorrhizae have been described in the roots of at least 12 genera (169–172). The supportive role of mycorrhizal infections in mineral absorption by nodulated plants is currently the subject of much interest. For example, among actinorhizal species, ectomycorrhizal roots of *Alnus viridis* have been shown to absorb phosphate five times more rapidly than non-mycorrhizal roots (170) while increases in nodule number and weight and elevated levels of N, Ca and P have been demonstrated in *Ceanothus velutinus*, infected with both vesicular-arbuscular mycorrhizae and *Frankia*, (172).

Of the trace elements, a larger requirement for molybdenum by nodulated plants is well established (e.g. 173, 174). It is an essential component of nitrogenase, so that molecules are particularly rich in molybdenum but its occurrence in nitrate reductase does not seem to pose such large demands in plants reliant on combined nitrogen for growth. In both legumes and non-legumes, deficiency can result in the formation of larger numbers of nodules distributed over the root systems, which appears to be a mechanism for partly compensating for the reduced efficiency of the nodules in nitrogen fixation. The effects of molybdenum deficiency on the growth of *Myrica gale* is illustrated in Fig. 7a.

Cobalt appears to be required specifically for satisfactory nodulation and again is present in root nodules at concentrations several times that of adjacent root tissues. Its requirement for growth of *Alnus glutinosa* is shown in Fig. 7b. The element is a specific requirement for vitamin B_{12} co-enzymes which are involved in such processes as propionate oxidation and DNA synthesis in both

Fig. 7. Effect of micronutrient deficiency on the growth of actinorhizal nodulated plants in water culture. a) *Myrica gale*, 3 month growth in N-free water culture showing effect of molybdenum deficiency. b) *Alnus glutinosa*, 6 month growth in N-free water culture showing effect of cobalt deficiency. In each case, a plant supplied with the complete range of nutrients is compared with the mineral deficient plant. Photographs reprinted by permission from Nature, Vol. 190, p. 133 and Vol. 183, p 319. Copyright © 1961 and 1965 Macmillan Journals Limited.

free living and symbiotic Rhizobium (176–178). In deficiency, the leghaemoglobin content and the bacteroid density of the nodules is decreased so that the larger mass of nodules which is produced on the lateral roots, functions in nitrogen fixation at only 5–13% of the efficiency of cobalt sufficient plants (179–181).

Substantial levels of vitamin B_{12} analogues are also present in the nodules of actinorhizal plants where they are necessary for proper functioning of the nodules (183–185). The synthesis of vitamin B_{12} by actinomycetes when cobalt is added to the growth medium is well known (186). The precise role of vitamin B_{12} in the metabolism of *Frankia* has not been elucidated as yet. However, it is conceivable that a methyl malonyl CoA mutase-catalysed pathway which oxidises propionate to succinyl CoA (187), could be the means by which propionate is utilised to support growth of the *Frankia* isolate from *Alnus viridis* (15).

Combined nitrogen

Inhibition of the nitrogen -fixing process by combined nitrogen has been recognised for over a 100 years (188) with effects on virtually all stages of the process from infection and nodulation to the expression of nitrogenase activity and nodule senescence. Inhibitory effects differ both between strains of *Rhizobium* and between different host species, with fast growing plants generally being most tolerant (189, 190). Determination of tolerance of different species to combined nitrogen requires special precautions to prevent depletion of low concentrations of nitrogenous salts in the growth medium, particularly by fast growing species, to levels which may be non-inhibitory or even stimulatory of nodulation and to control pH change consequent upon the uptake and assimilation of nitrogenous ions.

Several types of experiment suggest that inhibitory effects are localised primarily at the region of ion uptake e.g. in split root culture of both legumes (191) and actinorhizal plants (192), inhibition is confined to that portion of the root system fed combined nitrogen, while the other portion is unaffected. Nitrate bathing excised roots of *Phaseolus* inhibits nodulation but has little effect when fed through the cut end of the root (193). A most convincing demonstration has been obtained with the tropical legume *Sesbania rostrata* which bears nodules on both roots and stems. When ammonium nitrate is fed via the roots, root nodulation is suppressed but stem nodulation and nitrogenase activity actually increases (194).

The inhibitory effects of combined nitrogen on the nitrogen fixing process undoubtedly stem from a complex of events, all of which may not operate at every stage of the nitrogen fixing process. Diversion of photosynthates away from the nodules to support uptake of nitrogenous ions by the root system has been demonstrated and may contribute to reduced nodule activity (195). That deprivation of metabolites is not the sole cause of decreased nitrogen fixation during uptake of combined nitrogen is suggested from experiments with cultured roots, where supplying carbohydrate via the cut end of the excised root system only partially relieves the inhibition of nodulation shown when nitrate is taken up through the absorbing surface of the root (196). Other experiments with whole plants failed to reverse nitrate inhibition of nitrogen fixation when the rate of shoot photosynthesis was increased e.g. by enhancement of atmospheric CO_2 (191) although such results are not too surprising in view of the complex nature of competition for photosynthates by different 'sinks' in the whole plant. Nitrate only has a minor effect on sugar accumulation in nodules. In soybean plants where a 70% reduction in nodule weight followed feeding NO_3^-, levels of soluble sugars in the nodules were reduced by only 12% (198). However, the decline in glucose levels was greater than that in other sugars and it was suggested that NO_3^- may interfere with carbohydrate catabolism in the nodules.

Nitrate is generally viewed as being more inhibitory to nodulation than other sources of combined nitrogen (199, 200) with concentrations as low as 0.2 mM nitrate inhibiting root hair production and curling and infection thread formation in *Medicago sativa* (201, 202). Nitrite, produced by the reduction of nitrate by nodule or microbial nitrate reductase, has a variety of metabolic effects on components of the nitrogen fixing system, including inhibition of nitrogenase, inactivation of leghaemoglobin; and oxidation of indole-3-acetic acid produced by both the microsymbiont and the host plant and suggested to be involved in infection and nodulation (200, 203–206). It has proved difficult to establish that toxic effects of nitrate are due primarily to the reactivity of nitrite. Depression of nitrogen fixation and nodulation of *Macroptilium atropurpureum* and *Trifolium subterraneum* by nitrate, observed when plants were inoculated with rhizobial mutants which lacked nitrate reductase or with the parent strain (207), suggested that nitrite production by the microsymbiont is not involved in toxicity. Additionally, nitrite was not detected in nodule extracts, suggesting that toxicity was not due to the activity of host plant nitrite reductase either. However, this last possibility cannot be eliminated entirely since the reactivity of nitrite may render its existence in the nodule quite transient, even under conditions where it is being generated continually from nitrate. The possibility that uptake of nitrogenous ions may affect the proton motive gradient across the cytoplasmic membranes of the microsymbiont or organelles of the host plant cell, perhaps dissipating the driving force for nitrogen fixation or for ATP synthesis, are areas of current investigation which may help provide answers to these questions (208, 209).

In the field, except in extremely nitrogen-deficient soils, nodulated plants will utilize both combined nitrogen and symbiotically fixed nitrogen and, if soil nitrogen levels are sufficiently high this will be the main source of nitrogen for growth e.g. in soybeans symbiotically fixed nitrogen decreased from 48 to 10% of total shoot nitrogen as fertiliser nitrogen application was increased from 0 to 448 kg ha^{-1} (210). In conditions of low soil nitrate, the nodules themselves may be of importance as a site of nitrate reduction for the host plant since ^{15}N from $Na^{15}NO_3$ has been detected in the reduced nitrogen fraction of stem exudates of hydroponically grown soybeans, showing no detectable nitrate reductase activity in the roots (211). The general occurrence of some available nitrogen in the form of nitrogenous salts in most soils indicates the importance of selecting microsymbionts whose nitrogen fixation is least sensitive to the effects of external sources of combined nitrogen in order to achieve, for field application, symbiotic associations most efficient in nitrogen fixation.

Glasshouse experiments with actinorhizal plants suggest that the effect of combined nitrogen fixing process is similar in its complexity to that shown by leguminous plants, with considerable differences in response between different species. Sensitivity to NH_4^+ among the species studied increases in the order *Alnus glutinosa*, *Myrica gale*, *Casuarina cunninghamiana*, *Ceanothus velutinus*, *Hippophae rhamnoides* (212). In the least sensitive species, *Alnus glutinosa*, nodule dry weight per plant was not diminished with 5.55 mM NH_4^+, although at this concentration numbers of nodules decreased to 39% of O-NH_4^+ controls and nitrogen fixed formed 24% of the total nitrogen uptake of the plant (213). In a more sensitive species such as *Ceanothus velutinus* (214), nodule dry weight formed only 20% of O-NH_4^+ control plants at 2.8 mM NH_4^+ and nitrogen fixed formed 24% of the total nitrogen uptake of the plant. The relative toxicity of NH_4^+ and NO_3^- for growth of actinorhizal species still requires critical examination. However, the data which is currently available, all obtained under rather different experimental conditions, suggests slightly greater susceptibility to NO_3^- compared with the susceptibility shown to NH_4-N. Thus, in an experiment in which NO_3^- was fed to the nodulated half of *Alnus glutinosa*, the number of nodules on this portion of the root decreased by 74% (192), compared with the 61% inhibition of nodulation of plants fed 5.5 mM NH_4^+, noted above.

Although combined nitrogen above certain low levels generally depresses nitrogen fixation, the growth of the plant supplied with adequate combined nitrogen normally is much increased (e.g. 213) compared with plants solely reliant on nitrogen fixation. This presumably reflects, firstly, the additional demands placed on the plant for energy for nodulation and nitrogen fixation, compared with the requirements for assimilation of combined nitrogen. Secondly, the constraints on the supply of reduced nitrogen from nodules distributed in a relatively localised manner over the root system compared with nitrogenous ion uptake by the large absorbing surface represented by the root hair zone of the roots.

In rapidly growing plants, some of the inhibitory effects on nitrogen fixation of additional mineral nitrogen can be avoided if it is supplied in small doses, increasing in concentration to match current consumption, instead of as a concentration maintained at constant levels over the experimental period (215). In *Alnus incana*, nitrogen fixation by young plants is stimulated by the addition of ammonium nitrate up to an incremental rate of addition which increased by 12% daily. Under these experimental conditions, rates of nitrogen fixation were obtained which at their best were 55% higher than the rates shown by plants receiving no combined nitrogen. At daily incremental rates in mineral nitrogen above 12%, rates of nitrogen fixation decreased rapidly. It is not known whether these relationships hold for field situations, but in any event regular and frequent return to a plantation to supply additional fertiliser nitrogen to enhance symbiotic fixation of nitrogen is unlikely to be a practical proposition. In these experiments, additional nitrogen which inhibited nitrogen fixation nevertheless stimulated further growth of the plant, so again it appears that maximum growth may only be achieved by supply of fertiliser nitrogen at levels which largely suppress nitrogen fixation.

Despite its widespread use as a forest fertiliser, little is known of the effects of urea on nodulation and nitrogen fixation in the forest environment. The ability to utilise urea for growth differs between species e.g. growth of soybeans is similar with both nitrate and urea as nitrogen source, while growth of peas is less with urea than with nitrate (216). It is not surprising, therefore, that the effects of urea on nitrogen fixation in different plant species varies. Thus, in soybeans grown hydroponically with careful control of medium pH, nodulation was little affected by up to 18 mM urea, whereas 2 mM nitrate effectively inhibited nodulation (217). However, both ammonium nitrate and urea at 25 kg N $\cdot$ ha^{-1} caused degeneration of *Vigna unguiculata* nodules (218). Urea, while stimulating plant growth at 0.25 mM inhibited nodulation of *Alnus rubra* growing in pots of sandy soil whereas similar levels of nitrate stimulated both growth and nodulation (219). Urea may have advantages as a fertiliser for some nodulated species in some conditions, therefore, but further investigation under field conditions are required to establish the full range of its effects on nitrogen fixation in different species.

Moisture

Detailed investigations of effects of water stress on the nitrogen fixing process are mainly confined to a few crop legumes. Droughting has been studied in rather more detail than waterlogging, where detrimental effects on nodule development and function have been ascribed particularly to restrictions on the diffusion of the respiratory gases, O_2 and CO_2, into and away from the nodulated root system (220). As well as more direct effects on host plant nodulation, both waterlogged and dry conditions affect the migration, multiplication and survival of soil rhizobia. Following elevation of soil moisture content to toxic levels, the rate of decline of *Rhizobium* numbers is initially rapid, and then slows (221). Protozoa which feed on soil bacteria increase at higher moisture levels and accumulation of organic acids under conditions of high soil carbohydrate content stimulates the accumulation of organic acids with a concomitant decrease in soil pH.

In soil undergoing droughting, an initial rapid decline in numbers of *Rhizobium* occurs during the time of major water loss. In a comparison of six *Rhizobium* species, including *Rhizobium leguminosarum*, *Rhizobium japonicum* and a cowpea *Rhizobium*, no differences in susceptibility to drying were found between the fast and slow growing rhizobia (222). However, variation in response of different rhizobia to droughting undoubtedly does occur under natural conditions e.g. a greater drought tolerance in Australian soils of *Rhizobium trifolii* compared with *Rhizobium lupini* (223).

The nodules of several leguminous species show a sigmoidal relationship between nodule water content and nitrogenase activity, with reversible effects on nitrogen fixation with water losses of up to about 25% and more severe effects beyond this (224). Severe desiccation of soybean nodules produces collapse of nodule cortical cells and breakage of plasmodesmata (225) while in white clover, stress effects vary with cell age and differentiation with meristematic cells showing most resistance and those from the bacteroid zone accelerated senescence (220). These structural effects are reflected in decreased oxygen uptake by, and flow of reduced nitrogen from, the nodules (226, 227).

In the field, it is obvious that deep rooted species, able to bear nodules at great depth in the soil, should be most resistant to drought. Similarly, in those species whose root systems can withstand water logging, the bearing of nodules close to the soil surface should aid aeration necessary for continued nodular respiration, the extreme case of this being shown by some legumes tolerant of water-logged soils, such as *Aeschyomene indica* and *Sesbania rostrata*, which can bear nodules on both roots and stems (228, 194).

Some actinorhizal species show morphological or anatomical adaptations which suit them particularly well for growth and nodule function in water logged soils e.g. the presence of aerenchymatous tissue in the roots of *Alnus glutinosa* (229) or the development of negatively geotropic nodule roots, which develop profusely in some water culture plants and have been shown clearly in *Myrica gale* to aid nodule aeration (33). In dry condition, the possession of a corky periderm (230) may help to render the nodules more resistant to droughting than those of many legumes.

Water stress can affect nitrogen fixation both through its influence on essential aspects of host plant metabolism such as photosynthesis or translocation (231, 232), as well as by more direct effects on the structure and function of the nodules. In a study (233) of the recovery of soybeans from severe water stress following reduction of leaf water potential to −24 bars during a four-day drying cycle, the adenylate energy charge of the nodules recovered as rapidly as leaf water potential when plants were watered but acetylene reducing activity lagged behind the recovery of leaf water potential (Fig. 8). Adenylate energy charge provides a measure of ATP availability in the nodules (234) which would be expected to recover to normal levels when photosynthate supply to the nodules is restored. It is evident, therefore, from the lag in recovery of nodule nitrogenase activity that restoration of full nodule activity is dependent on more than restoration of photosynthate supply to the nodules, although the additional mechanisms which may be involved are as yet unknown.

Detailed studies of the effects of water stress on actinorhizal nodules, similar to those described above for leguminous species, have not yet been carried out but droughting has been suggested as a cause of lack of response of nitrogenase activity in *Alnus glutinosa* to fluctuations in temperature or solar radiation during the dry summer months (125). More detailed information concerning relationships between nitrogen fixation and plant water stress is available for *Purshia tridentata*, a xerophytic shrub of areas in N.W. America subject to summer drought. Measurement of xylem water potential of the host plant with a Scholander-type pressure bomb showed that potentials below −24 bars were associated with a substantial decrease of acetylene reducing activity (Fig. 9) both in field and glasshouse grown plants (133), presumably as a result of effects both on host plant processes such as photosynthesis and stomatal closure as well as

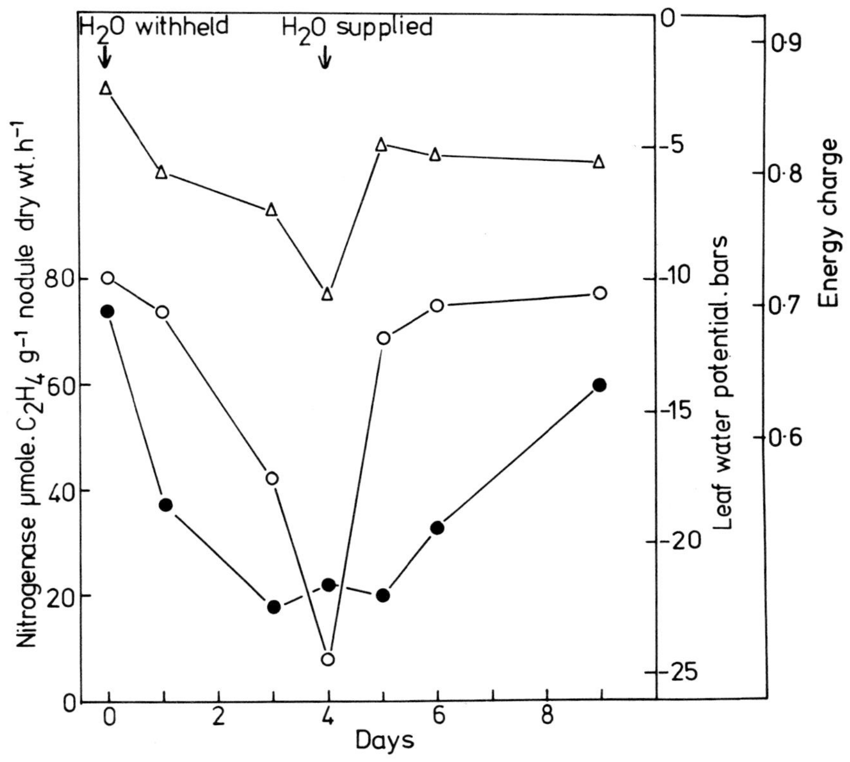

Fig. 8. Effect of moisture stress on nodulated soybean plants. Response and recovery in nitrogenase activity ●—●, leaf water potential ○—○ and nodule adenylate charge △—△ to applied moisture stress. Plants grown at 26° C day and 18° C night. Redrawn from ref. 231.

nodular metabolism. Unfortunately, accurate comparison of the levels of water potential which produce large inhibitions of nitrogenase activity in different species is currently not possible due to the variety of conditions under which such results have been obtained e.g. different parts of plant, different times of day or year etc. In those legumes where daytime measurements of leaf moisture stress have been made, nitrogenase activity appears to fall rapidly when moisture stress exceeds around −5 bars (233, 235, 236). Similarly, measurements made on broom plants in the field at different times of the year, showed that in the autumn activity dropped close to zero at predawn xylem pressure potentials more negative than −5 bars (237). Standardisation of techniques employed for study of water stress effects in nodulated plants would be of considerable value by permitting confident comparison of this sort of data.

The gaseous composition of the root environment

The composition of soil air is dependent largely on the porosity of the soil, its

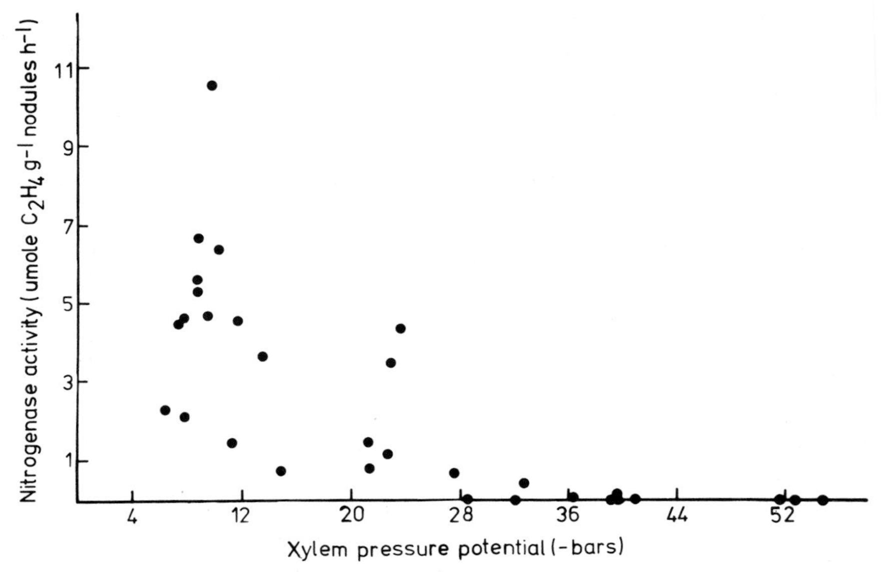

Fig. 9. Effect of moisture stress on the nitrogenase activity of *Purshia tridentata*. Field plants were transplanted into the greenhouse and subjected to different watering regimes. Measurements of xylem pressure potential were made by a pressure bomb technique. Reproduced from ref. 133 with permission.

water content, the biological activity of soil organisms and the rate of gaseous interchange with the atmosphere. Within the soil, the availability to plant life of soil gases depends on their partitioning between the soil pores, the soil water and the absorptive surface of the soil particles. Solute concentration in the soil solution and ambient temperature are major factors which affect the equilibrium between liquid and gaseous phases. Comparison of soil air composition for well aerated top soils often shows relatively little difference from atmospheric air, except for CO_2 content which can be elevated by an order of magnitude or more due to the respiratory activity of soil organisms (238). However, although the gas content of soil pores which connect with the atmosphere may remain relatively constant, because of the slow rate of diffusion of gases in water (some 10,000 times slower than in air), the water content of the soil plays a major part in modulating the rate at which soil gases can be supplied to the root (239). A combination of these environmental factors control the changes in gas composition which are seen with soil depth, water content and with season (favourable seasonal combinations of temperature and of water availability greater influence soil biological activity and hence CO_2 levels). In addition to the major atmospheric gases, a number of other gases may be present in the soil atmosphere the occurrence of which are most influenced by soil anaerobiosis e.g. hydrogen, hydrogen sulphide, ammonia, methane, ethylene, nitrous oxide (239).

Under the majority of soil conditions, changes in oxygen levels are most likely to have greatest effect on the activity of plant root systems. Concentrations of oxygen required for good root growth varies greatly with species and with the conditions of growth (240). The requirement for oxygen for respiration to support such activities as cell division and salt uptake suggests that the distal portions of the root will be most sensitive to changes in oxygen supply (241). The studies of Boynton *et al.* (242) on apple trees are frequently cited to illustrate root responses to changed oxygen supplies. The roots of this species required at least 3% oxygen in the soil air for survival while 5–10% was required to maintain growth. New root growth and normal mineral uptake required 12–15% oxygen.

These measurements of gas phase oxygen concentrations do not accurately describe oxygen flux in the immediate environment of the root since the rate of oxygen supply can only be calculated if the geometry of the soil and water around the root is known. An oxygen diffusion rate of 6 to 12 nmole cm^{-2} min^{-1} is necessary to support root growth of many plant species. For a root of 0.23 mm radius, requiring oxygen diffusion rates (ODR) within the above limits for growth and growing in a soil of porosity 0.5 in which the pores are in equilibrium with a gas phase containing 20% oxygen, it has been calculated that the water film around the root must not be more than 0.40 mm for root growth to occur (243). Increased root radius, decreased soil porosity and decreased oxygen content of the gas phase decrease the thickness of the water film which can be tolerated for root growth to be maintained. It is interesting to note that nodulation of *Trifolium subterraneum* in soils with oxygen diffusion rates of 2.5 nmole cm^{-2} min^{-1} was poor, while nodulation was satisfactory in soils with an ODR of 6 nmole cm^{-2} min^{-1} (244), which is the minimum diffusion rate commonly required for satisfactory root growth, as noted above.

Water culture experiments, in which gases of various concentrations are bubbled through the solution, do not provide a realistic assessment of soil gas levels which may limit growth since the diffusion layers, which develop around the roots under such conditions are likely to be quite different from those across a water film in soil. Indeed under conditions of high aeration where air is bubbled through the culture solution, diffusion layers may be virtually absent. Such experiments can provide useful comparative data but do not have great relevance to the soil condition. Reduction in oxygen levels below 20% have been noted to lower nitrogen fixation in water culture soybean (245) and red clover (246). Comparisons have also been made of the oxygen requirements of actinorhizal species and a legume, using detached nodules (actinorhizal plants) or nodulated roots (pea) immersed in nutrient solution, the surface of which was exposed to gas mixtures of desired oxygen concentration. With pea and *Myrica gale*, maximum fixation was found close to 20% oxygen with substantial depression of fixation at 40% oxygen, while in *Hippophae* and *Alnus* maximum fixation was observed at about 12% oxygen (247). Nitrogen fixation in soybean nodules, on the other hand, is stimulated by up to 50% oxygen (248). As discussed above, these differences between species in response to external oxygen concentration probably reflect the extent to which oxygen diffusion is restricted by anatomical barriers in the different nodules.

Carbon dioxide is the other soil gas which most commonly might be expected to affect nodule activity. Its concentration in the soil can vary widely, depending on the biological activity of the soil in question, and additionally substantial CO_2 will be produced within the nodules by respiration of host plant cells and microsymbiont. The high rate of respiration found in nodules active in nitrogen fixation conceivably may yield a major part of the CO_2 which is used as substrate in the pathways of dark carbon dioxide fixation to generate organic acids as a supplementary supply for amino acid synthesis in the nodules (249). Experiments designed to examine the effects of carbon dioxide supply to the root on nodule functions have produced somewhat conflicting results. A requirement for CO_2 for growth of *Rhizobium* has been clearly established (250). Supplementation of the air with 2% to 6% CO_2 has been found to inhibit nodulation (251) and on other occasions to increase nodulation of cultured plants (252). Further experiments using lower CO_2 concentrations would be worthwhile since 0.5% CO_2 stimulated growth of non-nodulated pea seedlings (253). In detached soybean nodules, CO_2 has an effect only at low pO_2 and at higher O_2 levels increased nodule respiration could well remove CO_2 limitations (254). On the other hand, too high an internal [CO_2] would be expected to inhibit cell respiration and in legumes perhaps inhibit the oxygen carrying functions of leghaemoglobin (255). Undoubtedly, apart from species differences, effects of CO_2 concentration are greatly affected by the pH of the growth medium through effects on the ionisation of carbonic acid, produced by solution of CO_2 in water. Despite these rather variable responses, some investigators have suggested that microbial decay of straw or culmiferous manures applied to soils where legumes grow may increase soil CO_2 evolution, which is then at least partly responsible for increased yield and nitrogen content of these crops. Obviously, the process of microbial decay of such low-nitrogen residues will itself consume substantial soil nitrogen for the saprophytic growth of the soil micro-organisms. Further experiments are required to show the applicability of, and reasons for, such treatments of nodulated species under a variety of soil and environmental conditions.

GENETIC IMPROVEMENT OF THE SYMBIOTIC ASSOCIATION

It is clear from the above discussion that many of the factors which affect the development and expression of the symbiotic process in agricultural legumes are now well described, even although the mechanisms by which such effects are brought about are often much less clear. However, translation of these findings into practice by the production and introduction into agriculture of improved strains of rhizobia for nodulation of grain and forage legumes is still by no means a world wide practice. It is perhaps not surprising therefore, that manipulation of the symbiotic association by improvement of the host plant, with its relatively slow growth and attendant requirement for lengthy experimentation, is still very much an experimental area which has not as yet been widely put to commercial advantage by growers. Several excellent reviews are available which cover the genetics of symbiotic nitrogen fixation in rhizobia (256–261) and in host legumes (262–265). It is to be hoped that the broad base of information which is developing for the rhizobium-legume association will be drawn upon early to provide guidelines for the genetic improvement of the stock available for use in forestry.

Desirable genetic factors in the free-living microsymbiont are those which best fit it for survival and multiplication in a particular soil environment and which at the same time allow expression of a high level of infectivity toward the host plant. Genetic factors of both the bacterium and of the host plant interact to determine infectiveness, however, and also subsequently to control the effectiveness of the symbiosis in nitrogen fixation. Rhizobia show wide natural variability in different environments, providing the opportunity for selection of naturally occurring strains which can be matched with a suitable variety of host plant to improve nodulation. Considerable care is required to ensure that greenhouse tests accurately predict strain performance when used in crop inoculation and it is important that strain selection should be undertaken in conditions which match as closely as possible those prevailing in the soil in the area where crops are to be grown. It is also necessary that the introduced *Rhizobium* is able to compete successfully with the native population of rhizobia for survival in the soil and for infection of the host plant. For example, it has been shown for *Rhizobium japonicum* that a strain which gave satisfactory nodulation of greenhouse plants, when introduced into the field, despite being numerically the major soil *Rhizobium*, was unable to compete with less effective native rhizobia, which came to inhabit virtually all the nodules (262). Greenhouse tests with inoculum prepared by mixing effective *Rhizobium* strains with ineffective strains in different proportions have been used to evaluate the competitiveness for nodulation of different strains and these results have been applied successfully to introduce competitive strains into the field (266, 267). Another solution to the competition problem might be to develop cultivars of host plant which are resistant to infection by the ineffective, native strains while retaining susceptibility to the introduced, effective strains (262).

In addition to strain selection which draws on the range of variability naturally present in populations, variation may also be induced by mutagenesis with chemicals or radiation which can be used to add to or remove useful functions from the microsymbiont. Mutations for resistance to antibiotics such as streptomycin or spectinomycin are useful as genetic markers for re-isolation of introduced rhizobium during ecological studies (268). Such mutation can also be achieved by transfer of sequences of DNA (transposons), carrying genes for the desired attribute, into the bacterium under study and this technique has proved particularly useful for analysis of gene function. Gram negative bacteria, such as rhizobia, contain plasmids (extra-chromosomal DNA) which are able to incorporate 'foreign' DNA to form cointegrate plasmids. Transfer of plasmids, and hence of genes between species of genera may occur during mating and transfer of a linked marker such as antibiotic resistance can be used to monitor plasmid transfer between bacteria (260). *Rhizobium* carries a number of plasmids which may be of large size, so that around 20% of the total genetic information is plasmid encoded (269). Such plasmids can carry the information for nitrogen fixation (270, 271). The plasmid pRLIJI from *Rhizobium leguminosarum* which transfers nodulation genes at high frequency, also specifies bacteriocin production (bacteriocins are substances produced by some strains of bacteria which inhibit other strains) which thus forms a natural marker during transconjugation experiments. This plasmid will tranfer nodulating ability for peas from *Rhizobium leguminosarum* to a non-nodulating strain of the same species and to *Rhizobium trifolii* and *Rhizobium phaseoli* (272). In *Rhizobium meliloti*, it has been shown that the genes for nodule formation and the nitrogenase structural genes are located on the same plasmid (273). Plasmid transconjugation offers considerable opportunities for future genetic improvement of the *Rhizobium*-legume symbiosis, as well as providing an important tool for studying gene location and function.

Hereditary host factors in legumes which affect the symbiosis include non-nodulation, the time interval for nodule initiation, nodule abundance and morphology and various factors which affect the development of nodule function (263). In both *Trifolium pratense* and *Vicia faba* early flowering cultivars are the most effective in nitrogen fixation. By contrast in *Phaseolus vulgaris* and in *Glycine max* the more effective cultivars are those which flower later, with more nitrogen fixed at highest rates during the vegetative phase (263, 128). In keeping with the requirements of symbiotic nitrogen fixation for host plant photosynthates better nitrogen fixation in late flowering soybean genotypes is correlated with the larger leaf area per plant developed during the longer vegetative phase (113).

The relatively short time in which suitable isolation techniques have been available and the slow growth of *Frankia* isolates have so far precluded genetic analysis of nitrogen fixation in this group of organisms. Cross inoculation studies of host plants with crushed nodules of different actinorhizal species (274) and host specificity studies of *Frankia* isolates have shown considerable differences between *Frankia* strains (275, 276). On the basis of serological and host compatibility studies, actinomycetes from the genera *Alnus* and *Myrica* have been found to be very similar, while those from *Eleagnus* are quite different from other strains tested (275). Protoplast fusion techniques have been applied successfully for DNA recombination studies with streptomyces by Hopwood and colleagues and Beringer (260) suggests that this technique (which requires that the experimental species can withstand cell wall removal by enzymic digestion and, further that the fused protoplast is able to regenerate cell walls) may have great potential for genetic analysis and improvement of *Frankia*.

Again, genetic improvement of actinorhizal plants for symbiotic nitrogen fixation is only just being considered. Some genera of actinorhizal plants are of wide geographic distribution and taxonomic diversity with correspondingly large genetic variation which may be used for species improvement. Important traits in *Alnus* species which might be utilised to breed for tolerance of particular environmental conditions include (277) water economy (*A. jorullensis*), resistance to water logging (*A. maritima*), tolerance of cool conditions (*A. viridis*), shrubby growth (*A. serrulata*, *A. viridis*, *A. rugosa*) or tree form (*A. rubra*, *A. incana*, *A. glutinosa*, *A. cordata*). Alders hybridise fairly readily, with many of the hybrids showing good growth e.g. hybrids between *A. incana* and *A. glutinosa* have a greater tree volume than the parental species (278). However, breeding for improved nitrogen fixation to accompany desired growth characteristics has not been attempted on any large scale. That such scope exists is shown by cross inoculation studies between clones of three different species of *Alnus* and *Frankia* isolates from *Comptonia peregrina* and *Alnus crispa*. Both the *Frankia* isolate and host plant species contributed significantly to the growth of plants in nitrogen free culture (276). A positive correlation between photosynthetic capacity and nitrogen fixation has also been shown for different clones of *Alnus glutinosa*, inoculated with the same endophyte source (279).

Vegetative reproduction from cuttings or from tissue culture has been used for several alder species (212, 280) and may be suitable for large scale multiplication of other plants showing characteristics of advantage in forestry or land improvement. Natural variation within the genus *Alnus* alone provides enormous scope for species selection and improvement to fit a wide variety of cultural practices and environmental situations. When considered together with the other known genera of actinorhizal plants, the prospective matching of *Frankia* and actinorhizal host plants offers almost limitless opportunities for improvement of growth habit and symbiotic nitrogen fixation in these associations in the future.

REFERENCES

1. Yates MG: Biochemistry fixation. In: *The Biochemistry of plants. 5 Amino acids and derivatives*. Miflin BJ (ed), New York, Academic Press, 1980, p 1–64.
2. Zumft WG, Mortensen LE, Palmer G: Electron-paramagnetic-resonance studies on nitrogenase. *Eur J Biochem* 46: 525–535, 1974.

3. Burris RH, Orme-Johnson WH: Mechanism of nitrogen fixation. In: *1st International symposium of nitrogen fixation.* Newton W, Nyman CJ (eds), Pullman, Washington State University Press, 1976, p 208–233.
4. Smith BE, Eady RR, Thorneley RNF, Yates MG, Postgate JR: Some aspects of the mechanism of nitrogenase. In: *Recent developments of nitrogen fixation.* Newton W, Postgate JR, Rodriguez-Barrueco C (eds), New York, Academic Press, 1977, p 191–204.
5. Burris RH: Fixation by free-living microorganisms: Enzymology. In: *The chemistry and biochemistry of nitrogen fixation.* Postgate JR (ed), London, Plenum Press, 1971, p 106–160.
6. Bisseling T, Van Staveren W, Van Kammen A: The effect of waterlogging on the synthesis of the nitrogenase components in bacteroids of *Rhizobium leguminosarum* in root nodules of *Pisum sativum. Biochem Biophys Res Comm* 93: 687–693, 1980.
7. Walker CC, Partridge CDP, Yates MG: The effect of nutrient limitation on hydrogen production by nitrogenase in continuous cultures of *Azotobacter chroococcum. J gen Microbiol* 124: 317–327, 1981.
8. Dixon ROD: Relationships between nitrogenase systems and ATP-yielding processes. In: *Nitrogen fixation by free-living micro-organisms.* Stewart WDP (ed), Cambridge, Cambridge University Press, 1975, p 421–436.
9. Scherings G, Haaker H, Veeger C: Regulation of nitrogen fixation by Fe-S protein II in *Azotobacter vinelandii. Eur J Biochem* 77: 621–630, 1977.
10. Carter KR, Rawlings J, Orme-Johnson WH, Becker RR, Evans HJ: Purification and characterization of a ferrodoxin from *Rhizobium japonicum* bacteroids. *J Biol Chem* 255: 4213–4233, 1980.
11. Veeger C, Laane C, Scherings G, Matz L, Haaker H, Van Zeeland-Wolberts: Membrane energisation and nitrogen fixation in *Azotobacter vinelandii* and *Rhizobium leguminosarum* In: *Nitrogen fixation 1 Free living systems and chemical models.* Newton WE, Orme-Johnson WH (eds), Baltimore, University Park Press, 1980, p 111–138.
12. Haaker H, Veeger C: Involvement of the cytoplasmic membrane in nitrogen fixation by *Azotobacter vinelandii. Eur J Biochem* 77: 1–10, 1977.
13. Aleem MJH, Lees H, Nicholas DJD: Adenosine triphosphate-dependent reduction of nicotinamide adenine dinucleotide by ferrocytochrome *c* in chemoautotrophic bacteria. *Nature* 200: 759–761, 1963.
14. Trinchant JC, Birot AM, Rigaud J: Oxygen supply and energy-yielding substrates for nitrogen fixation (acetylene reduction) by bacteroid preparations. *J gen Microbiol* 125: 159–165, 1981.
15. Blom J, Harkink R: Metabolic pathways for gluconeogenesis and energy generation in *Frankia* Avc Il. *FEMS Microbiol Lett* 11: 221–224, 1981.
16. Baker D, Torrey JG, Kidd GH: Isolation by sucrose-density fractionation and cultivation *in vitro* of actinomycetes from nitrogen-fixing root nodules. *Nature* 281: 76–78, 1979.
17. Tjepkema JD, Ormerod W, Torrey JG: Vesicle formation and acetylene reduction in *Frankia* sp CPI1 cultured in refined media. *Nature* 287: 633–635, 1980.
18. Ziegler H: Untersuchungen über die Leitung und Sekretion der Assimilate. *Planta* 47: 447–500, 1956.
19. Tjepkema JD, Yoccum CS: Respiration and oxygen transport in soybean nodules. *Planta* 115: 59–72, 1973.
20. Tjepkema JD, Yoccum CS: Measurement of oxygen partial pressure within soybean nodules by oxygen microelectrodes. *Planta* 119: 351–360, 1974.
21. Sinclair TR, Goudriaan J: Physical and morphological constraints on transport in nodules. *Plant Physiol* 67: 143–145, 1981.
22. Dixon ROD, Blunden EAG, Searle JW: Intercellular space and hydrogen diffusion in pea and lupin root nodules. *Plant Sci Lett* 23: 109–116, 1981.
23. Bergersen FJ, Goodchild DJ: Aeration pathways in soybean root nodules. *Aust J Biol Sci* 26: 729–740, 1973.
24. Ellfolk N: Leghaemoglobin, a plant haemoglobin. *Endeavor* 31: 139–142, 1972.
25. Jensen E, Paludan K, Hyldig-Nielson JJ, Jørgensen P, Marcker KA: The structure of a chromosomal leghaemoglobin gene from soybean. *Nature* 291: 677–679, 1981.
26. Dilworth MJ: The plant as the genetic determinant of leghaemoglobin production in the legume root nodule. *Biochim Biophys Acta* 184: 432–441, 1969.

27. Sievers G, Muhtala ML, Ellfolk N: The primary structure of soybean (*Glycine max*) leghaemoglobin. *Acta Chem Scand* B 32: 380–286, 1978.
28. Bergersen FJ, Goodchild DJ: Cellular location and concentration of leghaemoglobin in soybean root nodules. *Aust J Biol Sci* 26: 741–756, 1973.
29. Nadler KD, Avissar YJ: Heme synthesis in soybean root nodules. *Plant Physiol* 60: 433–436, 1977.
30. Verma DPS, Bal AK: Intracellular site of synthesis and localisation of leghaemoglobin in root nodules. *Proc Nat Acad Soc* 73: 3843–3847, 1976.
31. Bergersen FJ: Leghaemoglobin, oxygen supply and nitrogen fixation: Studies with soybean nodules. In: *Nitrogen Fixation*. Stewart WDP, Gallon JR (eds), New York, Academic Press, 1980, p 139–160.
32. Wheeler CT, Gordon JC, Ching TM: The oxygen relations of the root nodules of *Alnus rubra* Bong. *New Phytol* 82: 449–457, 1979.
33. Tjepkema J: Oxygen relations in leguminous and actinorhizal nodules. In: *Symbiotic nitrogen fixation in the management of temperature forests*. Gordon JC, Wheeler CT, Perry DA (eds), Corvallis, Oregon State University, 1979, p 175–816.
34. Phelps AS, Wilson PW: Occurrence of hydrogenase in nitrogen-fixing organisms. *Proc Soc Exp Biol* (NY) 47: 473–476, 1941.
35. Dixon ROD: Hydrogen uptake and exchange by pea root nodules. *Ann Bot* NS 31: 179–188, 1967.
36. Dixon ROD: Hydrogenase in pea root nodule bacteroids: *Arch Mikrobiol* 62: 272–283, 1968.
37. Dixon ROD: Hydrogenase in legume root nodule bacteroids: Occurrence and properties. *Arch Mikrobiol* 85: 193–201, 1972.
38. Schubert KR, Evans HJ: Hydrogen evolution: A major factor affecting the efficiency of nitrogen fixation in nodulated symbionts. *Proc Nat Acad Sci* USA 73: 1207–1211, 1976.
39. Evans HJ, Emerich DW, Ruiz-Argüeso T, Maier RJ, Albrecht SL: Hydrogen Metabolism in legume-Rhizobium symbiosis. In: *Nitrogen fixation*. Newton WE, Orme-Johnson WH (eds), Baltimore University Park Press, p 69–86, 1980.
40. Evans HJ, Emerich DW, Maier RJ, Hanus FJ, Russell SA: Hydrogen cycling within the nodules of legumes and non-legumes and its role in nitrogen fixation. In: *Symbiotic nitrogen fixation in the management of temperate forests*. Gordon JC, Wheeler CT, Perry DA (eds), Corvallis Oregon State University, 1979, p 196–206.
41. Bethnelfalvay GJ, Phillips DA: Ontogenetic interactions between photosynthesis and symbiotic nitrogen fixation in legumes. *Plant Physiol* 60: 419–421, 1977.
42. Phillips DA: Efficiency of symbiotic nitrogen fixation in legumes. *Ann Rev Plant Physiol* 31: 29–49, 1980.
43. Dixon ROD, Blunden EAG: unpublished.
44. Roelofson W, Akkermans ADL: Uptake and evolution of H_2 and reduction of C_2H_2 by root nodules and nodule homogenates of *Alnus glutinosa*. *Plant and Soil* 52: 571–578, 1979.
45. Laane C, Krone W, Konings W, Hasker H, Veeger C: Short term effect of ammonium chloride on nitrogen fixation by *Azotobacter vinelandii* and by bacteroids of *Rhizobium leguminosarum*. *Eur J Biochem* 103: 39–46, 1980.
46. Bergersen FJ, Turner GL: Nitrogen fixation by the bacteroid fraction of breis of soybean nodules. *Biochim Biophys Acta* 141: 507–515, 1967.
47. O'Neal D, Joy KW: Glutamine synthetase of pea leaves. *Plant Physiol* 54: 773–779, 1974.
48. King J, Yung-Fan Wu W: Partial purification and kinetic properties of glutamic dehydrogenase form soybean cotyledons. *Phytochemistry* 10: 915–928, 1971.
49. Brown CM, Dilworth MJ: Ammonia assimilation by *Rhizobium* cultures and bacteroids. *J gen Microbiol* 86: 39–48, 1975
50. Robertson JG, Warburton MP, Farnden KJF: Induction of glutamate synthase during nodule development in lupin. *FEBS Lett* 55: 33–37, 1975.
51. Akkermans ADL, Roelofson W, Blom J: Dinitrogen fixation and ammonia assimilation in actinomycetous root nodules of *Alnus glutinosa* In: *Symbiotic nitrogen fixation in the management of temperate forests*. Gordon JC, Wheeler CT, Perry DA (eds), Corvallis, Oregon State University, p 160–174, 1979.

52. Scott DB, Farnden KJF, Robertson JG: Ammonia assimilation in lupin nodules. *Nature* 263: 703–705, 1976.
53. Pate JS, Atkins CA, White ST, Rainbird RM, Woo KC: Nitrogen fixation and xylem transport in ureide-producing grain legumes. *Plant Physiol* 65: 961–965, 1980.
54. Thomas RJ, Schrader LE: Ureide metabolism in higher plants. *Phytochem.* 20: 361–371, 1981.
55. Triplett EW, Blerins DG, Randall DD: Allantoic acid synthesis in soybean root nodule cytosol via xanthine dehydrogenase. *Plant Physiol* 65: 1203–1206, 1980.
56. Hanks JF, Tolbert NE, Schubert KR: Localisation of enzymes of ureide biosynthesis in peroxisomes and microsomes in nodules. *Plant Physiol* 68: 65–69, 1981.
57. Kapoor M, Waygood ER: Biosynthesis of nucleotides in wheat. *Biochem Biophys Res Comm* 9: 7–11, 1962.
58. Robern H, Wang D, Waygood ER: Initial steps of purine biosynthesis in wheat germ. *Can J Biochem* 43: 225–235, 1965.
59. Iwai K, Nakegawas S, Osamu O: Isolation and identification of glycinamide ribonucleotide accumulated in pea seedlings in a folate-deficient state. *Biochim Biophys Acta* 68: 152–156, 1965.
60. Leaf G, Gardner IC, Bond G: Observations on the composition and metabolism of the nitrogen-fixing root nodules of *Alnus*. *J Exp Bot* 9: 320–331, 1958.
61. Leaf G, Gardner IC, Bond G: Observations on the composition and metabolism of the nitrogen-fixing root nodules of *Myrica*. *Biochem J* 72: 662–667, 1959.
62. Wheeler CT, Bond G: The amino acids of non-legume root nodules. *Phytochemistry* 9: 705–708, 1970.
63. Wheeler CT: Unpublished.
64. O'Neal TD, Taylor AW: Partial purification and properties of carbamoyl phosphate synthetase of Alaska pea. (*Pisum sativum* L cultivar Alaska). *Biochem J* 113: 271–279, 1979.
65. Schubert KR, Ryle GJA: The energy requirements for nitrogen fixation in nodulated legumes. In: *Advances in legume science*. Summerfield RJ, Bunting AH (eds), Kew, Royal Botanic Gardens, 1980, p 85–96.
66. Lespinat PA, Gerster R, Berlier YM: Direct mass-spectrometric determination of the relationship between respiration, hydrogenase and nitrogenase activities in *Azotobacter chroococcum*. Biochemie 60: 339–341, 1978.
67. Dixon ROD, Berlier YM, Lespinat PA: Respiration and nitrogen fixation in nodulated roots of soya bean and pea. *Plant Soil* 61: 135–143, 1981.
68. Schubert KR, Jennings NT, Evans HJ: Hydrogen reactions of nodulated plants. II Effects on dry matter accumulation and nitrogen fixation. *Plant Physiol* 61: 398–401, 1978.
69. Nutman PS: IBP field experiments on nitrogen fixation by nodulated legumes. In: *Symbiotic nitrogen fixation in plants*. Nutman PS (ed), Cambridge, Cambridge University Press, 1976, p 211–237.
70. Vincent JM: *A manual for the practical study of root nodule bacteria*. IBP Handbook 15, Oxford, Blackwell Scientific Publications, 1970.
71. Dalton H: The cultivation of diazotrophic microorganisms. In: *Methods for evaluating biological nitrogen fixation*. Bergersen FJ (ed), Chichester, Wiley, 1980, p 65–110.
72. Wheeler CT, Henson IE: Hormones in plants bearing nitrogen-fixing root nodules: the nodule as a source of cytokinins in *Alnus glutinosa* (L.) Gaertn. *New Phytol* 80: 557–565, 1978.
73. Gadgil RL: The nutritional role of *Lupinus arboreus* in coastal sand dune forestry. I: The potential influence of undamaged lupin plants on nitrogen uptake by *Pinus radiata*. *Pl Soil* 34: 357–367, 1971.
74. Moiroud A, Capellano A, Bärtschi H: Fixation d'azote chez les espèces ligneuses symbiotiques. I. Ultrastructure des nodules, mycorhizes à vésicules et à arbuscules et activité réductrice de C_2H_2 de jeunes plants de *Robinia pseudoacacia* cultivés au laboratoire. *Can J Bot* 59: 481–490, 1981.
75. Akkermans A: Symbiotic nitrogen fixers available for use in temperate forestry. In: *Symbiotic nitrogen fixation in the management of temperate forests*. Gordon JC, Wheeler CT, Perry DA (eds), Corvallis, School of Forestry, Oregon State University, 1979, p 23–37.
76. Ek-Jander J, Fahraeus G: Adaptation of *Rhizobium* to subarctic environment in Scandinavia. *Pl Soil*, Special Volume: 129–138, 1971.

77. Eaglesham A, Seaman B, Ahmad H, Hassouna S, Ayanaba A, Mulongoy K: High temperature tolerant 'cowpea' rhizobia. In: *Current perspectives in nitrogen fixation.* Gibson AH, Newton WE (eds), Canberra, Aust Acad Sci, 1981, p 436.
78. Ranga Rao V: Effect of root temperature on the infection process and nodulation in *Lotus* and *Stylosanthes. J Exp Bot* 28: 241–259, 1977.
79. Gibson AH: Factors in the physical and biological environment affecting nodulation and nitrogen fixation by legumes. *Pl Soil*, Special Volume: 139–152, 1971.
80. Kumarasinghe RMK, Nutman PS: The influence of temperature on root hair infection of *Trifolium parviflorum* and *Trifolium glomeratum. J Exp Bot* 30: 503–515, 1979.
81. Lie TA: Environmental effects on nodulation and symbiotic nitrogen fixation. In: *The biology of nitrogen fixation.* Quispel A (ed), Amsterdam, North Holland/Elsevier, 1974.
82. Roughley RJ, Dart PJ: Root temperature and root hair infection of *Trifolium subterraneum* L. cv. Cranmore. *Pl Soil* 32: 518–20, 1970.
83. Roughley RJ, Dart PJ: Growth of *Trifolium subterraneum*, selected for spare and abundant nodulation as affected by root temperature and *Rhizobium* strain. *J Exp Bot* 21: 776–786, 1970.
84. Gibson AH: Physical environment and symbiotic nitrogen fixation. VI. Nitrogen retention within the nodules of *Trifolium subterraneum* L. *Aust J Biol Sci* 22: 829–838, 1969.
85. Roughley RJ: The influence of root temperature, *Rhizobium* strain and host selection in the structure and nitrogen-fixing efficiency of the root nodules of *Trifolium subterraneum. Ann Bot* 34: 631–646, 1970.
86. Gibson AH: Recovery and compensation by nodulated legumes to environmental stress. In: *Symbiotic nitrogen fixation in plants.* Nutman PS (ed), Cambridge, Cambridge University Press, 1976, 385–404.
87. Roughley RJ, Dart PJ, Day JM: The structure and development of *Trifolium subterraneum* L. root nodules. II. In plants grown at sub-optimal root temperatures. *J Exp Bot* 27: 431–440, 1976.
88. Fyson A: Effects of low temperature on the development and functioning of the root nodules of *Vicia faba* L. PhD Thesis, University of Dundee, 1981.
89. Pankhurst CE, Gibson AH: *Rhizobium* strain influences and disruption of clover nodule development at high root temperature. *J Gen Microbiol* 74: 219–231, 1973.
90. Wollum AG, Youngberg CT: Effect of soil temperature on nodulation of *Ceanothus velutinus. Proc Soil Sci Soc Amer* 33: 801–803, 1969.
91. Waughman GJ: The effect of temperature on nitrogenase activity. *J Exp Bot* 28: 949–960, 1977.
92. Wheeler CT: The causation of the diurnal fluctuation in nitrogen fixation in *Alnus glutinosa. New Phytol* 70: 487–496, 1971.
93. Hensley DL, Carpenter PL: The effect of temperature on N_2 fixation (C_2H_2 reduction) by nodules of legumes and actinomycete-nodulated woody species. *Bot Gaz* 140: 558–564, 1979.
94. Bond G, Mackintosh AH: Diurnal changes in nitrogen fixation in the root nodules of *Casuarina. Proc Roy Soc B* 192: 1–12, 1975.
95. Granhall V, Lid-Torsvik V: Nitrogen fixation by bacteria and free living blue green algae in tundra area. In: *Fennoscandian Tundra Ecosystems Pt. 1. Plants and microorganisms.* Wielgolaski FE (ed), Ecological studies 16, Berlin, Springer Verlag, p 305–315.
96. Akkermans ADL: Nitrogen fixation and nodulation of *Alnus* and *Hippophae* under natural conditions. PhD Thesis, University of Leiden, 1971.
97. Wheeler CT, Perry DA, Helgerson O, Gordon JC: Winter fixation of nitrogen in scotch broom (*Cytisus scoparius* L.). *New Phytol* 82: 697–701, 1979.
98. Wheeler CT, McLaughlin ME, Steele P: A comparison of symbiotic nitrogen fixation in Scotland in *Alnus glutinosa* and *Alnus rubra. Pl Soil* 61: 169–188, 1981.
99. Gibson AH: Root temperature and symbiotic nitrogen fixation. *Nature, London* 191: 1080–1081, 1961.
100. Ching TM, Hedtke S, Russell SA, Evans HJ: Energy state and dinitrogen fixation in soybean nodules of dark grown plants. *Pl Physiol* 55: 796–798, 1975.
101. Lie TA: Environmental effects on nodulation and symbiotic nitrogen fixation. In: *The biology of nitrogen fixation.* Quispel A (ed), Amsterdam, North-Holland, 1974, p 555–582.

102. Sprent JI, Silvester WB: Nitrogen fixation by *Lupinus arboreus* grown in the open and under different aged stands of *Pinus radiata. New Phytol* 72: 991–1003, 1973.
103. Vezina PE, Boulter DKW: The spectral composition of near ultraviolet and visible radiation beneath forest canopies. *Can J Bot* 44: 1267–1284, 1966.
104. Atzet T, Waring RH: Selective filtering of light by coniferous forests and minimum light energy requirements for regeneration. *Can J Bot* 48: 2163–2167, 1970.
105. Virtanen AI, Moisio T, Burris RH: Fixation of nitrogen by nodules excised from illuminated and darkened pea plants. *Acta Chem Scand* 9: 184–193, 1955.
106. Roponen IE: The effect of darkness on the leghaemoglobin content and amino-acid levels in the root nodules of pea plants. *Physiol Plant* 23: 452–460, 1970.
107. Wheeler CT, Bowes BG: Effects of light and darkening upon nitrogen fixation in root nodules of *Alnus glutinosa* in relation to their cytology. *Z Pflanzenphysiol* 71: 71–75, 1974.
108. Lawn RJ, Brun WA: Symbiotic nitrogen fixation in soybeans. I. Effect of photosynthetic source-link manipulations. *Crop Sci* 14: 11–16, 1974.
109. Bethlenfalvay GJ, Phillips DA: Effect of light intensity on efficiency of carbon dioxide and nitrogen reduction in *Pisum sativum* L. *Plant Physiol* 60: 868–871, 1977.
110. Wilson PW, Fred EB, Salmon MR: Relation between carbon dioxide and elemental nitrogen assimilation in leguminous plants. *Soil Sci* 35: 145–165, 1933.
111. Phillips DA, Newell KD, Hassell SA, Felling CE: Effect of CO_2 enrichment on root nodule development and symbiotic N_2 reduction in *Pisum sativum* L. *Am J Bot* 63: 356–362, 1976.
112. Hardy RWF, Havelka UD: Photosynthate as a major factor limiting nitrogen fixation by field-grown legumes with emphasis on soybeans. In: *Symbiotic nitrogen fixation in plants*, Nutman PS (ed), IBP 7, Cambridge, Cambridge University Press, 1976, p 421–442.
113. Lawn RJ, Fischer KS, Brun WA: Symbiotic nitrogen fixation in soybeans. II. Inter-relationship between carbon and nitrogen assimilation. *Crop Sci* 14: 17–22, 1974.
114. De Jong TM, Phillips DA: Nitrogen stress and apparent photosynthesis in symbiotically grown *Pisum sativum* L. *Pl Physiol* 68: 309–313, 1981.
115. Nátr L: Influence of mineral nutrition on photosynthesis and the use of assimilates. In: *Photosynthesis and productivity in different environments.* Cooper JP (ed), Cambridge, Cambridge University Press, 1975, p 537–555.
116. Masterson CL, Murphy PM: Application of the acetylene reduction technique to the study of nitrogen fixation by white clover in the field. In: *Symbiotic nitrogen fixation in plants.* Nutman PS (ed), IBP 7, Cambridge, Cambridge University Press, 1976, p 299–318.
117. Eckart JF, Raguse CA: Effects of diurnal variation in light and temperature on the acetylene reduction activity (nitrogen fixation) of subterranean clover. *Agron J* 72: 519–523.
118. Mague TH, Burris RH: Reduction of acetylene and nitrogen by field-grown soybeans. *New Phytol* 71: 275–286, 1972.
119. Trinick MJ, Dilworth MJ, Grounds M: Factors affecting the reduction of acetylene by root nodules of *Lupinus* species. *New Phytol* 77: 359–370, 1976.
120. Minchin FR, Pate JS: Diurnal functioning of the legume root nodule. *J Exp Bot* 25: 295–308, 1974.
121. Johnsrud SC: Nitrogen fixation by root nodules of *Alnus incana* in a Norwegian forest ecosystem. *Oikos* 30: 475–479, 1978.
122. Tripp LN, Bezdirek D, Heilman PE: Seasonal and diurnal patterns and rates of nitrogen fixation by young red alder. *Forest Sci.* 25: 371–380, 1979.
123. Wheeler CT: The diurnal fluctuation in nitrogen fixation in the nodules of *Alnus glutinosa* and *Myrica gale. New Phytol* 68: 675–682, 1969.
124. Wheeler CT, Lawrie AC: Nitrogen fixation in root nodules of alder and pea in relation to the supply of photosynthetic assimilates. In: *Symbiotic nitrogen fixation in plants.* Nutman PS (ed), IBP 7, Cambridge, Cambridge University Press, 1976, p 497–509.
125. McNiel RE, Carpenter PL: The effect of temperature, solar radiation and moisture on the reduction of acetylene by excised root nodules from *Alnus glutinosa. New Phytol* 82: 459–465, 1979.
126. Sprent JI, Silvester WB: Nitrogen fixation by *Lupinus arboreus* grown in the open and under different aged stands of *Pinus radiata. New Phytol* 72: 991–1004, 1973.
127. Lawrie AC, Wheeler CT: Nitrogen fixation in the root nodules of *Vicia faba* L. in relation to

the assimilation of carbon. I. Plant growth and metabolism of photosynthetic assimilates. *New Phytol* 74: 429–436,1975.

128. Hardy RWF, Burns RC, Herbert RR, Holsten RD, Jackson EK: Biological nitrogen fixation: a key to world protein. *Pl Soil*, Special Volume: 561–590, 1971.
129. Sprent JI: Nitrogen fixation by legumes subjected to water and light stresses. In: *Symbiotic nitrogen fixation in plants*. Nutman PS (ed), IBP 7, Cambridge, Cambridge University Press, 1976, p 405–420.
130. Sprent JI, Bradford AM, Norton C: Seasonal growth patterns in field beans (*Vicia faba*) as affected by population density, shading and its relationships with soil moisture. *J Agric Sci*, Camb. 88: 293–301, 1977.
131. Pizelle G: Variations saisonnières de l'activité nitrogénasique des nodules d'*Alnus glutinosa* (L.) Gaertn., d'*Alnus incana* L. Moench et d'*Alnus cordata* (Lois.) Desf *CR Acad Sc Paris*, D 281: 1829–1832, 1975.
132. Sprent JI, Scott R, Perry KM: The nitrogen economy of *Myrica gale* in the field. *J Ecol* 66: 657–668, 1978.
133. Dalton DA, Zobel DB: Ecological aspects of nitrogen fixation by *Purshia tridentata*. *Pl Soil* 48: 57–80, 1977.
134. Stewart WDP, Pearson MC: Nodulation and nitrogen fixation by *Hippophae rhamnoides* in the field. *Pl Soil* 26: 348–360, 1967.
135. Vincent JM: *Rhizobium*: general microbiology. In: *A treatise on biological nitrogen fixation. III. Biology*. Hardy RWF, Gibson AH (eds), New York, Wiley, 1977, p 277–366.
136. Mulder EG, van Veen WL: Effect of pH and organic compounds on nitrogen fixation by red clover. *Pl Soil* 13: 91–113, 1960.
137. Munns DN: Mineral nutrition and the legume symbiosis. In: *A treatise on dinitrogen fixation. IV. Agronomy and ecology*. Hardy RWF, Gibson AH (eds), New York, Wiley, 1977, p 353–391.
138. Mulder EG, Lie TA, Houwers A: The importance of legumes under temperate conditions. In: *A treatise on dinitrogen fixation. IV. Agronomy and ecology*. Hardy RWF, Gibson AH (eds), New York, Wiley, 1977, p 221–242.
139. Lie TA: Effect of low pH on different phases of nodule formation in pea plants. *Pl Soil* 31: 391–406, 1969.
140. Munns DN: Nodulation of *Medicago sativa* in solution culture. I. Acid-sensitive steps. *Pl Soil* 28: 129–146, 1968.
141. Andrew CS: Effect of Ca, pH and nitrogen on growth and chemical composition of tropical and temperate pasture legumes. I. Nodulation and growth. *Aust J Agr Res* 29: 611–623, 1976.
142. Munns DN, Fox RL, Koch BL: Influence of lime on nitrogen fixation by legumes. *Pl Soil* 46: 591–601, 1977.
143. Bond G: The fixation of nitrogen associated with the root nodules of *Myrica gale* L. with special reference to its pH relation and ecological significance. *Ann Bot* 25: 447–459, 1951.
144. Canizo A, Miguel C, Rodriguez-Barrueco C: The effect of pH on nodulation and growth of *Coriaria myrtifolia* L. *Pl Soil* 94: 195–198, 1978.
145. Bond G, Fletcher WW, Ferguson TP: The development and function of the root nodules of *Alnus*, *Myrica* and *Hippophae*. *Pl Soil* 5: 309–323, 1954.
146. Ferguson TP, Bond G: Observations on the formation and function of the root nodules of *Alnus glutinosa* (L.) Gaertn. *Ann Bot* 17: 175–188, 1953.
147. Bond G: The development and significance of the root nodules of *Casuarina*. *Ann Bot* 21: 373–380, 1957.
148. Bond G, MacConnell JT, McCallum AH: The nitrogen nutrition of *Hippophae rhamnoides*. *Ann Bot* 20: 501–512, 1956.
149. Gardner IC, Bond G: Observations on the root nodules of *Shepherdia*. *Can J Bot* 35: 305–314, 1957.
150. Bond G: Symbiosis of leguminous plants and nodule bacteria. III. Observations on the growth of soya beans in water culture. *Ann Bot* 14: 245–261, 1950.
151. Nyatsanga T, Pierre WH: Effect of nitrogen fixation by legumes on soil acidity. *Agron J* 65: 936–940, 1973.
152. Franklin JF, Dyrness CT, Moore DG, Tarrant RF: Chemical soil properties under coastal

Oregon stands of alder and conifers. In: *Biology of alder*. Trappe JM, Franklin JF, Tarrant RF, Hansen GM (eds), Portland, Oregon Pacific Northwest Forest and Range Experiment Station, USDA, 1968, p 157–172.

153. Israel DW, Jackson WA: The influence of nitrogen nutrition on ion uptake and translocation by leguminous plants. In: *Mineral nutrition of legumes in tropical and sub-tropical soils*. Andrew CS, Kamprath EJ (eds), Melbourne, CSIRO, 1978, 112–129.
154. Crocker RL, Major J: Soil development in relation to vegetation and surface age at Glacier Bay, Alaska. *J Ecol* 43: 427–448, 1955.
155. Mitchell WW: On the ecology of sitka alder in the subalpine zone of South-central Alaska. In: *Biology of alder*. Trappe JM, Franklin JF, Tarrant RF, Hansan GM (eds), Portland, Oregon, Pacific North West Forest and Range Experimental Station, USDA, 1968, p 45–56.
156. Robson AD: Mineral nutrients limiting nitrogen fixation in legumes. In: *Mineral nutrition of legumes in tropical and sub-tropical soils*. Andrew CS, Kamprath EJ (eds), Melbourne, CSIRO, 1978, 277–294.
157. Dart PJ, Pate JS: Nodulation studies in legumes. III. The effects of delaying inoculation on the seedling symbiosis of barrel medic, *Medicago tribuloides* Desr. *Aust J Biol Sci* 12: 427–441, 1959.
158. Holding AJ, Lowe JF: Some effects of acidity and heavy metals on the Rhizobium leguminous plant association. *Pl Soil*, Special Volume: 153–166, 1971.
159. Andrew CS: Mineral characterisation of tropical forage legume. In: *Mineral nutrition of legumes in tropical and sub-tropical soils*. Andrew CS, Kamprath EJ (eds), Melbourne, CSIRO, 1978, 93–112.
160. Andrew CS: Effect of calcium, pH and nitrogen on the growth and chemical composition of some tropical and temperature pasture legume species. *Aust J Agric Res* 27: 611–623, 1976.
161. Lowther WL, Lonerogan JF: Calcium and nodulation in subterraneum clover (*Trifolium subterraneum* L.). *Pl Physiol* 43: 1362–1366, 1968.
162. Munns DN: Nodulation of *Medicago sativa* in solution culture. V. Calcium and pH requirements during infection. *Pl Soil* 32: 92–102, 1970.
163. Ovington JD: The calcium and magnesium contents of tree species grown in close stands. *New Phytol* 58: 164–175, 1959.
164. Youngberg CT, Wollum AG: Nitrogen accretion in developing *Ceanothus velutinus* soils. *Soil Sci Soc Amer J* 40: 109–112, 1976.
165. Zarovy MG, Munns DN: Nodulation and growth of *Lablab purpureus* in relation to *Rhizobium* strain, liming and phosphorus. *Pl Soil* 53: 329–337, 1979.
166. Smith SE, Daft MJ: Interactions between growth, phosphate content and nitrogen fixation in mycorrhizal and non-mycorrhizal *Medicago sativa*. *Aust J Pl Physiol* 4: 403–413, 1977.
167. Daft MJ: Nitrogen fixation in nodulated and mycorrhizal crop plants. *Ann Appl Biol* 88: 461–2, 1978.
168. Munns DN, Mosse B: Mineral nutrition of legume crops. In: *Advances in legume science*. Summerfield RJ, Bunting AH (eds), Kew, Royal Botanic Gardens, 1980.
169. Riffle JW: First report of vesicular arbuscular mycorrhizae in *Elaeagnus augustifolia*. *Mycologia* 69: 1200–1203, 1977.
170. Mejstrik V, Benecke U: The ectotrophic mycorrhizae of *Alnus viridis* (Chaix) DC and their significance in respect to phosphorus uptake. *New Phytol* 68: 141–149, 1969.
171. Molina R: Ectomycorrhizal specificity in the genus *Alnus*. *Can J Bot* 57: 325–334, 1979.
172. Rose SL, Youngberg CT: Tripartite associations in snowbrush (*Ceanothus velutinus*): effect of vesicular-arbuscular mycorrhizae on growth, nodulation and nitrogen fixation. *Can J Bot* 59: 34–49, 1981.
173. Evans HJ, Russell SA: Physiological chemistry of symbiotic nitrogen fixation by legumes. In: *The chemistry and biochemistry of nitrogen fixation*. Postgate JR (ed), London, Plenum Press, 1971, 191–245.
174. Becking JH: Root nodules in non-legumes. In: *The development and function of roots*. Torrey JG, Clarkson DT (eds), London Academic Press, 1976, p 507–566.
175. Bond G, Hewitt EJ: Molybdenum and the fixation of nitrogen in *Myrica gale* root nodules. *Nature* 190: 1033–1034, 1961.
176. De Hertogh AA, Mayeux PA, Evans HJ: The relationship of cobalt requirement to propionate

metabolism in *Rhizobium*. *J Biol Chem* 239: 2446–2453, 1964.
177. De Hertogh AA, Mayeux PA, Evans HJ: Effect of cobalt on the oxidation of propionate by *Rhizobium meliloti*. *J Bacteriol* 87: 746–747, 1964.
178. Cowles JR, Evans HJ, Russell SA: B_{12} coenzyme-dependent ribonucleotide reductase in *Rhizobium* species and the effect of cobalt deficiency on the activity of the enzyme. *J Bacteriol* 97: 1460–1465, 1969.
179. Gladstones JS, Loneragan JF, Goodchild NA: Field responses to cobalt and molybdenum by different legume species, with inferences on the role of cobalt in legume growth. *Aust J Agric Res* 28: 619–628, 1977.
180. Chatel DL, Robson AD, Gartrell JW, Dilworth MJ: The effect of inoculation and cobalt application on the growth and nitrogen fixation of sweet lupin. *Aust J Agr Res* 29: 1191–1202, 1978.
181. Dilworth MJ, Robson AD, Chatel DL: Cobalt and nitrogen fixation in *Lupinus angustifolius* L. II. Nodule formation and function. *New Phytol* 83: 63–79, 1979.
182. Robson AD, Dilworth MJ, Chatel DL: Cobalt and nitrogen fixation in *Lupinus angustifolius* L. I. Growth, nitrogen concentration and cobalt distribution. *New Phytol* 83: 53–62, 1979.
183. Bond G, Adams JF, Kennedy EH: Vitamin B_{12} analogues in non-legume root nodules. *Nature*, 207: 319–320, 1965.
184. Bond G, Hewitt EJ: Cobalt and the fixation of nitrogen by root nodules of *Alnus* and *Casuarina*. *Nature*, Lond 195: 94–95, 1962.
185. Hewitt EJ, Bond G: The cobalt requirement of non-legume root nodule plants. *J Exp Bot* 17: 480–491, 1966.
186. Waksman SA: The actinomycetes. New York, Ronald Press, 1967, 168–169.
187. Metzler DE: Biochemistry. New York, Academic Press, 1977, p 539.
188. Wilson PW: The biochemistry of symbiotic nitrogen. Madison, University of Wisconsin Press, 1940.
180. Dart PJ, Wildon DC: Nodulation and nitrogen fixation by *Vigna sinensis* and *Vicia atropurpurea*: the influence of concentration, form and site of application of combined nitrogen. *Aust J Agric Res* 21: 45–56, 1970.
190. Allos HF, Bartholomew WV: Replacement of symbiotic fixation by available nitrogen. *Soil Sci* 87: 61–66, 1959.
191. Virtanen AI, Jorma J, Linkola H, Linnasalmi A: On the relation between nitrogen fixation and leghaemoglobin content of leguminous root nodules. *Acta Chem Scand* 1: 90–111, 1947.
192. Pizelle G: L'azote minéral et la nodulation de l'aune glutineux (*Alnus glutinosa*). II. Observations sur l'action inhibitrice de l'azote minéral a l'égard de la nodulation. *Ann. Inst. Pasteur* 111: 259–264, 1966.
193. Cartwright PM: The effect of combined nitrogen on the growth and nodulation of excised roots of *Phaseolus vulgaris* L. *Ann Bot* 31: 309–321, 1967.
194. Dreyfus BL, Dommergues YR: Non-inhibition of nitrogen fixation by combined nitrogen in a stem bearing nodules in the legume *Sesbania rostrata*. CR Acad Sci Ser D 291: 767–770, 1980.
195. Small JGC, Leonard AO: Translocation of C^{14}-labelled photosynthate in nodulated legumes as influenced by nitrate nitrogen. *Amer J Bot* 56: 187–194, 1969.
196. Raggio M, Raggio N, Torrey JG: The interaction of nitrate and carbohydrates in rhizobial root nodule formation. *Pl Physiol* 40: 601–606, 1965.
197. Chen P-C, Phillips DA: Induction of nodule senescence by combined nitrogen in *Pisum sativum*. *Pl Physiol* 59: 440–442, 1977.
198. Streeter JG: Effect of nitrate in the rooting medium on carbohydrate composition of soybean nodules. *Pl Physiol* 68: 840–844, 1981.
199. Darbyshire JF: Studies on the physiology of nodule formation. IX. The influence of combined nitrogen, glucose, light intensity and dry length on root-hair infection in clover. *Ann Bot* 30: 623–638, 1966.
200. Dixon ROD: Rhizobia (with particular reference to relationships with host plants). *Ann Rev Microbiol* 23: 137–158, 1969.
201. Munns DN: Nodulation of *Medicago sativa* in solution culture. II. Compensating effects of nitrate and of prior nodulation. *Pl Soil* 28: 246–257, 1968.

202. Munns DN: Nodulation of *Medicago sativa* in solution culture III. Effects of nitrate on root hairs and infection. *Pl Soil* 29: 33–47, 1968.
203. Kennedy IR, Rigaud J, Trinchant JC: Nitrate reductase from bacteroids of *Rhizobium japonicum*: enzyme characteristics and possible interactions with nitrogen fixation. *Biochim Biophys Acta* 397: 24–35, 1975.
204. Tanner JW, Anderson IC: An external effect of inorganic nitrogen in root nodulation. *Nature* 198: 303–4, 1963.
205. Dart PJ: The infection process. In: *The biology of nitrogen fixation*. Quispel A (ed), Amsterdam, North Holland, 1974, p 381–429.
206. Libbenga KR, Bogers RJ: Root nodule morphogenesis. In: *The biology of nitrogen fixation*. Quispel A (ed), Amsterdam, North Holland, 1974, p 430–472.
207. Gibson AH, Pagan JD: Nitrate effects on the nodulation of legumes inoculated with nitrate-reductase-deficient mutants of *Rhizobium*. *Planta* 134: 17–22, 1977.
208. Veeger C, Haaker H, Laane C: Energy transduction and nitrogen fixation. In: *Current perspectives in nitrogen fixation*. Gibson AH, Newton WE (eds), Canberra, Aust Acad Sci, 1981, p 101–104.
209. Salminen SO: Effect of NH_4^+ on nitrogenase activity in nodule breis and bacteroids from *Pisum sativum* L. *Biochim Biophys Acta* 658: 1–9, 1981.
210. Johnson JW, Welch LF, Kurtz LT: Environmental implications of N fixation by soybeans. *J Environ Qual* 40: 303–306, 1975.
211. Randall DD, Russell WJ, Johnson DR: Nodule nitrate reductase as a source of reduced nitrogen in soybean, *Glycine max Pl Physiol* 44: 325–328, 1978.
212. Wheeler CT, McLaughlin ME: Environmental nodulation of nitrogen fixation in actinomycete nodulated plants. In: *Symbiotic nitrogen fixation in the management of temperate forests*. Gordon JC, Wheeler CT, Perry DA (eds), Corvallis, School of Forestry, Oregon State University, 1978, p 124–142.
213. Stewart WDP, Bond G: The effect of ammonium nitrogen on fixation of elemental nitrogen in *Alnus* and *Myrica*. *Pl Soil* 14: 347–359, 1961.
214. Rodriguez-Barrueco D, Mackintosh AH, Bond G: Some effects of combined nitrogen on the nodule symbioses of *Casuarina* and *Ceanothus*. *Pl Soil* 33: 129–139, 1970.
215. Ingestad T: Growth, nutrition and nitrogen fixation in grey alder at varied rates of nitrogen addition. *Physiol Plant* 50: 353–364, 1980.
216. Lahair E, Harper JE, Hageman RH: Improved soybean growth in urea with pH buffered by a carboxy resin. *Crop Sci* 16: 325–328, 1976.
217. Vigue JT, Harper JE, Hageman RH, Peters DB: Nodulation of soybeans grown hydroponically on urea. *Crop Sci* 17: 169–172, 1976.
218. Eaglesham ARJ, Day J, Dart PJ: Rep. Rothamstead Exp. Sta., 85, 1973.
219. Zavitkovski J, Newton M: Effect of organic matterand combined nitrogen on nodulation and nitrogen fixation in red alder. In: *Biology of alder*. Trappe JM, Franklin JF, Tarrant RF, Hansen GF (eds), Portland, Oregon, Pacific North West Forest and Range Experimental Station, USDA, 1968, p 157–172.
220. Sprent JI: Water deficits and nitrogen-fixing root nodules. In: *Water deficits and plant growth* IV. New York, Academic Press, 1976, p 291–315.
221. Osa-afiana LO, Alexander M: Effect of moisture on the survival of *Rhizobium* in soil. *Soil Sci Soc Amer J* 43: 925–930, 1979.
222. Pena-cabriales JJ, Alexander M: Survival of *Rhizobium* in soils undergoing drying. *Soil Sci Soc Amer J* 43: 962–966, 1979.
223. Chatel DL, Parker CA: Survival of field-grown Rhizobia over the dry summer period in Western Australia. *Soil Biol Biochem* 5: 415–423, 1973.
224. Sprent JI: Nitrogen fixation by legumes subjected to water and light stresses. In: *Symbiotic nitrogen fixation in plants*. Nutman PS (ed), IBP 7, Cambridge, Cambridge University Press, 1976, p 405–420.
225. Sprent JI: The effects of water stress on nitrogen fixing root nodules. I. Effects on the physiology of detached soybean nodules. *New Phytol* 70: 9–17, 1971.
226. Pankhurst CE, Sprent JI: Effects of water stress on the respiratory and nitrogen-fixing activity of soybean root nodules. *J Exp Bot* 26: 287–304, 1975.

227. Minchin FR, Pate JS: Effect of water, aeration and salt regime on nitrogen fixation in a nodulated legume-definition of an optimum root environment. *J Exp Bot* 26: 60–69, 1975.
228. Tatazawa M, Yoshida S: Stem nodules in *Aeschynomene indica* and their capacity of nitrogen fixation. *Physiol Plant* 45: 293–295, 1979.
229. McVean DN: Ecology of *Alnus glutinosa* (L.) Gaertn. IV. Root system. *J Ecol* 43: 219–225, 1955.
230. Bond G: Root-nodule symbioses with actinomycete-like organisms. In: Quispel A (ed), *The Biology of nitrogen fixation* North Holland, Amsterdam, 1974, p 342–380.
231. Huang C-T, Boyer JS, Vanderhoeff LN: Acetylene reduction (nitrogen fixation) and metabolic activities of soybean having various leaf and nodule water potentials. *Pl Physiol* 56: 222–227, 1975.
232. Huang C-T, Boyer JS, Vanderhoeff LN: Limitation of acetylene reduction (nitrogen fixation) and metabolic activities of soybean having various leaf and nodule water potentials. *Pl Physiol* 56: 228–232, 1975.
233. Patterson P, Raper DC, Gross HD: Growth and specific nodule activity of soybeans during application and recovery of a leaf moisture stress. *Pl Physiol* 64: 551–556, 1979.
234. Ching TM, Hedtke S, Russell SA, Evans HJ: Energy state and dinitrogen fixation in soybean nodules of dark grown plants. *Pl Physiol* 55: 796–798, 1975.
235. Sprent JI: Growth and nitrogen fixation in *Lupinus arboreus* as affected by shading and water supply. *New Phytol* 72: 1005–1022, 1973.
236. Albrecht SL, Bennett JM, Quesenberry KH: Growth and nitrogen fixation of *Aeschynomene* under water stressed conditions. *Pl Soil* 60: 309–315, 1981.
237. Helgerson OT, Wheeler CT, Perry DA, Gordon JC: Annual nitrogen fixation in Scotch Broom (*Cytisus scoparius* L.). In: Symbiotic nitrogen fixation in the management of temperate forests. Gordon JC, Wheeler CT, Perry DA (eds), Corvallis, School of Forestry, Oregon State University, 1979, p 477–478.
238. Buckman HO, Brady NC: The nature and properties of soils. Toronto, Macmillan (7th edition 1969), p 246.
239. Stolzy LH: Soil atmosphere. *The plant root and its environment*. Carson EW (ed), Charlottesville, University Press of Virginia, 1974, p 335–362.
240. Loneragan JF: The interface in relation to root function and growth. In: *The soil-root interface*. Harley JL, Scott Russell R (eds), London, Academic Press, 1979, p 351–368.
241. Amoore JE: Dependence of mitosis and respiration in roots upon oxygen tension. *Proc Roy Soc (Lond)* B 154: 109–129, 1961.
242. Boynton D, De Villiers JE, Reuther W: Are there different critical oxygen concentrations for the different phases of root activity? *Science* 88: 569–570, 1938.
243. Letey J, Stolzy LH: Limiting distances between root and gas phase for adequate oxygen supply. *Soil Sci* 103: 404–409, 1967.
244. Loveday J: Influence of oxygen diffusion rate in (*Rhizobium*) nodulation of subterranean clover. *Aust J Sci* 26: 90–91, 1963.
245. Bond G: Symbiosis of leguminous plants and nodule bacteria. IV. The importance of the oxygen factor in nodule formation and function. *Ann Bot* 15: 95–108, 1950.
246. Ferguson TP, Bond G: Symbiosis of leguminous plants and nodule bacteria. V. The growth of red clover at different oxygen tensions. *Ann Bot* 18: 385–396, 1954.
247. Bond G: The oxygen relation of nitrogen fixation in root nodules. *Zeitsch Allg Mikrobiol* 12: 93–99, 1960.
248. Burris RH, Magee WE, Bach MK: The pN_2 and the pO_2 function for nitrogen fixation by excised soybean nodules. *Ann Acad Sci Fenn* 60: 190–199, 1955.
249. Rawsthorne S, Minchin FR, Summerfield RJ, Cookson C, Coombs J: Carbon and nitrogen metabolism in legume root nodules. *Phytochem* 19: 341–355, 1980.
250. Lowe RH, Evans HJ: Carbon dioxide requirement for growth of legume nodule bacteria. *Soil Sci* 94: 351–356, 1962.
251. Grobelaar N, Clarke B, Hough MC: The nodulation and nitrogen fixation of *Phaseolus vulgaris* L. III. The effect of CO_2 and C_2H_4. *Pl Soil*, Special Volume: 215–223, 1971.
252. Mulder EC, van Veen WL: The influence of carbon dioxide in symbiotic nitrogen fixation. *Pl Soil* 13: 265–278, 1960.

253. Stolwijk JAJ, Thimann KV: On the uptake of carbon dioxide and bicarbonate by roots and its influence on growth. *Pl Physiol* 32: 513–520, 1957.
254. Bergersen FJ: Biochemistry of symbiotic nitrogen fixation in legumes. *Ann Rev Pl Physiol* 22: 121–140, 1971.
255. Wheeler CT: Carbon dioxide fixation in the legume root nodule. *Ann Appl Biol* 481–484, 1978.
256. Schwinghamer EA: Genetic aspects of nodulation and dinitrogen fixation by legumes: the microsymbiont. In: *A treatise on dinitrogen fixation.* III. *Biology* Hardy RWF, Silver WS (eds), New York, Wiley, 1977, p 577–622.
257. Brill WJ: Biochemical genetics of nitrogen fixation. Microbiol. Rev. 44: 449–467, 1980.
258. Brewin NJ, Johnston AWB, Beringer JE: The genetics of *Rhizobium.* In: *Nitrogen fixation.* Stewart WDP, Gallon JR (eds), London, Academic Press, 1980, p 365–374.
259. Beringer JE, Brewin NJ, Johnston AWB: The genetic analysis of *Rhizobium* in relation to symbiotic nitrogen fixation. *Heredity* 45: 161–186, 1980.
260. Beringer JE: Genetics of symbiotic nitrogen-fixing microorganisms. In: *Current perspectives in nitrogen fixation.* Gibson AH, Newton WE (eds), Canberra, Aust Acad Sci, 1981, p 131–136.
261. Postgate JR, Cannon FC: The molecular and genetic manipulation of nitrogen fixation. *Phil Trans Roy Soc B* 292: 589–599, 1981.
262. Caldwell BE, Vest HG: Genetic aspects of nodulation and dinitrogen fixation by legumes: the macrosymbiont. In: *A treatise on dinitrogen fixation.* III. *Biology.* Hardy RWF, Silver WS (eds), New York, Wiley, 1977, p 557–576.
263. Nutman PS: Genetics of symbiosis and nitrogen fixation in legumes. *Proc Roy Soc B* 172: 417–438, 1969.
264. La Rue TA: Host plant genetics and enhancing symbiotic nitrogen fixation. In: *Nitrogen fixation.* Stewart WDP, Gallon JR (eds), London, Academic Press, 1980, p 355–364.
265. Nutman PS: Hereditary host factors affecting nodulation and nitrogen fixation. In: *Current perspectives in nitrogen fixation.* Gibson AH, Newton WE (eds), Canberra, Aust Acad Sci, 1981, p 194–204.
266. Amarger N: Selection of *Rhizobium* strains on their competitive ability for nodulation. *Soil Biol Biochem* 13: 481–486, 1981.
267. Amarger N: Competitiveness for nodule formation between effective and ineffective strains of *Rhizobium meliloti. Soil Biol Biochem* 13: 475–480, 1981.
268. Obaton M: Influence de la composition chimiques du sol sur l'utilitée de l'inoculation des graines de luzerne avec *Rhizobium meliloti. Pl Soil,* Special Volume, 273–285, 1971.
269. Nuti MP: The large plasmids of Rhizobia: current status and their involvement in the expression of symbiotic functions. In: *Current perspectives in nitrogen fixation.* Gibson AH, Newton WE (eds), Canberra, Aust Acad Sci, 1981, p 167–168.
270. Dunican LK, O'Gora F, Tierney AB: Plasmid control of effectiveness in *Rhizobium*: transfer of nitrogen-fixing genes on a plasmid from *Rhizobium trifolin* to *Klebsiella aerogenes.* In: *Symbiotic nitrogen fixation in plants.* IBP 7, Nutman PS (ed), Cambridge, Cambridge University Press, 1976, p 77–90.
271. Nuti MP, Lepidi AM, Prakash RK, Schilperoot RA, Cannon FC: Evidence for nitrogen fixation (nif) genes on indigenous *Rhizobium* plasmids. *Nature* 282: 533–534, 1979.
272. Johnston AWB, Beynon JL, Buchanan-Wollaston AV, Setchell SM, Hirsch PR, Beringer JE: High frequency transfer of nodulating ability between strains and species of *Rhizobium. Nature* 276: 634–636, 1978.
273. Dénarié J, Rosenberg C, Boistard P, Truchet G, Casse-Delbart F: Plasmid control of symbiotic properties in *Rhizobium meliloti.* In: *Current perspectives in nitrogen fixation.* Gibson AH, Newton WE (eds), Canberra, Aust Acad Sci, 1981, p 137–141.
274. Rodriguez-Barrueco R, Miguel C, Canizo A: Host plant endophyte specificity in actinorhizal plants. In: *Current perspectives in nitrogen fixation.* Gibson AH, Newton WE (eds), Canberra, Aust Acad Sci, 1981, p 476.
275. Baker D, Lechevalier MP, Dillon JT: Strain analysis of actinorhizal microsymbionts (Genus: *Frankia*). In: *Current perspectives in nitrogen fixation.* Gibson AH, Newton WE (eds), Canberra, Aust Acad Sci, 1981, p 479.
276. Dawson JO, Soon-Hwa Sun: The effect of *Frankia* isolates from *Comptonia peregrina* and

Alnus crispa on the growth of *Alnus glutinosa*, *A. cordata* and *A. incana* clones. *Can J For Res*, in press.

277. Hall RB, Maynard CA: Considerations in the genetic improvement of alder. In: *Symbiotic nitrogen fixation in the management of temperate forests*. Gordon JC, Wheeler CT, Perry DA (eds), Corvallis, School of Forestry, Oregon State University, 1979, p 322–344.
278. Mejnartowicz L: Genetyka. In: *Olsze – Alnus Mill.* Bialokok S (ed), Warsaw, Polska Akad. Nauk Inst. Dendrologii, 1980, p 201-228.
279. Gordon JC, Wheeler CT: Whole plant studies on photosynthesis and acetylene reduction in *Alnus glutinosa*. *New Phytol.* 80: 179–186, 1978.
280. Perinet P, Lalonde M: In vitro propagation and nodulation of the actinorhizal host plant *Alnus glutinosa* (L.) Gaertn. *Pl. Sci. Letters* 29: 9–17, 1983.

6. Analysis of nitrogen fixation

WARWICK B. SILVESTER

Department of Biological Sciences, University of Waikato, Hamilton, New Zealand

INTRODUCTION

General

Unlike pastoral and cropping systems where the nitrogen fixing plants are normally managed and cropped, the agents of biological nitrogen fixation in forests are normally subordinate to the cropped system. Thus the primary management procedures in forest concern the forest crop and measurement of nitrogen fixation within a stand, whether managed or not, will be subject to the effects of the dominant species. While all field analysis is beset with methodological and fundamental errors and problems, these problems are enhanced in many forest systems because of the lack of homogeneity both in vegetation and environment. The lack of homogeneity increases when going from plantation to natural managed to virgin forests and leads to extreme temporal and spatial sampling problems.

While the agents of nitrogen fixation are dealt with in detail elsewhere it is convenient to categorise them at this point according to their actual and potential contribution to forest ecosystems. The range of measured biological fixation in forest covers over three orders of magnitude (Fig. 1) ranging from less than 1 kg ha^{-1} a^{-1} for many forests having sparse lichens and heterotrophic bacteria to over 300 kg ha^{-1} a^{-1} in dense stands of nodulated tree crops such as *Alnus*, *Casuarina* or *Acacia* (1, 2).

Analysis of nitrogen fixation involves a number of steps which can be set out in sequence.

1. Identification of sites and/or agents responsible for N_2 fixation.
2. Measurement of the spatial and temporal distribution of the N_2 fixing agents.
3. Choice of method(s) for measuring N_2 fixation.
4. Estimation of specific activity of organisms, or organs (i.e. rate per unit weight, area, number per unit time).
5. Investigation of specific activity and the temporal and spatial distribution of specific activity.
6. Calculation of total N inputs related to time (e.g. diurnal, seasonal) space (e.g. soil profile, stand) environment (e.g. light penetration, moisture).

Gordon, J.C. and Wheeler, C.T. (eds.). Biological nitrogen fixation in forest ecosystems: foundations and applications

ISBN 90-247-2849-5. Printed in The Netherlands

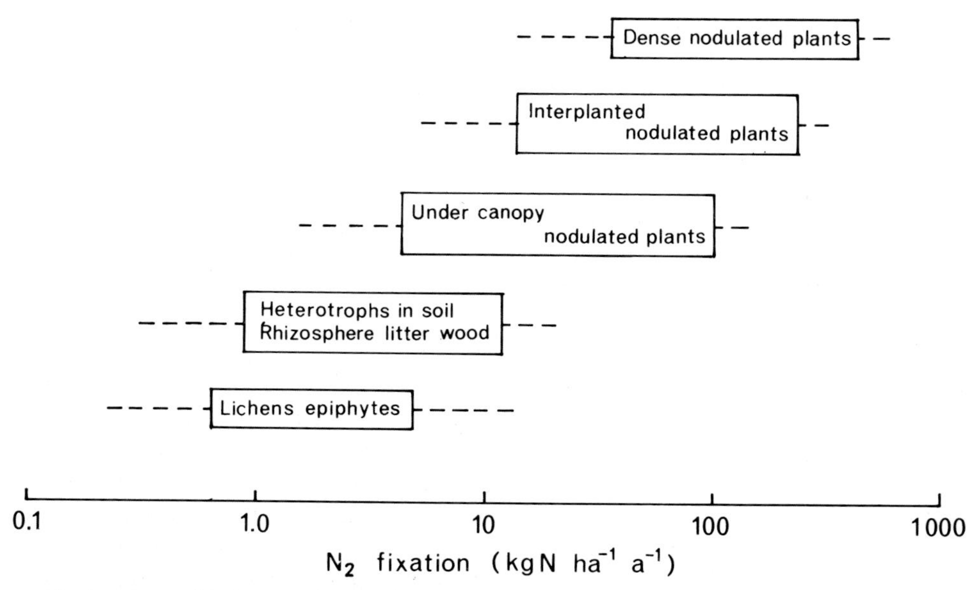

Fig. 1. Average N_2 fixing activity and extreme ranges for the major biological inputs into forest ecosystems.

7. Measurement of response in forest system in terms of productivity, whether total biomass or roundwood.

While it is not necessary to include all of the above steps, the eventual aim is to quantify N_2 fixation in absolute or relative terms and, to be of ultimate value, to show whether this results in significantly increased biomass production in the tree crop.

Basic terminology

N_2 fixation and N_2ase activity. With the advent of the acetylene reduction technique during the 1960s, those studying nitrogen fixation were given a powerful tool. While there was almost instant and widespread adoption of the technique as a measure of N fixation there is also a growing awareness of its limitations especially in field and ecological studies. The acetylene reduction assay, and other related assays (3) rely on the ability of the nitrogenase enzyme system to reduce a number of substrates other than N_2. Such substrates as nitrous oxide, acetylene, azide, cyanide and isocyanide are reduced by the enzyme and the product measured. By far the most common of these assays is the reduction of acetylene to ethylene. While the requirements for ATP and reductant are similar to those for nitrogen fixation there are some differences in the reaction (see pp. 200–205). Because the reactions do not measure nitrogen fixation,

they are better referred to as *nitrogenase assays* rather than nitrogen fixation assays. It is necessary therefore to differentiate between assays which measure nitrogen fixation and those which measure nitrogenase activity.

Direct and indirect methods. Direct methods assess actual nitrogen fixed as e.g. $^{15}N_2$ uptake or nitrogen accumulation in biomass. Indirect methods assess associated reactions e.g. acetylene reduction or effects of N fixing plants on associated species. Calibration of indirect measures to give absolute measures of nitrogen fixation require considerable effort and at times, faith and intuition.

Significant levels. The accepted measure of nitrogen fixation at the community and ecosystem level is kg N ha^{-1} a^{-1}. Authors often go to great lengths to fit their data into this format for comparative purposes. In some cases rates measured on very small samples may be blown up to hectare basis or sampling over a small time period related to an annual rate with use of some appropriate multiplication factors. Use of the term 'significant' in this context has three meanings which are seldom differentiated. Statistical significance is often easy to determine in relating one area to another, or in determining whether the rate is significantly above zero. Biological significance relates to the organism only, is the rate of fixation significant to the organism that has nitrogenase. Biological (agronomic) significance is probably the most difficult to determine as it involves an assessment of whether the input has any short or long term significance to the productivity of the ecosystem. In the forestry sense, this is obviously of greatest importance as it is ultimately the effect of added N on the growth of trees that will determine the so called significance of the event.

Errors

Spatial and temporal. Like all biological processes, nitrogen fixation is distributed in space and time and errors in analysis relate not only to the assay methods used but also to the degree to which experimenters can come to grips with space/time variation. While at the physiological and biochemical levels it is possible to measure rates with great accuracy over seconds to hours, field measurements pose difficulties because of the large medium (daily) and long term (annual) fluctuations in rate. Similarly, the distribution of both organisms and environment in a forest is seldom homogeneous, usually tending towards discontinuity and clumping and these also lead to particular and sometimes unique spatial sampling problems. Thus in order to obtain meaningful measures of nitrogen fixation in the field, care is needed to take account of and to measure both spatial and temporal variation. Measurements of nitrogen fixation taken at one time and place (spot readings) are most prone to the above errors and it is essential to calibrate these in space and time. However there are integrative techniques which can be used to integrate time and/or space variation to give measures which are free of the sampling errors mentioned.

Methodological. Methodological errors in measuring any biological system may creep in from application of methods that alter the system or fail to obtain a full sample for example. In all cases there is a requirement that the biology of the system be understood and that the methods that are applied do not either over or under represent the system or modify it so as to stimulate or supress normal activity. Examples of these two types of error are firstly, failure to excavate all nodules in a sample and secondly, use of a small container for assay such that the gas phase is altered during the assay and thus biological activity modified.

Classification of methods

All methods of assessing N_2 fixation can be classified as direct or indirect and as spot or integrative, and examples of each are shown in Table 1.

Overview

While many recent reviews and texts on nitrogen fixation include chapters on methods, they are not detailed and often do not address the real problems associated with field sampling of heterogeneous systems. The recent publication of a detailed field manual on the subject (4) edited by Bergersen has made the task of workers in this field much simpler, and the writing of this chapter much easier.

In the present account emphasis is placed on reliable and tested methods that can be used in the field and that have been used or show promise in forest ecosystems. Firstly qualitative methods that indicate N_2 fixing species or sites will be described. Secondly direct methods of N fixation assessment will be outlined, both those concerned with biomass analysis and with the N_2 fixation process. This is followed by indirect methods whereby N_2 fixation is assessed by measuring associated reactions such as acetylene reduction. Finally, the most difficult question of all will be addressed, how do we assess the impact of the added N on the system. Tree growth is doubtless the final arbiter of the quantity and quality of added N and effects of added N may be both direct on tree growth and indirect on stimulation of other processes such as mineralisation. As specific

Table 1. Classification of methods for assessing N_2 fixation inputs.

	Spot	Integrative
Direct	Spot $^{15}N_2$ uptake	Integrative N balance
Indirect	C_2H_2 reduction	^{15}N dilution Stimulation of associated vegetation

techniques are dealt with exhaustively in the Bergersen manual (4) they will not be treated in this chapter where the application of those techniques will be developed for assessing forest nitrogen fixation.

IDENTIFICATION OF N_2 FIXING SITES AND AGENTS

General

A necessary prerequisite to assessing N_2 fixation is the identification of the agents responsible and their distribution in space and time. N_2 fixing organisms can be classified as symbiotic where they exist in root nodules, lichens etc, associative as part of rhizosphere or phyllosphere, or asymbiotic where they are either heterotrophic bacteria or autotrophic cyanobacteria. In natural forest ecosystems all of the above types, except for asymbiotic cyanobacteria, have been shown to be of some importance.

Biological

Taxonomy. The taxonomy of known N_2 fixing organisms gives a primary indication of the likely agents and this is dealt with in chapters 2 and 3. Over the last two decades there has been an enormous increase in the number of identified N_2 fixing organisms, including the identification of legume type nodules on a non-legume (*Parasponia*) and actinorhizal nodules on a herbaceous genus (*Datisca*). The identification of *Cowania* as a nodulated species (5) and *Pseudocyphellaria* as a cyanobacterial lichen (6) illustrate recent discoveries of N_2 fixing species both present in forest communities.

Root nodules. Identification of specific structures such as nodules is a useful tool in determining likely agents of N_2 fixation. However, the possession of nodule like structures is not sufficient to prove N_2 fixation as they must be checked for N_2ase activity. Two examples of nodule misidentification illustrate this point. Coralloid roots on *Arctostaphylos* thought to be N_2 fixing structures have been identified as mycorrhizal roots (1). Similarly the short roots which form on many of the southern conifers in the Podocarpaceae, look very much like clover nodules but have now been shown to be simply short roots, which may become mycorrhizal and possess only low rhizosphere N_2 fixing activity (7, 8).

Ecological

Three particular ecological situations aid in the identification of nitrogen fixing species. Firstly on poor soils, especially young alluvial or volcanic material, primary succession is often initiated by N_2 fixing species (9). However, care must be taken to check such ecological evidence as the major limiting nutrient may be phosphorus and the exploiting plants strongly mycorrhizal as in the case of *Arctostaphylos*.

Secondly organic material with a high C/N ratio often provides a medium for N_2 fixing species and in forests both leafy and woody litter are capable of supporting heterotrophic N_2 fixing species.

Finally a large number of reports show the growth stimulation effect of an associated N_2 fixing plant, especially on poor soils. A dramatic effect of the presence of the actinorhizal plant *Coriaria* on *Pinus radiata* growth was demonstrated (10) on landing sites where a log-linear relationship was shown between height of pines and proximity to the nodulated plant.

Chemical

N content. As a broad generalisation most diazotrophs have over 2% nitrogen in their foliage while forest trees growing in similar situations normally have much lower N levels (Table 2).

Isotope ratios. Many soils are slightly enriched with ^{15}N compared with atmospheric N_2, and this is presumably due to the fractionation effect which occurs during denitrification. Plants which are able to utilise atmospheric N_2 will possess an enrichment somewhere between air and soil values according to the proportion of N_2 fixed to soil N assimilated. While earlier work (12) indicated little quantitative or qualitative value in isotope ratio methods, more recent work (13) has shown that there are significant differences in isotope ratio between N_2 fixing and non-fixing plants.

Identification of putative N_2 fixing organisms by the above techniques pro-

Table 2. Nitrogen concentration of foliage of N fixing species and associated species summarised from reviews (1, 2, 10, 11).

Nodulated species	Foliage %N	Associated tree	Foliage %N
Alnus incana	2.73*	*Tilia cordata*	1.16*
		Acer platanoides	0.63*
Alnus glutinosa	2.57*	*Corylus arellana*	1.30*
		Pinus sylvestris	0.49*
Alnus rubra	2.10*	*Pseudotsuga taxifolia*	0.50*
Cytisus scoparius	4.7**		
Ceanothus velutinus	1.78**		
Coriaria arborea	2.2–3.0**	*Pinus radiata*	1.19–1.68**

* fallen or senescent leaves
** fresh foliage

vides a first line of attach which is normally followed by more specific tests described in sections 3, 4 and 5 to quantify the N input and assess its importance.

Nitrogen balance techniques

General principles. The overall N balance of a system, whether individual plant, community or ecosystem is a function of inputs and losses and can be expressed in the simple equation

$$\Delta N = N_{in} - N_{out} \tag{1}$$

The N balance of a N_2 fixing crop is given by

$$\Delta N = Nf + Na - N_{out} \tag{2}$$

Where Nf is nitrogen fixation, Na is N assimilated from soil and N_{out} are the losses due to leaf fall, root turnover and leaching

When community or ecosystem N is analysed

$$\Delta N = Nf + P - D - S \tag{3}$$

Where P is precipitation (wet and dry deposition) of N, D is loss due to denitrification and S is solution and particulate loss (Fig. 2). Thus by rearrangement of equations 2 and 3 equations can be produced which allow estimates of nitrogen fixation.

Forest ecosystems are seldom at equilibrium, usually slowly accumulating nitrogen during their normal life cycle and suffering periodic large losses when vegetation is artificially or naturally removed (14, 15). Long term inputs of N are readily measured in forests during early stages of development as, when total ecosystem N is low, N inputs tend to be high both in absolute and relative terms. However in mature ecosystems, total ecosystem N is high and inputs both in absolute and relative terms tend to be low making measurement possible only over long time spans. These problems are illustrated in Fig. 2 which illustrates a hypothetical forest ecosystem in which 50 kg N ha^{-1} a^{-1} is being contributed by the N_2 fixation, which represents 1% of total ecosystem N, 5% of biomass N, 17% of diazotroph N and 25% of total litter fall N. The precision required in estimating inputs into such a system by mass balance requires that inputs be measured in a stratified way into specific components, or that they be measured over a long period.

The heterogeneous nature of forests demands that a large number of samples be taken to assess N changes adding a further restriction to the technique.

Despite the obvious difficulties of the N balance approach, it possesses the major advantage in being able to integrate the time/space variation with which ecological sampling is fraught producing a net value, which in terms of ecosystem function is probably the most significant. Further approaches to nutrient balance in forest ecosystems have been considered by Ovington (16).

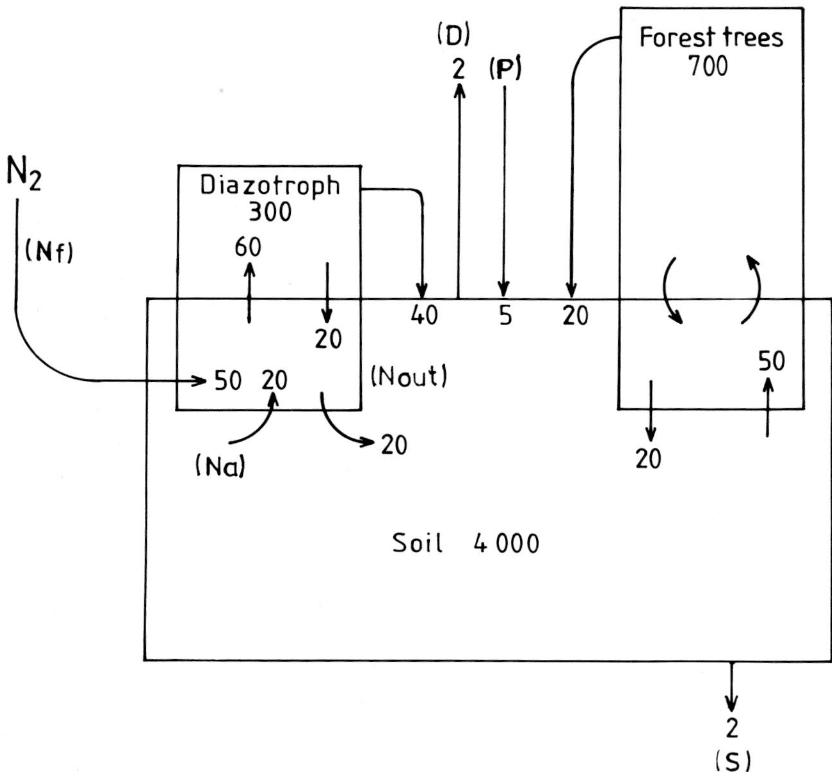

Fig. 2. Generalised N balance sheet for forest systems containing a nodulated plant. All values are kg N ha^{-1} for standing crop and kg N ha^{-1} a^{-1} for transfers. Nitrogen fixation (Nf), N assimilation (Na), N output (N_{out}), denitrification (D), precipitation (P) and solution loss (S).

Analysis of total nitrogen

The essence of nitrogen balance studies involve the sampling and harvesting of plant and soil material and estimation of total nitrogen, or of nitrogen fractions within it. The subject of nitrogen analysis has been dealt with in great detail in a large number of references and this material has been brought together in one volume recently (4).

Inputs into components

General. Reference to Fig. 2 will show that increased sensitivity of analysis is achieved if N fluxes are measured in components of the ecosystem. The further away from the site of fixation the measurement is taken the more likely it is that other losses and gains may influence the result. Rearrangement of equation 2 gives,

$$Nf = \Delta N - Na + N_{out} \quad (4)$$

which for a nodulated plant, N_2 fixation equals nitrogen increment over time, less nitrogen assimilated from soil (Na) plus nitrogen lost in leaf fall, root turnover etc. These principles are developed and examples given for various components of forest ecosystems.

Nodulated plants. Early work on both legumes and non-legumes (17) established that 90% or more of nitrogen fixed is rapidly translocated from the nodule firstly to the shoot and thence to other parts of the plant (18). The components of N balance in nodulated plants are fixation, soil uptake, leaf drop, root turnover/exudation. In short term studies it may be possible to ignore the small losses during active growth or to measure the litter fall component. However in longer term studies it is essential that some attempt is made to quantify these.

When soil nitrogen is low, the majority of ΔN is due to fixation, and many reports provide accurate measures of N_2 fixation simply as ΔN of nodulated plants. On recessional moraines Lawrence (19) made harvests of *Alnus crispa* on surfaces of increasing age and showed the plant was fixing 157 kg N ha^{-1} a^{-1}. Studies on dune forest in New Zealand (see chapter 12) have shown that tree lupin (*Lupinus arboreus*) is an important contributor of nitrogen (soil N level 0.008%). In a short term biomass study on lupin (20) four sequential harvests showed 1.7 g m^{-2} of N was incorporated into lupins over 10 weeks. Parallel C_2H_2 reduction assays (calibrated with $^{15}N_2$) indicated N_2ase activity at 1.5 g m^{-2}, showing that the biomass technique may be slightly overestimating nitrogen fixation due to uptake from soil.

Laboratory and field studies have shown that soil N supply markedly affects both the rate of N_2 fixation and the ratio of N in plants due to N_2 fixation. Thus ΔN is a poor estimator of Nf in rich soils. In general, nitrogen fixation may be stimulated by small additions of nitrogen but suppressed at high levels. For example, in the field pea (21) when plants were grown at 0–330 ppm NO_3-N, whole plant N was doubled but the contribution by fixation at the high NO_3 concentration was only 25%. This problem has been studied in some detail for *Alnus rubra* (22) where it was grown in the greenhouse in total soil nitrogen (TSN) levels ranging from 0.01–0.5% by addition of scotch broom litter to a low N cinder soil. Nitrogen fixation was assessed by total biomass and soil harvesting at the end of the experiment and nitrogen uptake into plants measured separately. Both plant weight and plant nitrogen increased with increasing soil nitrogen. However nitrogen fixation was stimulated at low TSN and inhibited at high TSN. At 0.5% TSN only 18% plant nitrogen originated from fixation with 82% coming from mineral nitrogen in soil. Similar results have been obtained in many nodulated plants (17) and a general response curve illustrating this phenomenon is shown in Fig. 3. Although the figure must be interpreted

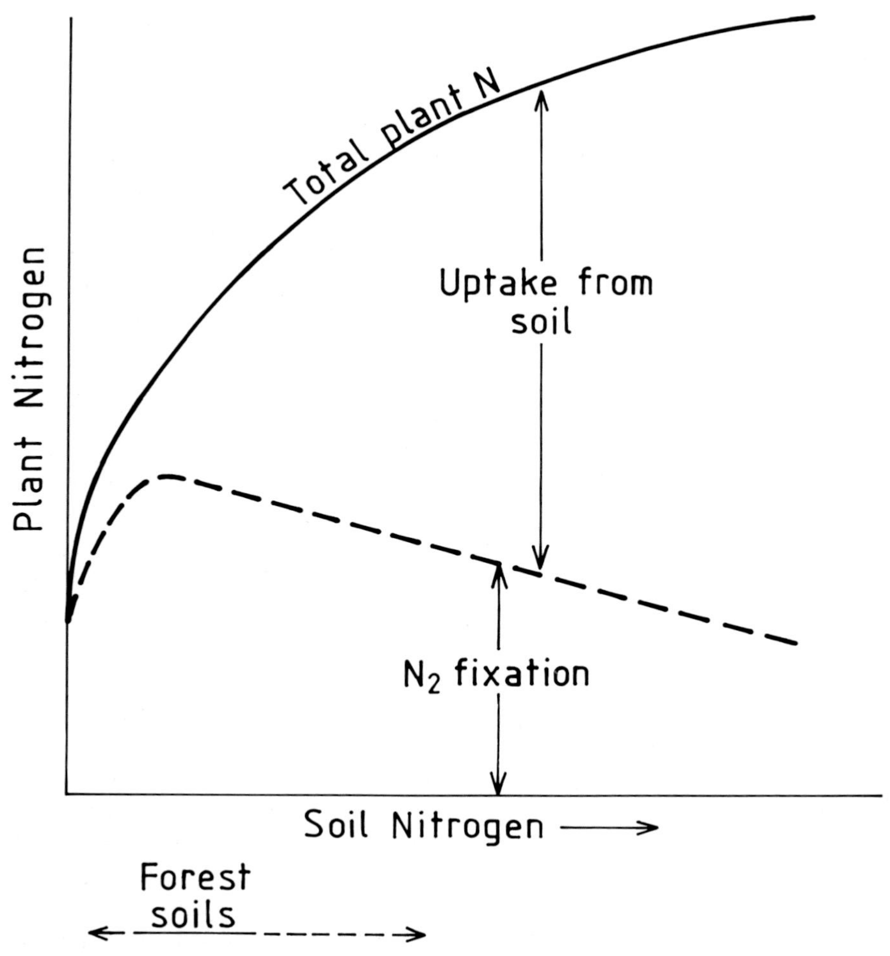

Fig. 3. Generalised response curve for nodulated plants showing proportion of nitrogen assimilated from soil and from N_2 under varying soil N levels.

only in the most general terms, one generalisation can be made that at low soil nitrogen levels (less than 0.1% TSN) the majority of nitrogen in diazotrophs originates from current fixation.

A further point made in the above work with *Alnus rubra* (22) is that when total N inputs were measured in plant and soil, whether in greenhouse or in field, a proportion of the increment could not be accounted for in the accumulated biomass. Thus in an ecosystem analysis of stands ranging from 2–14 years, it was shown that annual fixation was 300 kg N ha^{-1} but only 100 kg ha^{-1} was found in litter return and 35 kg N ha^{-1} in above ground parts. The authors' conclusion was that over 50% of the N accretion in the alder stand was under ground by '... secretions of nitrogen from nodules and roots into the soil and nitrogen fixation by free living organisms'. Recent evidence on free living heterotrophic fixation in soils indicates this process is minor in comparison with the above rates (7, 23) and that this could contribute at most 10 kg N to the missing 150 kg N ha^{-1}. It is my contention that the missing nitrogen in the above experiments is most likely due to small root and mycorrhizal turnover. Entry into these pools is rapid and turnover of small roots can be a very significant contributor to a nutrient budget. For example, an analysis of N turnover in a *Liriodendron* stand (24) showed that 52 kg N ha^{-1} was returned in leaf litter and 85 kg was turned over below ground. The difficulty of sampling this below ground 'litter' due to decay of small roots has led to a gross underestimation of nutrient turnover generally. A further point which may be at least as significant is the role of mycorrhizas. A large proportion of forest soil nutrients are tied up in mycorrhizal fungi (25), these turn over very rapidly and recent evidence in the author's laboratory has shown that plant nitrogen contributes significantly to mycorrhizal nitrogen.

Bearing all the above in mind it may be concluded that nitrogen increments in biomass of nodulated plants may give reasonable results on low N soils, but errors due to soil N input and loss from roots may lead to large errors in richer soils.

Litter fall. Leaf litter fall has been used as an indication of nitrogen fixation by nodulated plants. In a review of a number of non-legume plants often associated with forest it was shown (1) that annual litter fall in pure stands varied from 1–29 t ha^{-1} and a nitrogen content of 1.0–3.0%. For a perennial plant the nitrogen returned in litter may be a high proportion of that taken in by fixation and soil uptake, and as in the example (Fig. 2) leaf fall of 40 kg N is over half of the 70 kg N ha^{-1} a^{-1} assimilated. In low soil N conditions leaf drop may be used as a very rough estimator of Nf, but it neglects the N stored in the plant during growth and N turned over underground.

Epiphytes. Forest epiphytes often contain a proportion of nitrogen fixing species of lichens. The most detailed study of nitrogen inputs have been conducted in Oregon USA where it was shown that the total epiphyte load in old growth Douglas Fir forest is over one tonne per hectare (26). The nitrogen fixing lichen *Lobaria oregana* dominates the community and growth rate, biomass and litter-fall data have been used to estimate the nitrogen input by this plant (27, 28). Total biomass of *Lobaria* was measured by systematic sampling of one old growth tree giving 10–15 kg *Lobaria* per tree or 500–900 kg ha^{-1}.

It is illuminating at this stage to compare the four different estimates of fixation that have been calculated for this system from the above data.

(a) A growth rate of 0.30 per annum was measured, lichen N content of 2.1%, and assuming that all nitrogen in the lichen is due to fixation (27) = 3.15 kg N ha^{-1} a^{-1}.

(b) It has been shown that *Lobaria* litterfall grossly underestimates true production because of *in situ* decay (29) and that much of the enhanced throughfall N is due to N leaking from lichen thallus. Thus N fixation was estimated as: Total *Lobaria* litter N plus half throughfall N (28) = 3.5 kg N ha^{-1} a^{-1}.

(c) As some *Lobaria* N might reasonably be expected to originate from epiphyte 'soil' an alternative approach (28) is half throughfall N plus half litter N = 2.8 kg N ha^{-1} a^{-1}.

(d) Finally a growth rate of 0.34 per annum, biomass of 900 kg ha^{-1} and lichen N content of 1.73 (28) gives 5.3 kg N ha^{-1} a^{-1}.

What is perhaps the most surprising result from the above, is that despite the wide array of assumptions and errors in the approaches the rates of fixation are remarkably alike, especially when it is realised that (d) above uses a biomass figure of 900 kg ha^{-1} which is right at the upper boundary of the range of biomass figures.

Free living heterotrophs. Probably the first N balance study of nitrogen fixation in forests was reported by Henry in 1897 (30) and it is of both historical and methodological interest to describe these and later experiments in detail. Henry incubated fallen leaves of oak and hornbeam in boxes outside and analysed for nitrogen at the beginning and end of the experiment. His results showed an increase in N of 36% and 82% for oak and hornbeam leaves respectively. Later experiments of Hornberger in 1905 (31) cast considerable doubt on these results, especially in that being conducted outside, other inputs of nitrogen from dust or rain were not accounted for.

Olsen in 1932 (32) took up the problem again and in a series of well conducted experiments showed that decaying leaves of oak and beech do indeed support nitrogen fixation while hornbeam leaves showed no increase in nitrogen over the 337 day incubation.

In Olsen's experiments, leaf material was cut and carefully homogenised to give very even samples and replicated samples taken throughout the experiment. Because the natural heterogeneity of the material was eliminated precision of better than 1% was achievable and in fact increases of 27–131 mg N per 100 g sample were detected in beech leaves and 1–170 mg N per 100 g in oak leaves. In these experiments great care was taken to eliminate extraneous N inputs and to prevent leaching and thus $\Delta N = Nf$. The techniques used by Olsen confirmed the earlier results of Henry and further work since then (7, 23) using other techniques have established that free living N fixation is a common phenomenon in forest litter.

Ecosystem and soil N balance

General principles. The pioneering work of Jenny (33) on soil development laid much of the foundation for ecosystem nutrient balance work. Jenny proposed that soil development is a function of climate, organisms, relief, parent material and time and that the effect of any one of these on soil development can be studied by finding a situation where all other factors are constant. In the present context the effect of time and organisms are the pertinent factors and changes in soil and ecosystem nitrogen with time are described in many chronosequence studies in which the presence or absence of the nitrogen fixing organism determines the rate of N accretion.

Many of the principles relating to the estimation of nitrogen fixation in nutrient budget studies have been covered in other reviews of this subject (17, 34, 35). However, little emphasis has been given to the variety of techniques, the relative errors or to the theoretical basis of such techniques. The following account attempts to develop a theoretical framework for such studies and to develop simple models to illustrate the five major ways in which N budgets can be used to determine N accretion and to isolate a N_2 fixation component.

Reference to Fig. 2 shows that ecosystem N has two major inputs; biological fixation (Nf) and precipitation (P) (dry and wet deposition). Losses are also twofold as denitrification (D) and particulate and solution (S) losses in streams. Managed forests may also have harvest losses and fertiliser gains but these are ignored in the present discussion. A simple statement of the total ecosystem nitrogen change with time N given by equation 3 may be rewritten

$$Nf = \Delta N - P + D + S \quad (5)$$

It is apparent that if ΔN is known e.g. from successive ecosystem sampling for nitrogen then something can be said about nitrogen fixation but only if something is known about P, D and S or if they can otherwise be eliminated.

Precipitation input into forest has been measured in many forested areas and shown to be in the range 2–20 kg N ha^{-1} a^{-1} (15, 16). Of the components of this, the soluble inputs and particulate matter are well accounted for, however gaseous loss and uptake by leaves has been studied only recently and the extent to which ammonia for example is assimilated and recycled in forests is largely conjectural. Most inputs in precipitation fall in the range 3–10 kg N ha^{-1} with only forests near heavy industry receiving more than 10 kg N ha^{-1}.

Mature forests lose little nitrogen in solution and this and particulate N loss becomes large only when the system is damaged (37). Losses of the order of 1–5 kg ha^{-1} would span the range for most forests. Denitrification is perhaps the least studied of all the processes, especially in forests. Figures for deciduous forest in eastern United States show that there is potential for denitrification but there is insufficient data to place any real figures on it as yet (15).

Most forest ecosystems, even in the absence of nitrogen fixation are accumulating nitrogen and the generalised values of Fig. 2 indicate an annual net accretion of 1 kg N ha^{-1}.

Mass balance. As has been aptly stated by Knowles (34) 'Precise estimates of N fixation are exceedingly difficult to obtain by means of overall N balance studies ... the proper interpretation of any N balance data requires the precise quantitation of as many as possible of the N inputs and outputs ...' Despite the above some very detailed N mass balance studies have been made e.g. for European deciduous forest (38) eastern US deciduous forest (39) and in western US conifer forest (28).

When all parameters of a mass balance are measured independently, any single unknowns in the equation may be calculated by difference. Thus for Hubbard Brook deciduous forest (15, 40) N_2 fixation has been estimated by measuring ΔN as the change in vegetation N over time of + 17, precipitation of 7 and losses both in solution and denitrification of 4. When substituted in equation 5

$$Nf = \Delta N - P + (D + S)$$
$$Nf = 17 - 7 + 4 = 14$$

In this work the premise was apparently made that soil N stock remains constant. An alternative approach to this problem (15) incorporates all parts of the system

$$\Delta N_{veg} + \Delta N_{floor} + \Delta N_{soil} - \text{net N flux} = 0 \qquad (6)$$

and it is postulated that the increase in vegetation N is derived from the soil. Subsequent work (23) has shown that N_2 fixation in these forests are less than 1 kg N ha^{-1} a^{-1}.

A much more complete N mass balance has been undertaken for old growth Douglas fir forest (28) in which vegetation is losing 2.8 kg N ha^{-1} a^{-1} and detritus gaining at 7.8 kg N ha^{-1} a^{-1}. This detailed account estimates that 2.8 kg N ha^{-1} are being added by lichen fixation (see p. 183) and that there still remains a 1.7 kg ha^{-1} annual input unaccounted for. Subsequent work has shown (41) that wood and litter fixation may well account for this deficit.

In conclusion, careful N mass balance can lead to information on the general nature of unidentified N inputs and in fact the precision obtained in such studies (28) may allow precise predictions to be made. However it is important to ensure that all parameters are measured and that the net balance of the system is accurately gauged.

Soil N accumulation. As soil represents the largest store of nitrogen in an ecosystem and because it is easier to sample than biomass, increase in soil nitrogen is often used as a measure of nitrogen accretion. The general case can be written as

$$\Delta N_s = \frac{{}_2N_s - {}_1N_s}{{}_2T - {}_1T} \tag{7}$$

Where ${}_2N_s$ and ${}_1N_s$ are soil N quantities at times 2 and 1 respectively giving a rate of increase in soil N per unit time. Four separate approaches to estimating Nf from ΔN_s can be identified.

A. In which $\Delta N \simeq \Delta N_s$ and neither P, D, S nor vegetation N are measured. In this case, total soil nitrogen is measured at various time intervals and it is assumed that vegetation is in equilibrium and the equation

$$Nf = \Delta N_s \tag{8}$$

is assumed. Perhaps the best example of this approach is the measurement of N accretion on moraines left after glacial retreat at Glacier Bay in Alaska. The detailed account (9) shows that nitrogen accumulates rapidly in the presence of *Alnus crispa* and other N fixing species and over a 40 year moraine history soil and forest floor nitrogen increase from 16 g m^{-2} to 290 g m^{-2} giving a mean rate of N accretion of 68 kg N ha^{-1} a^{-1}. During the 40 year period nitrogen was also accumulating in above ground vegetation and this was not included in the analysis so that the accumulation rate is minimal. This approach has also been used in assessing N inputs over very long periods of time where there is probably some justification in assuming that vegetation is in equilibrium. The conifer forests of the Pacific northwest have been shown to have accumulated soil N at a rate of 0.2–0.4 kg ha^{-1} a^{-1} over the past 12,000 years (42). Such measures integrate all losses and gains of N and where Nf is low and of the same order of magnitude as P, D and S it is obvious that a measure of Nf is unobtainable without independent estimates of P, D and S.

B. When early stages of succession are being studied, N accumulation in vegetation can be a significant N sink and can be measured independently. In this case it is recognised that $\Delta N \neq \Delta N_s$ and that some measure of N change in vegetation (ΔN_v) is required, especially N accretion in the N_2 fixing species.

$$Nf = \Delta N_s + \Delta N_v \tag{9}$$

This approach is illustrated by the work of Youngberg (43, 44) measuring the contribution of *Ceanothus* to the N economy of pine and Douglas fir forests in Oregon. Soil N accretion was measured by soil sampling in 1962, 1972 and 1977 and gave ΔN_s of 859 kg N ha^{-1} and 424 kg N ha^{-1} in stands after pine and Douglas fir respectively for the 15 year period. Using the example of the Douglas fir clearcut, 859 kg N ha^{-1} of nitrogen was the total increment of which 15 kg N ha^{-1} was derived from P, and total N in *Ceanothus* biomass and litter was 417 kg N ha^{-1} for the 15 year period. Nf can be derived as

$$\mathrm{Nf} = \frac{859 - 15 + 417}{15} = 86 \text{ kg N ha}^{-1}\text{ a}^{-1}$$

C. In the above examples P, D and S may be measured independently but more often are treated as unknowns or insignificant quantities.

A simple expedient used to overcome the problems of making independent measures of P, D and S is to measure N increment in an area of control forest lacking the N fixing species. The basic equation (7) then becomes

$$\Delta N_s = \frac{{}_2N_s - {}_1N_s}{{}_2T - {}_1T} - \left[\frac{{}_2N_s - {}_1N_s}{{}_2T - {}_1T} \right] \quad (10)$$

where the quantity in brackets is nitrogen change in soil lacking the N_2 fixing species or differing in some known way. In the above case it is usual for an area of forest to naturally or artificially have a N_2 fixing species removed or inserted so that at ${}_1T$ the quantities ${}_1N_s$ and $({}_1N_s)$ are the same. It is not then necessary to measure ${}_1N_s$ as simple rearrangement of equation 10 gives

$$\Delta N_s = \frac{{}_2N_s - [{}_2N_s]}{{}_2T - {}_1T} \quad (11)$$

such that all that is required is a measurement of soil nitrogen in control and experimental blocks. In this case the formula for Nf is similar to equation 9. It is not necessary to make independent measurements of P, D and S as these are automatically subtracted from both experimental and control. It may be argued that while precipitation inputs would be similar for each plot differences in vegetation between experimental and control may affect relative values for D and S, but it is likely that the differences are marginal. Similarly the need to measure ΔN_v is diminished as both control and experimental plots are taking N up from the soil; nevertheless, omission of such measurements will always underestimate N fixed since both acummulation of N in the nitrogen fixing species and additional N in the associated vegetation resulting from enhancement of soil fertility by nitrogen fixation are not taken into account.

D. Another way to circumvent the need for measurement of P, D and S is to measure soil nitrogen increments in an area both before and after the addition or subtraction of a N fixing species, when

$$\Delta N_s = \frac{{}_3N_s - {}_2N_s}{{}_3T - {}_2T} - \frac{{}_2N_s - {}_1N_s}{{}_2T - {}_1T} \quad (12)$$

In this case if a legume was inserted into a plot at ${}_2T$, ΔN_s expresses the rate of nitrogen increment due to the legume and $\mathrm{Nf} \simeq \Delta N_s$. The approaches described in C and D above probably provide the most commonly used methods of measuring forest nitrogen inputs. One of the most regularly quoted papers is that of Tarrant and Miller (45) who measured soil nitrogen increment under Douglas fir forest on the Wind River burn in South Western Washington USA. The burn was replanted with trees in 1929 and a strip planted with *Alnus rubra* in 1933. No initial soil sampling was done but control and alder plots were measured in 1959 and the soils shown to have 2833 and 3771 lb N $acre^{-1}$ respectively representing a net gain ΔN_s of 938 lb acre. This gave a conservative estimate of nitrogen fixation of 36 lb $acre^{-1}$ a^{-1} (40 kg N ha^{-1} a^{-1}).

Ecosystem N accumulation. The most accurate measurements of N accretion are those that include all ecosystem components in deriving N and the general case is given by

$$\Delta Ne = \frac{{}_2Ne - {}_1Ne}{{}_2T - {}_1T} \tag{13}$$

where Ne is ecosystem N. Using N derived in this way Nf can be calculated as

$$Nf = \Delta Ne - P + (D + S) \tag{14}$$

This approach is analogous to the one described in B above and a large number of examples could be given. An early succession example is of *Alnus incana* in Alaska (46) where total ecosystem N analyses were made on 0, 5, 10 and 20 year old stands. Maximum N accretion occured over the 0–5 year interval the total increment being 362 kg N ha^{-1} a^{-1} made up of 48 kg in vegetation and 314 kg in soil. In this case Nf $\simeq \Delta Ne$.

An analogous model to that of equations 10 and 11 can be constructed for ecosystem nitrogen where nitrogen in control and experimental plots is measured thus

$$\Delta Ne = \frac{{}_2Ne - ({}_2Ne)}{{}_2T - {}_1T} \tag{15}$$

This technique is ideal for measuring smaller changes in N increment in older stands where P, D, and S may be significant. However because of the difficulty in adequately sampling total ecosystem N few examples of this approach are available. Because of the significance of this method of estimating Nf two studies are cited. Firstly the underplanting of *Myrica* in pine forests in Japan (47). The planting, like that of the measurements cited by Tarrant and Miller for *Alnus* (45) was not set up as an experiment on nitrogen fixation and no initial measurements of ecosystem nitrogen were made. However, when final measurements were made after 12 years, on the assumption that the two sites were equivalent at planting the results are unequivocal.

Two plots containing both pine and *Myrica* contained 363 and 232 kg N ha^{-1} in biomass while the pine only contained 58 kg N ha^{-1}. Soils for the mixed plots contained 3089 and 3163 kg N ha^{-1} and for the pine only plot 2848. Using means and equation 12 and subtracting 55 kg N ha^{-1} added to mixed plots in an initial mulch

$$\Delta Ne = 3423 - (2096 - 55) = 30 \text{ kg N ha}^{-1} \text{ a}^{-1}$$

Despite the almost total lack of experimental design and statistical sampling this nitrogen addition by fixation can reliably be attributed to nitrogen fixation by *Myrica* as all other losses and gains may be assumed to be similar for both areas.

A similar approach was made by Dommergues (48) in his study of N increment under *Casuarina equisetifolia* forest on Cape Verde Island off the west coast of North Africa. Measurements made under 13 year old *Casuarina* and on an equivalent non vegetated area gave a net difference of 760 kg N ha^{-1} and an increment of 58 kg N ha^{-1} a^{-1}.

In conclusion a variety of different techniques can be used to distinguish nitrogen fixation by N budget techniques, the relative merits have seldom been addressed in the past, and as can be seen the method given in equation 15 represents the simplest and least equivocal measure of nitrogen fixation. The problem of finding suitable control and experimental sites which have both had identical prior history and have similar topography, climate, etc. make it a difficult one to realise.

ISOTOPE TECHNIQUES

General

Nitrogen has two isotopes that have been used for measurements of biological nitrogen fixation. A radioactive isotope ^{13}N can be generated using a cyclotron or accelerator but has a half life of only c. 10 minutes and this is of little use in any field experiments, although it has been used in the laboratory for rapid biochemical investigations. The heavy stable isotope ^{15}N is the only really useful isotope of nitrogen, and it has been widely used since 1941 in the investigation of nitrogen fixation.

^{15}N occurs naturally in all materials containing N. The natural abundance of ^{15}N in air is 0.366 atoms % which means that the majority of molecules will be ^{14}N ^{14}N (mass 28) and c. 0.7% will be ^{14}N ^{15}N (mass 29). Equilibrium samples derived from air N_2 should contain the same proportion of ^{15}N, but some mass discrimination does occur in both biological and chemical reactions resulting in materials which are depleted or enriched in ^{15}N. Thus some soils may be enriched by as much as 0.004 atom% ^{15}N (12) while some chemicals may be considerably depleted.

Early experiments established that there was no measurable isotope effect in the biochemical reactions of nitrogen. More recently however it has been established that all the major reactions discriminate against the uptake of ^{15}N (49). The level of discrimination is extremely small and can be measured with only the most sophisticated of equipment. It is greatest in denitrification and this leaves most soils slightly enriched in ^{15}N.

There are basically two ways in which ^{15}N can be used in N_2 fixation studies, firstly direct methods whereby the uptake of $^{15}N_2$ is measured in the organism or system and secondly by dilution where the organism is enriched with ^{15}N, usually via the soil, and the dilution of that material by N_2 uptake gives an indirect measure of N_2 fixation. The details of many of these techniques have been reviewed recently (50); they will be explained in outline here and reference made to specific forest applications and examples.

$^{15}N_2$ *assimilation*

General. With the advent of the acetylene reduction technique (see p. 200) in 1966 the use of $^{15}N_2$ as a direct measure of N_2 fixation has diminished. However it still represents the most reliable and accurate technique for short term assays of biological N_2 fixation. In essence, the technique involves the exposure of a N_2 fixing system such as root nodules, lichen or soil to a mixture containing an enriched $^{15}N_2$ atmosphere. The incorporated ^{15}N is then extracted by standard Kjeldahl digestion and the resulting NH_3 is converted back to N_2 for analysis by mass spectrometry or optical emission spectroscopy.

While the technique reliably estimates N_2 fixation during the incubation period it suffers from two basic defects which must always be addressed, measured and minimised. Firstly, the removal of the plant material into an unnatural contained environment, its excision and incubation all change the functioning of the system. At the very least there are changes in temperature, humidity, gas composition and light resulting from enclosing the system. It is extremely important both to minimise these changes and to take measurements of the important environmental effects. Secondly, exposure of samples for a short time of several hours represents a sampling of one point in space and time. It is therefore important to cover this variation both by spatial sampling within the environment and also along daily and seasonal gradients to ensure that integrated values for activity are measured.

$^{15}N_2$ uptake, because it is expensive and time consuming, is not used for routine N_2 fixation analysis but is now most often used to calibrate the acetylene reduction technique. The following account outlines the important steps in the procedure; it does not attempt to provide an exhaustive account of the many variations but is intended to highlight possible pitfalls and technical difficulties. Further details are to be found in some of the more exhaustive accounts of the subjct (17, 35, 50).

Incubation of material. ^{15}N may be obtained from a number of chemical supply companies (4, 5) in a wide range of forms including N_2, NH_4, NO_3 etc. at enrichments from 1–99.9 atom % ^{15}N. It is prepared by continuous distillation techniques and until quite recently was relatively expensive. Many workers generate N_2 gas from labelled NH_4 compounds but $^{15}N_2$ is now freely available and only marginally more expensive than equivalent $^{15}N_2$ quantities as NH_4. Bergersen (50) gives details of vacuum and manometer techniques for preparing gas mixtures containing $^{15}N_2$ and advocates the preparation of $^{15}N_2$ from NH_4 salts by oxidation with copper at 600° C. The following account illustrates a simple approach to the problem of $^{15}N_2$ use which is adaptable to field experiments.

$^{15}N_2$ gas is obtained commercially in partially evacuated glass containers with a break seal inside a glass tube. It is recommended that the highest enrichment possible is used for such experiments as the marginal reduction in cost by using lower enriched isotope is not warranted and results in a reduction in precision and accuracy. A brass slug is introduced into the neck of the flask while carefully lying it on its side, a thick septum is used to plug the glass tube (Fig. 4a) and a vacuum is pulled on the dead space through a 24 g syringe needle passing through the septum. When a good vacuum is obtained, the flask is shaken so the slug breaks the glass seal, taking care not to let it crash through to break the flask. At this stage the gas in the flask is under reduced pressure and this is relieved by immersing the neck of the flask into water.* This initial solution should also be mildly acidified with H_2SO_4 to mop up any traces of NH_3 which may be an impurity in the gas. While under water a syringe needle is passed through the septum and water moves into the flask to regain atmospheric pressure (Fig. 4b). All subsequent gas transfers are made using plastic disposable syringes, taking care to ensure that syringe dead space volume in the needle and hub is filled with evacuated water at each transfer. Experience in our lab has shown that the accuracy and gas tightness of most plastic syringes is equal to the task and volume measurements on 1 cm^3 syringes shows they do not deviate by more than 1%. An extremely useful syringe for some transfers is a 40 unit insulin syringe (Becton-Dickinson) in which the 26 gauge needle is fused into the syringe and contains a minimum dead space.

Gas mixtures for aerobic N_2 fixation experiments are made by mixing $^{15}N_2$ and O_2 in appropriate proportions with syringes which obviates the requirement for gas mixing lines, manometers, vacuum gauges etc. For relative work it is expedient to supply only 0.10 atm of $^{15}N_2$ to the fixing system with the balance made up with an inert gas such as argon. However because nitrogen fixing sites in some organisms may not be saturated with less than c. 0.5 atm N_2 it is essential in

* All water used for transfers should be recently boiled and/or evacuated to remove dissolved air, and also saturated with K_2SO_4 to reduce N_2 solubility. Alternatively and more efficiently, K_2SO_4 saturated water can be continuously sparged with argon in a narrow necked flask, this effectively eliminates N_2 in 5 minutes.

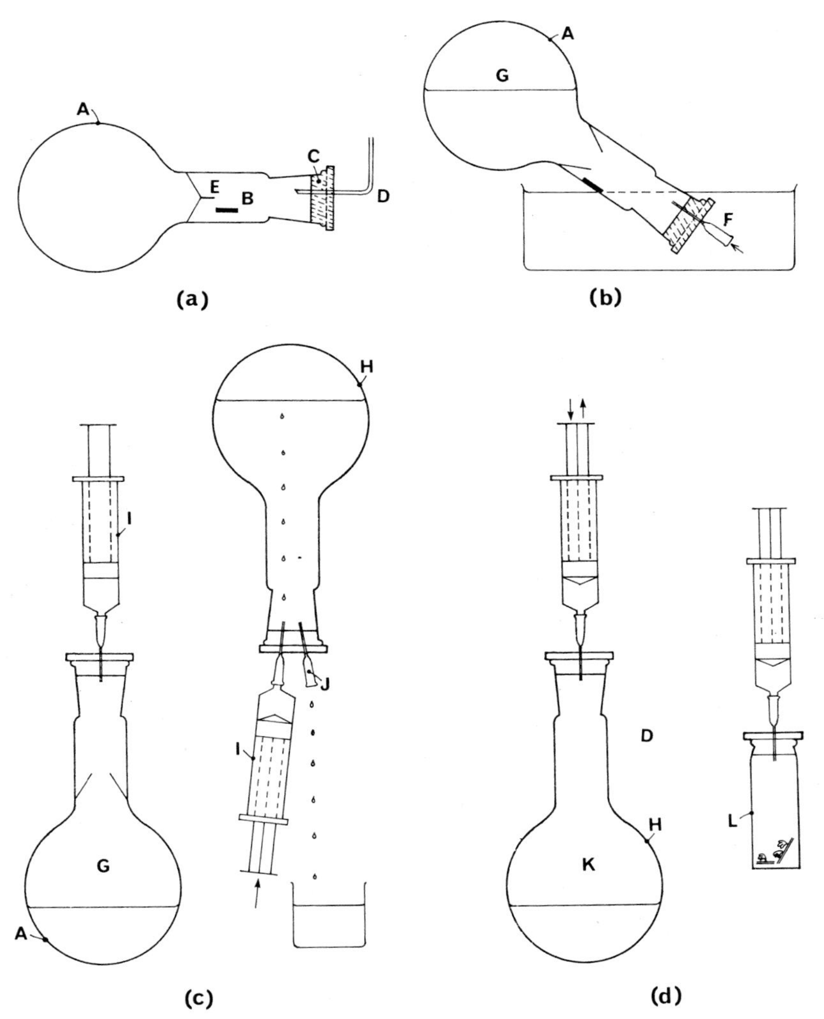

Fig. 4. Simple $^{15}N_2$ gas mixing and incubation procedures. (a) $^{15}N_2$ gas is obtained in glass breakseal flask (A) under vacuum. A brass slug or magnetic stirrer bar (B) is inserted in neck and rubber septum (C) put on top. The flask neck is evacuated through a 24g (D) needle and then the flask shaken to break glass seal (E). (b) The flask is immersed in water and a needle inserted (F) so that water enters to leave the $^{15}N_2$ (G) in the flask at near atmospheric pressure. (c) Gas mixing is done in a separate flask (H) which is filled with water. $^{15}N_2$ is withdrawn from G by injecting a volume of water with syringe (I) into (G) then immediately withdrawing slightly less $^{15}N_2$ out to leave a small positive pressure. Quickly transfer this into (H) by displacing water out needle J. This is done for all gases required in the mixture (K). (d) Gas mixture (K) is then transferred from H into evacuated incubation vessels (L).

all experiments where absolute values of N_2 fixation are required to use the normal pN_2 of c. 0.80 atm.

To make 50 cm^3 of a normal gas mixture the following procedures are followed. A flask similar to the $^{15}N_2$ flask is totally filled with water and stoppered with a septum seal.

Draw 3 cm^3 of evacuated water into a 50 cm^3 syringe, quickly turn needle up and expel bubbles and water leaving the total dead space (hub and needle, 24 g) filled with water. Now draw up 42 cm^3 of evacuated water ensuring that no bubbles enter the syringe. Immediately insert the needle into the $^{15}N_2$ reservoir, inject the 42 cm^3 of water into the reservoir and draw out 40 cm^3 of gas. In this way the reservoir and syringe are slightly over atmospheric pressure and any gas movement during transfer will be by mass flow out of the syringe. The syringe is rapidly inserted into the inverted gas mixing flask (Fig. 4c) along with a 24 g exit syringe needle. As the gas is forced into the flask, water is displaced until atmospheric pressure is realised.

A similar exercise is done with 10 cm^3 O_2 in which the syringe is carefully sparged with oxygen to remove any N_2 from the atmosphere.

This gives a final gas mixture of 80:20 v/v with a minimum of contamination by air. Further elaborations of the technique can be made by sparging all solutions with a continuous stream of argon to remove N_2 and/or doing transfers in an argon purged glove bag.

Incubation vessels may be any septum sealed container which has a vacuum tight seal. N_2 fixing material is placed in the incubation chamber and evacuated either with a rotary pump or a hand vacuum pump. Most material can withstand evacuation for short periods; the major damages that occur are dehydration of tissue and removal of the equilibrium gas phase in the tissue which may or may not be optimal for the tissue or plant. Having predetermined the incubation vessel size, a known volume of gas mixture is transferred to the chamber to bring it to atmospheric pressure (Fig. 4d), ensuring that syringe dead space is water filled as before.

A number of precautions need to be borne in mind when carrying out the operations.

1. The ratio of gas phase to tissue volume must be large enough to avoid significant oxygen depletion during the incubation.

2. Evacuation of tissue should be for no longer than 30 s, preferably shorter. It should be noted that evacuation changes internal gas composition and dehydrates tissues which in some cases may drastically affect activity.

3. Tissue should not be washed or chilled prior to incubation as both these operations inhibit activity in many tissues.

4. Samples of gas phase need to be taken at the end of the incubation to determine the $^{15}N_2$ content for later calculations. This is done conveniently using a one cm^3 insulin syringe filled with evacuated water which is injected into

the incubation vessel to give a positive pressure, the syringe pumped at least 10 times to free water and mix the gas. Then a 0.1 cm^3 sample is drawn out of the vessel and immediately transferred to a septum port on the inlet of the mass spectrometer. As the pressure in the syringe is above atmospheric while the syringe is removed all gas movement is mass flow from the needle in the 1–2 s required for the transfer. If rigorous attention to detail is maintained, excellent repeatability can be achieved. Samples may be stored in syringes by inserting the needle into the rubber bung and submerging them in evacuated water or inert atmosphere whereby any gas diffusion in or out of the syringe does not affect isotope ratio.

5. It is essential that all apparatus, and especially rubber stoppers be clean and have NEVER been used for acetylene assays. Acetylene is such an effective competitor with N as a N_2ase substrate that small traces will interfere with $^{15}N_2$ uptake.

6. Incubation conditions should reflect as closely as possible the environment of the original material, especially temperature, which markedly affects N_2ase. In field analyses, it is convenient to bury the incubation vessels in soil or incubate them in an insulated box at field temperature.

Harvest and analysis. These topics have previously been covered well (50, 51, 52, 53) and will be dealt with in outline here. After $^{15}N_2$ exposure material is exposed to air and dried and milled for weighing. At this point it has been a common practice to extract a soluble (3N HCl) component from the system in order to isolate the recent products of nitrogen fixation and thus produce a more highly enriched sample. While this procedure may have been expedient when highly enriched $^{15}N_2$ was very expensive it is not to be encouraged as it may leave substantial ^{15}N in the insoluble material, especially if longer term $^{15}N_2$ incubations are used. After drying and weighing, samples are digested, distilled and prepared for mass spectrometry.

It should be emphasised at this point that analysis of stable isotopes as done either by mass spectrometry or optical emission, produce results which are ratios of the relevant masses. Thus ^{15}N is never measured in analytical procedures as an absolute amount but as a percentage of the total nitrogen or atom % and the natural abundance of ^{15}N is given as 0.366 atom %. In the mass spectrometer nitrogen gas is ionised and the two masses 28 due to ^{14}N ^{14}N and 29 due to ^{14}N ^{15}N are measured to give a ratio. It can be seen then that any fixation of $^{15}N_2$ into a nodule or soil results in a dilution of the $^{15}N_2$ into a large mass of already existing ^{14}N and the amount of preexisting N determines to a large extent the sensitivity of the technique.

Samples collected after distillation as NH_4 are slightly acidified to stabilise the NH_4 then are regenerated to N_2 by reaction with lithium or sodium hypobromite (NaOBr) either in Rittenberg tubes (50) or by freeze layer technique (54).

It is essential that control material, not exposed to $^{15}N_2$, be processed for ^{15}N analysis at the same time and that all enrichments be referred to control natural abundance. Enrichments of ^{15}N are normally referred to as atoms % excess, in which the percentage of ^{15}N above that of natural material is expressed. If natural abundance is 0.365 atom % and enriched material is 0.412 atom % then excess ^{15}N is 0.047 atom % excess ^{15}N.

Designs and calculations. Before $^{15}N_2$ experiment is run it is necessary to make preliminary calculations to ensure that an expected rate of fixation can be detected with the isotope. The isotope technique is strongly limited by both the rate of $^{15}N_2$ uptake and by the amount of ^{14}N already existing in the system. The following calculations show how firstly, an experiment is designed to determine the length of exposure of ^{15}N and the level of isotope enrichment to use and secondly, to illustrate how N_2 fixation rate is calculated from isotope ratios.

In considering an experiment to estimate nitrogen fixation in nodules of *Alnus*, we can make two basic assumptions; that root nodules of both legumes and non-legumes have a normal range of activities ranging from 1–60 mg N g^{-1} day^{-1} (17) or 0.04–2.5 mg N g^{-1} $hour^{-1}$ and that they contain between 2 and 5% N.

Using conservative estimates of 0.2 mg N g^{-1} h^{-1} for N_2 fixation and nodule N level of 3% N and assuming the nodules are in pure $^{15}N_2$ then the trial calculations would be for a sample of lg dry weight of nodules

$$\text{total nodule N} = 1000 \times \frac{3}{100} = 30 \text{ mg N}$$

$$\text{fixation of } N_2 \text{ in 1 hour} = 0.2 \times 1.0 = 0.2 \text{ mg N}$$

$$\text{therefore } \% \ ^{15}N \text{ in sample} = \frac{0.2}{30} \times 100 = 0.66 \text{ atom } \% \ ^{15}N \text{ excess}$$

If 50 atom % excess ^{15}N were used the result would be halved to 0.33 atom % ^{15}N.

As the minimum detectable level can be set arbitrarily at e.g. 0.01 atom % excess it can be seen that such an experiment is easily analysed under the above conditions. In general uptake of $^{15}N_2$ into very active tissues such as root nodules and cultures of bacteria poses little problem even for $^{15}N_2$ exposures of only a few minutes. However when very slow rates of activity such as lichens (55) or decaying wood (41) is analysed long exposure times and high $^{15}N_2$ enrichment must be used.

Following the experiment a series of simple calculations establishes the rate of nitrogen fixation.

e.g. 0.1 g nodules
3 mg N extracted

1 hour exposure
48 atom % ^{15}N excess in gas phase
^{15}N enrichment in sample – 0.382 atom % excess ^{15}N
Initially, the absolute amount of $^{15}N_2$ assimilated is calculated by

$$\frac{0.382}{100} \times 3 = 0.0115 \text{ mg } ^{15}\text{N excess}$$

As this originated from a 48 atom % excess gas phase – total N uptake is given by

$$0.0115 \times \frac{100}{48} = 0.024 \text{ mg N}$$

$$\text{therefore fixation rate} = \frac{0.024}{0.1} = 0.24 \text{ mg g}^{-1} \text{ h}^{-1}$$

Limitations. The obvious advantages of $^{15}N_2$ uptake are the direct measurement of the process and analysis of products of reaction. These advantages make it the preferred technique for sensitive nitrogen fixation assays. Some of the more apparent limitations are the cost of isotope, and the cost and complexity of analytical procedures. As has been indicated, the production of kilogram quantities of ^{15}N is now possible and the real cost of stable isotopes has been continually reduced over the past 10 years. While analytical equipment is still expensive, the advent of automated apparatus that both processes samples and calculates results has meant that at least one laboratory can now analyse in excess of 200 samples per day and can provide an analytical service at reasonable cost. (Dr B.B. McInteer, Los Alamos Scientific Laboratory, New Mexico, U.S.A.).

While the $^{15}N_2$ technique is ideal for relative activities and biochemical testing, its use for estimation of fixation in the field has been limited. The space/time variations discussed in the introduction require that many samples need to be taken to cover diurnal and seasonal patterns of activity and extrapolation from the inevitably few $^{15}N_2$ incubations that can be made do not lead to reliable estimations of field nitrogen inputs. A much more common application of the technique is in the calibration of acetylene reduction.

Applications. Major applications of $^{15}N_2$ techniques have been in the confirmation of N_2 fixing activity of organisms (1) and in biochemical studies (50). However some attempts have been made to extrapolate from $^{15}N_2$ uptake to total N accumulation. Specific activity of *Ceanothus* nodules has been estimated (56) and extrapolation over a nominal growing season gave an estimated input of 60 kg N ha^{-1} a^{-1}. Exposure to $^{15}N_2$ of *Hippophae* nodules on coastal dunes (57) gave values of nitrogen fixation of 0.10–3.93 mg N m^{-2} for a two hour exposure. Crude multiplication of these figures to a 4 month growing season gave N accumulation figures of 0.15–5.75 g N m^{-2} a^{-1}, figures which are considerably less than those the same authors obtained from biomass analysis.

^{15}N *dilution*

Principles. The measurement of nitrogen fixation by isotope dilution was first reported in 1959. Since then, it has found application in pasture and crop systems (58, 59) and has been shown to give results very comparable with N harvest data (60). The method is indirect and relies upon using natural or artificially ^{15}N enriched soil. Non nitrogen fixing plants growing in the soil will possess isotopic ratios identical to the soil, while N_2 fixing plants will dilute the soil N with atmospheric N_2. The method has two very significant advantages, firstly it is an integrative technique that can indicate the cumulative N_2 fixation over a period. Secondly the difference in isotope ratios allows not only a measure of N_2 fixed but, by difference, a measure of N taken up from soil into the N_2 fixing plant. The technique is extremely powerful, with great potential, and suffers perhaps only in the difficulty of uniformly labelling soil organic matter. The techniques have so far been restricted to pastures and annual crops because of the necessity to start with bare ground, but in long term studies there is great potential in forest systems.

There are basically two variations to the technique, one in which natural abundance variations in soil N are utilised and the other in which ^{15}N is added to the soil to artificially enrich it.

Natural abundance dilution. The measurement of differences in natural abundance requires that isotope effects be measured and while these were once thought to be insignificant they can now be measured using the new generation of mass spectrometry. Natural abundance differences are normally measured in delta units, one delta unit being natural abundance per 1000, δ-$= \frac{0.366}{1000} = 0.000366$ atom % ^{15}N. A modern mass spectrometer can measure with confidence $\delta\ ^{15}N$ values as low as 0.5 parts per 1000.

A detailed appraisal of the possible points at which isotope discrimination may occur (12) has shown that all reactions result in some selection for the lighter isotope. The discrimination factor for N_2 fixation is extremely small $\beta = 1.004$ while the discrimination during denitrification is relatively large $\beta = 1.02$ for the instantaneous rate. It must be remembered that if there is a finite substrate and the reaction goes to completion no fractionation will be observed, but at various times during a process a range of fractionation effects will be realised. Without looking at the details of the processes it is now apparent that fractionation effects lead to a soil and therefore biosphere N pool that is variously enriched in ^{15}N when compared with the atmosphere. It may be assumed, but it is by no means proven, that the degree of enrichment may reflect the extent of denitrification that occurs, as opposed to other N losses. Since the early 1970's when it was shown that soils may be significantly enriched, a large amount of work has been done to establish the extent of this.

A number of recent studies on forest soils are summarised in Table 3. As can be seen forest soils, like grassland and agricultural soils, show an enrichment of ^{15}N by as much as 10 per mil. Unfortunately workers in this field draw two quite opposite conclusions from their data. On the one hand a number of workers with wide experience in the field (61, 62, 64) report consistent results with small variation and their results give rise to the conclusion (64) '... within the Chernozemic order, spatial variability would not be expected to unduly limit use of natural variation in ^{15}N abundance to study specific agronomic processes such as biological nitrogen fixation, denitrification under field conditions, and N cycling in general'. On the other hand an equally reputable group of workers (63, 65) have shown that soils show very large variations both horizontally and vertically (Table 3) and have recently concluded (63) '... it is clear that the spatial variation in distribution of $\delta\, ^{15}N$ values is of the same order of magnitude as the difference between soil and atmospheric N. The data support the conclusion reached by others that it is not feasible to trace biological events in soil by means of natural abundances of ^{15}N'. Such widely discrepant conclusions obviously leave a lot more work to be done, and leave only one unequivocal conclusion, that soils are generally enriched in ^{15}N.

Despite the reservations expressed above natural abundance measurements have been used to assess both qualitative and quantitative N_2 fixation. A study of the legume *Prosopis* growing in the Sonoran desert (66) showed that plants had a $\delta\, ^{15}N$ of $+\,3.0$ (soil $= +6{\cdot}0$) even though nodules were not found. The results extended to other legumes show that N_2 fixing plants consistently contain ^{15}N levels close to atmospheric levels while growing in soils that are significantly

Table 3. Variation in natural abundance of soils under forest. All values are in δ Units (see text).

Location	Depth cm	$\delta^{15}N$	Standard deviation	Ref.
Hubbard Brook	–	+ 9.6	2.1	61
Various	–	+ 7.01	2.19	62
Tahoe	0–15	+ 2.8	4.6*	63
	15–30	+ 2.4	4.7*	63
	30–45	+ 7.1	3.9*	63
	45–60	+ 5.8	3.3*	63
	60–75	+ 6.0	5.3*	63
	75–90	+ 9.5	7.2*	63
	90–105	+ 10.2	7.6*	63
	105–120	+ 9.2	12.2	63

* 95% confidence limits

enriched. *Prosopis* was presumed to develop nodules on deep roots which are not normally harvested. Quantitative results on nitrogen fixation have also been attempted (13, 67) but in pasture systems it has been shown to be impossible because of the many fractionation processes associated with animals (68). If consistent soil ^{15}N enrichment levels are confirmed in forest environments then the technique promises to be a useful one for the identification of N_2 fixation and also for the transfer of symbiotically fixed N to associated forest plants.

^{15}*N enrichment*. The standard procedures for isotope dilution techniques have been applied to N_2 fixation with considerable success. This technique requires that ^{15}N (as NO_3^- or NH_4^+) is applied to the soil and comparison is made between the isotope content of N_2 fixing plants and control plants in the same soil. Considerable success has been claimed for the technique (59, 60, 68) and it has all of the hallmarks of a useful integrating method of assessing N_2 fixation.

Application of this technique to forest systems is problematical as it requires that the soil be enriched with ^{15}N prior to any plant growth. However there are very great benefits to be obtained by long term experiments in which the soil is enriched prior to planting out nodulated plants and forest trees, for example to assess not only total N_2 fixation but also the transfer of N from the N_2 fixer to the trees. An excellent assessment of the relative advantages and disadvantages of the various dilution techniques has been made by Knowles (35). The advantages he cites are that no enclosure need be used for nodules and the integrating nature of the technique which gives a cumulative measure of activity in which diurnal and seasonal patterns of activity can be ignored. It is interesting to note that many of the six disadvantages or problems noted by Knowles, would not be so damaging to the technique in a long term experiment designed to test N_2 fixation inputs into forestry. Not only does the dilution technique give a measure of N_2 fixation, but in the context of a mixed stand of e.g. legume and tree it would also give a measure of the contribution of the legume to tree growth, which after all is the real test of N_2 fixation.

ACETYLENE REDUCTION

General principles

The nitrogen fixing enzyme system N_2ase (nitrogenase) of all organisms so far studied has the property of reducing a large number of substrates other than N_2. Most useful amongst these is the reduction of acetylene to ethylene which was first reported in 1966 (69) and since then has been utilised as a sensitive quantitative assay for N_2ase activity (53, 70, 71, 72) and with suitable correction as a measure of nitrogen fixation (73, 73).

Acetylene reduction has gained wide acceptance because it is simple, sensitive and inexpensive in operation. In general outline the technique relies on the ability of all N_2 fixing organisms to reduce acetylene (C_2H_2) to ethylene (C_2H_4) in a reaction which has the same reductant and ATP requirements as N_2 fixation. In practice all that is required is to enclose a N_2 fixing system in a container, add a saturating concentration of acetylene (0.10 atmospheres normally) and measure the product by H_2 flame ionisation gas chromatography. As C_2H_2 is so much more soluble than N_2, N_2 fixation is effectively blocked and it is not necessary to remove N_2 from the assay chamber. Under conditions where there is no limitation of substrate, energy or reductant supply the molar ratio of the reactants for an equal number of electrons used in substrate reduction should be 3:1 according to the reactions.

$$3C_2H_2 + 6H^+ + 6e = 3C_2H_4$$

$$N_2 + 6H^+ + 6e = 2NH_3$$

While values close to this theoretical value have been obtained, considerable variation is shown for living systems, where a range from 1.7–25 has been reported for a wide variety of conditions (70, 73). A significant complication that contributes to the wide range in ratio is the fact that during N_2 fixation a proportion of electron flow is channelled into H_2 production (see p. 204 and Chapter 5, p. 111) and some organisms can utilise the H_2 produced as a reductant supply in a H_2 uptake mechanism. In the presence of C_2H_2, H_2 production is suppressed and all electron flow is into C_2H_2 reduction thus widening the ratio of C_2H_2:N_2.

Measurements of N_2 fixation obtained from C_2H_2 reduction are often designated N_2 (C_2H_2) fixation (70) and increasingly it is being show that conversion factors need to be determined for the specific conditions of the experiment rather than the blanket application of 3:1 or any other predetermined ratio.

Methodology

Acetylene reduction is now firmly established as the standard method for assessing N_2ase activity and after calibration also of N_2 fixation. Various techniques for its use have been described in excellent detail recently (53, 71, 72) and will not be dealt with again. However attention will be drawn here to some details that require special attention.

Acetylene gas, of commercial grade or generated from CaC_2 contains impurities of acetone, PH_3 and H_2S amongst others. It has recently been shown (74) that these impurities may lead to inhibition of N_2ase and that unpurified commercial C_2H_2 may inhibit C_2H_2 reduction significantly. It is recommended that careful attention be paid to purification procedures if absolute values for C_2H_2 reduction are required.

Most sources of C_2H_2 also contain C_2H_4 at low concentrations and while this is usually so low as not to affect many assays, it does become significant when low activities are being detected. As C_2H_4 is much less soluble in acetone this background drops during the life of a cylinder and for low background work the last portion of a cylinder only should be used. Checks for spontaneous ethylene production should also be made as this may be significant in some soils and tissues.

Assessment of specific activity from ethylene production has been dealt with in detail in recent reviews (71, 72). Ethylene may be measured as the product in an absolute fashion by calibration with pure C_2H_4 or by using an internal standard which may be C_2H_2 or another non-metabolised hydrocarbon such as propane against which the C_2H_4 peak is measured. Use of C_2H_2 in the field poses a number of problems both in incubation and in storing samples. A recent paper (75) outlines a number of techniques that simplify field techniques and exemplify use of the internal standard in calculating activity.

A major problem associated with incubations in acetylene relate to the time of incubation and sensitivity of assays. It is normal and desirable to run a time based activity to establish the degree of linearity and then to run assays over a period when linearity is established. The problem relates mainly to the fact that removal of nodules, soil, etc. into a closed container is bound to upset *in situ* activity. At least five different situations exist which may over or under estimate N_2ase activity, and while some of these relate to any assay made in enclosed space, others are specific to C_2H_2 reduction

(a) Size of material. Pieces of nodules or even whole intact nodules give lower rates and a shorter period of linearity than intact nodules attached to a section of root. This phenomenon was studied in *Alnus* (76) and shown to create a significant reduction in N_2ase, whereas nodules still attached or excised with a root fragment maintained high rates over a long time period. Similarly for legumes (73) N_2ase activity was reduced by 80–90% by excision from roots during incubations.

(b) Gas composition. During incubations in a fixed volume gas composition may change dramatically. In particular pO_2 will decrease and lead to possible stimulation of N_2ase in asymbiotic organisms, or reduction of activity in root nodules which have a high oxygen requirement (17).

(c) Inhibitors. Production of volatile inhibitors by nodules is suspected (77) and this phenomenon is eliminated by using a large gas space. A ratio of 300 cm^3 gas phase to 1g of tissue is recommended (34).

(d) Derepression. Products of normal nitrogen fixation control N_2ase enzyme levels by processes of repression (78). As acetylene eliminates the products of N_2 fixation this leads to enhanced enzyme synthesis (79). Long term incubations of organisms in C_2H_2 have been shown to produce affects that appear to be derepression as they are not mirrored by reductions in N_2 fixation (41, 80). It is essential therefore to study a time course of reaction and work only in the linear part of the reaction and to keep incubation times as short as possible.

(e) Growth. Early work with root associated (rhizosphere) bacteria showed a significant lag in induction of C_2H_2 reduction in soil cores (81). The normal procedure was to wait until linear rates of C_2H_2 reduction were obtained and treat this part of the curve as normal. It is now realised (17, 82) that the lag period represents a period of bacterial growth or enzyme synthesis and that the high rates originally obtained are not representative of *in situ* activity. All of the above precautions underline the need to make acetylene reduction assays as short as possible, to endeavour to duplicate the environment of the system being studied and to treat C_2H_2 reduction assays with some caution. A restatement of some concluding remarks of the recent review of Turner and Gibson (72) are appropriate '... the apparent simplicity of the assay can be deceptive. Apart from operational problems, interpretation of the results requires great care'. A recent paper (107) considers further limitations of this assay.

Applications

General. Reports of C_2H_2 reduction to assess N_2 fixation in forestry situations are numerous and examples given here illustrate the range of applications, broken down according to the types of organisms studied. Acetylene reduction has been used to measure N_2ase activity and more particularly to assess the physiological effects of shading, moisture nutrients etc, where it is a powerful tool. It has also been used with and without calibration to measure N (C_2H_2) fixation and examples of both will be given.

Nodulated plants. Perhaps the most dramatic and best authenticated account of legume fixation of proven significance to forests is the use of lupins in sand dune forestry in New Zealand (see Chapter 12). Biomass and soils analysis showed the importance of lupin to growth of *Pinus radiata* (83). Acetylene reduction, calibrated with $^{15}N_2$ uptake, was used to investigate the effects of canopy closure on lupin growth (84) and also in the lab on shading and water supply (85). In this work frequent samples of lupin nodules under various forest canopy conditions showed a linear relationship between C_2H_2 reduction activity and nodule biomass with an extinction at c. 20% relative light intensity.

One of the big problems of acetylene reduction assays in the field is the timing of analysis which may be abnormal in terms of diurnal and longer term variation. Thus sampling at one time of day may grossly overestimate the daily total activity of the plant. For alder seedlings (86) and pea plants (87) there is a close relationship between daily light intensity and N ase activity while in lupins (20, 88) and perhaps in some other large plants no detectable diurnal periodicity is shown by successive acetylene reduction assays. It is obvious that the major factors that require to be investigated in looking at N_2 fixation by legumes under forests are shading, moisture supply to the relatively shallow-rooted species, nutrients and coping with litter.

Reported work using acetylene reduction with non-legumes is much more extensive. In some ways this is because the plants themselves may be forest trees e.g. *Alnus* and *Casuarina* and also because many species often grow naturally in association with forest trees. Red alder has been looked at in great detail (89, 90, 91) and the relative seasonal and diurnal activity tudied. In addition C_2H_2 reduction has been used to assess total $N_2(C_2H_2)$ fixation (89).

Non symbiotic organisms. The early work on free living N fixation in leaf litter was assessed laboriously by total N assays. However recent work using C_2H_2 reduction has shown that there is significant N_2 fixation in many substrates in forest including leaf litter (92, 93, 94) decaying wood (23, 41, 94, 95) soil (96, 97) phyllosphere (98) and rhizosphere (7, 8).

Acetylene reduction because of its great sensitivity has been instrumental in solving a number of enigmas in non-symbiotic forest N_2 fixation. Firstly, several accounts of nitrogen fixation ($^{15}N_2$ uptake) associated with conifer roots (99, 100, 101) implicated either the 'nodule' like structures of podocarps or the ectomycorrhizae of pines while a similar number of negative results were obtained on conifers (102, 103). Using the much more sensitive and rapid C_2H_2 reduction it has been shown since that N_2 fixation is often associated with conifer roots (7, 8) but that the activity is in the rhizosphere and can be removed by washing and surface sterilisation (7).

Similarly the extremely low specific rates associated with organic decay of both leafy and woody forest residues was a much debated subject, which when reviewed in 1966 (104) and again in 1970 (105) left the view that this process is of little significance generally and especially in forest soils. However in the past decade a number of workers have shown, especially by use of C_2H_2 reduction, that N_2 fixation is widespread, and although the annual rates are low (1–10 kg N ha^{-1} a^{-1}) when considered against the lifetime of a forest, provide a significant input.

Calibration

While C_2H_2 reduction has enormous assets in speed and cost, it still suffers the disadvantages already discussed for $^{15}N_2$ uptake due to disturbance and to spot analyses. Over and above this however C_2H_2 reduction is an indirect measure of N_2 fixation and as such needs accurate calibration against $^{15}N_2$ or biomass analysis.

While the theoretical relationship of 3:1 is often closely adhered to, many departures from this have been found and many environmental variables have been closely looked at (73). However, since then it has been elucidated that the major reason for the departure from the theoretical ratio is that during N_2 reduction 25–30% of the electron flow is diverted into H_2 production and this does not happen during acetylene reduction. To further confuse this issue some organisms are able to recycle the H_2 through a hydrogenase system that oxidises H_2 in an ATP yielding process (106). Some 15% of rhizobial strains investigated show ability to recycle all H_2 produced. While the above has important implications for productivity and N_2 fixation it also means that the calibration of C_2H_2 reduction against some reliable method of assessment is essential if absolute measures of N_2 fixation are to be derived from C_2H_2 data. The majority of calibration checks with $^{15}N_2$ indicate that ratios of 3.5–4.0 are more normal but where accurate data are required it is not sufficient to use a ratio obtained by someone else for the same species.

Calibration of acetylene reduction with both $^{15}N_2$ uptake and biomass sampling by lupins in pine forest has been attempted (20). Having established there was no significant diurnal pattern in C_2H_2 reduction, parallel acetylene reduction and biomass sampling were conducted over 10 weeks. Total N accumulated in biomass was 1.69 g m^{-2} while (C_2H_2) data (adjusted for measured C_2H_2/N_2 ratio of 4.2) gave 1.51 g m^{-2}. These results and others indicate that C_2H_2 reduction carefully interpreted can give a close approximation to real activity.

EFFECTS ON ASSOCIATED SPECIES

Principles

Most forests are nitrogen stressed (89) i.e. respond quantitatively to added fertiliser N (107), however there is little evidence yet as to the extent that this N stress can be relieved by biological N inputs. While these inputs can be measured with some accuracy using the foregoing methods, in the final analysis it is forest yield that is the arbiter of whether that N input is significant or not.

Forest ecosystems contain a wide range of total N from 1–20 t N ha^{-1} and N stress has been recorded right across this broad spectrum (108). Normally less than one percent of the N is in available form, with the rest immobilised in organic form in soil, litter and living biomass. It is possible to divide arbitrarily N responses across this broad spectrum into two major types. Firstly, on very N deficient sites where total ecosystem N is below 2 t ha^{-1}, there is acute N deficiency and response to added N in any form is marked. Under these conditions N stress is at least 50 percent (i.e. plants growing at less than half maximum) and biological N inputs provide a very significant increase in growth. Examples are early succession stages on sand dunes, alluvium, eroded sites etc. (10, 17, 83, 84). Secondly, forests with a large supply of N often still display chronic N deficiency with stress in the order of 10–30 percent. In this case a number of factors control the rate of mineralisation of soil N which is the rate controlling step in productivity. In this situation the addition of new nitrogen as organic residues from leaves or roots of N_2 fixing plants may or may not relieve N stress and the addition of a N_2 fixing plant, may or may not be able to mirror the strategic application of fertiliser N.

Measuring N stress

N stress (S_N) is the quantitative estimation of the intensity of current nitrogen deficiency in a crop (107) and is expressed as a percentage of maximum growth attained in non limiting N supply.

$$S_N = 100 \frac{(\text{growth rate at } N_{max} - \text{growth rate at deficiency})}{\text{growth rate at } N_{max}}$$

The split plot technique is commonly used to assess S_N where one subplot is left untreated and the other given sufficient N to make N non-limiting. In forestry practice it is more normal to measure increase in yield as a percentage of unfertilised yield. Two important considerations have emerged from recent work on N stress in forest trees.

Firstly, N stress in forests is most commonly expressed during rapid tree growth and response to N addition is most dramatic either in early stages of growth or after releasing and thinning of established stands. Closed canopy and dense mature stands seldom respond to N additions. Secondly, and perhaps most importantly, N supply affects tree morphology. Will (109) has shown that *P radiata* trees under N stress have slender tapering trunks and light limbs and when N is added volume increment is most dramatic in changing stem form with no evidence of height increase. In the practical sense it was shown (110) that measurements of basal area and height showed no response to N addition in fast growing *P. radiata* but that when total stem volume was measured a 20 percent increase in volume gain was recorded from one application of urea.

N transfer

Nitrogen which is fixed is rapidly incorporated into the host protoplasm and is normally liberated on leaf fall and decomposition. Leaf litter return probably accounts for the greatest amount of N transfer in most systems but underground transfer by root and mycorrhizal turnover may be significant (22). The rate of litter decomposition will determine the rate of N recovery, and while most nodulated plants have high litter turnover rates, some e.g. *Ceanothus*, *Casuarina* and *Elaeagnus* have low foliage N and may not add significantly to mineral N supply.

The possibility of more direct N transfer between N fixing plants and associated trees is an intriguing one. Root grafts and mycorrhizal bridges are the most likely avenues for nitrogen transfer but as yet there is no evidence that N is transferred other than by the normal decay processes.

Other effects

The presence of a N_2 fixing plant may produce other both positive and negative effects on the tree crop. In many cases the N_2 fixing plant is vigorous and looked upon as a forest weed e.g. *Alnus* (89) *Coriaria* (20) and *Ceanothus* (1) are often looked upon as unwelcome invaders of forest sites because of their competitive effects. On the other hand there are good examples of these plants producing subsidiary benefits in disease prevention such as *Phellinus* in Douglas fir and *Fomes* in rubber (108).

N_2 fixing plants may play other roles in general site amelioration, increasing organic matter and soil tilth, decreasing bulk density and improving aeration, as well as stimulating nutrient turnover (45).

Conclusions

Documented examples of forest yield increase, directly attributable to N inputs from N_2 fixation, are scarce. The most dramatic effects are seen on low N soils and pioneer situations and these are well reviewed (10, 17, 89). It is likely that in high productivity stands the effects of biological N_2 fixation will be much more difficult to quantify and elucidate, and much of the present evidence is based on known N_2 fixation rates and extrapolation to known N responses.

It should be emphasised that N derived from N_2 fixation has both quantity and quality and the quantitative aspects have been emphasised. Quality of nitrogen is also important, especially C/N ratio and rate of decomposition. With increasing total N in the system, the quality of N additions becomes increasingly important, such that in very high producing forest stands, N response may be seen only by tactical use of readily available fertiliser N after thinning.

However, the use of nodulated plants as understory in high producing forests awaits careful evaluation by reliable productivity assessment.

REFERENCES

1. Silvester WB: Dinitrogen fixation by plant associations excluding legumes. In: *A Treatise on Dinitrogen Fixation*, Hardy RWF and Gibson AH (eds), John Wiley New York, 1977 pp 141–190.
2. Cromack K, Delwiche C, McNabb DH: Prospects and problems of nitrogen management using symbiotic nitrogen fixers. In: *Symbiotic Nitrogen Fixation in the Management of Tem-*

perate Forests, Gordon JC, Wheeler CT, Perry DA (eds), Corvallis Oregon State University, 1979, p 210–223.

3. Turner GL, Gibson AH: Measurements of nitrogen fixation by indirect means. In: *Methods for Evaluating Nitrogen Fixation*, Bergersen FJ, (ed), Chichester, John Wiley and Sons, 1980, p 111–138.
4. Bergersen FJ: *Methods for Evaluating Biological Nitrogen Fixation*, Chichester, John Wiley and Sons, 1980.
5. Rhigetti TL, Muns DN: Nodulation and nitrogen fixation in cliffrose (*Cowania mexicana* var. stansburiana) (Torr.) Jeps. *Plant Phys.* 65: 411–412, 1980.
6. Green TGA, Horstmann J, Bonnet H, Wilkins A, Silvester WB: Nitrogen fixation by members of the Stictaceae (Lichenes) of New Zealand. *New Phytologist* 84: 339–348, 1980.
7. Silvester WB, Bennett KJ: Acetylene reduction by roots and associated soil of New Zealand conifers. *Soil Biol Biochem* 5: 171–179, 1973.
8. Richards BN: Nitrogen fixation in the rhizosphere of conifers. *Soil Biol Biochem* 5: 149–152, 1973.
9. Crocker RL, Major J: Soil development in relation to vegetation and surface age at Glacier Bay, Alaska. *J. Ecology* 43: 427–448, 1955.
10. Silvester WB: Ecological and economic significance of the non-legume symbiosis. In: *Nitrogen Fixation*, Proceedings of the 1st International Symposium. Newton WE, Nyman J (eds), Washington State University Press. Pullman, 1976, pp 489–506.
11. Tarrant RF and Trappe JM: The role of *Alnus* in improving the forest environment. *Plant and Soil* Special Volume pp 335–348, 1971.
12. Delwiche CC, Steyn PL: Nitrogen isotope fractionation in soils and microbial reactions. *Environmental Science and Technology*, 4: 929–935, 1970.
13. Armager N, Mariotti A, Durr JC, Bourguignon C, Lagaeherie B: Estimate of symbiotically fixed nitrogen in field grown soy beans using variations in ^{15}N natural abundance. *Plant and Soil* 52: 269–280, 1979.
14. Gosz JR: Nitrogen cycling in coniferous ecosystems. In: *Terrestrial Nitrogen Cycles*, Clark FE, Rosswall T (eds), Ecol. Bull. (Stockholm) 33: 405–426, 1981.
15. Melillo JM: Nitrogen cycling in deciduous forests. In: *Terrestrial Nitrogen Cycles*, Clark FE, Rosswall T (eds), Ecol. Bull. 33: 427–442, 1981, (Stockholm).
16. Ovington JD: Quantitative ecology and the woodland ecosystem concept. Adv Ecol Res 1: 103–192, 1962.
17. Sprent JI: *The Biology of Nitrogen Fixing Organisms*, McGraw Hill London, 1979.
18. Date JS: Uptake assimilation and transport of nitrogen compounds by plants. *Soil Biol Biochem* 5: 109–119, 1973.
19. Lawrence DB: Glaciers and vegetation in south eastern Alaska. *American Scientist* 46: 89–122, 1958.
20. Silvester WB, Carter DA, Sprent JI: Nitrogen input by *Lupinus* and *Coriaria* in *Pinus radiata* forest in New Zealand. In: *Symbiotic Nitrogen Fixation in the Management of Temperate Forests* Gordon JC, Wheeler CT, Perry DA (eds), Corvallis, Oregon State University, p 253–265, 1979.
21. Oghoghorie CGO, Pate JS; The nitrate stress syndrome of the nodulated field pea (*Pisum satiuum L*), *Plant and Soil*, Special Volume 185–202, 1971.
22. Zavitkovski J, Newton M: Effect of organic matter and combined nitrogen on nodulation and nitrogen fixation in red alder. In: *Biology of Alder*, Trappe JM, Franklin JF, Tarrant RF, Hausen GM (eds), Portland, Oregon: US Forest Service, 1968.
23. Roskoski J: Nitrogen fixation in hardwood forests of the northeastern United States. *Plant and Soil* 54: 33–44, 1980.
24. Harris WF, Santantonio D, McGinty D: The dynamic below ground ecosystem. In: *Forests: Fresh Perspective from Ecosystem Analysis*, Waring NH, (ed) Corvallis, Oregon. Oregon State University Press, 1979.
25. Fogel R, Hunt G: Fungal and arboreal biomass in a western Oregon Douglas-fir ecosystem: distribution patterns and turnover. *Canadian J For Res* 1979: 245–256.
26. Pike LH, Tracy DM, Sherwood MA, Nidson D: Estimates of epiphytes from old growth Douglas-fir. In: *Research on Coniferous Forest Ecosystems*, First year progress in the Coni-

ferous Forest Biome, US/IBP, Franklin FJ, Dempster LJ, Waring RH (eds). Portland Oregon. US Forest Service, 1972.

27. Denison WC: *Lobaria oregana*, a nitrogen fixing lichen in oldgrowth Douglas-fir forests. In: *Symbiotic Nitrogen Fixation in the Management of Temperate Forests*, Gordon JC, Wheeler CT, Perry DA, (eds), Corvallis, Oregon State University, p 266–275, 1979.
28. Sollins P, Grier CC, McCorisen FM, Cromack K, Fogel R, Fredericksen RL: The internal element cycle of an old growth Douglas-fir ecosystem in Western Oregon. *Ecological Monographs* 50: 261–285, 1980.
29. Carroll GC: Forest canopies: complex and independent subsystems in forests: *Fresh Perspectives from Ecosystem Analysis*, Waring RH (ed), Corvallis, Oregon State University Press, 1979.
30. Henry E: L'azote et la vegetation forestiere. In: *Revue d'eaux et forets* 36: 641, 1897.
31. Hornberger: Streu und stickstoff, *Zeitschr f Forst- und Jagdwesen* 37: 71, 1905.
32. Olsen C: Studies of nitrogen fixation I. Nitrogen fixation in the dead leaves of forest beds. *Comptes-rendus du laboratoire Carlsberg* 19: 1–36, 1932.
33. Jenny H: *Factors of Soil Formation*, New York, McGraw-Hill, 1941.
34. Knowles R: Nitrogen fixation in natural plant communities. In: *Methods for Evaluating Nitrogen Fixation*, Bergersen FJ (ed) Chichester, John Wiley and Sons, p 557–582, 1980.
35. Knowles R: The measurement of nitrogen fixation. In: *Current Perspectives in Nitrogen Fixation*, Gibson AH, Newton WE (eds), Canberra Australian Academy of Science, 1981.
36. Soderlunk R: Dry and wet deposition of nitrogen compounds. In: *Terrestrial Nitrogen Cycles*, Clark FE, Rosswall T (eds), *Ecol Bull* 33; 123–130, 1981.
37. Vitousek PM, Gosz RJ, Grier CC, Melillo JM, Reiness WA, Todd RL: Nitrate losses from disturbed ecosystems. *Science* 204: 469–474, 1979.
38. Duvgneaud P, Denaeyer-DeSmet S: Biological cycling of minerals in temperate deciduous forests. In: *Analysis of Temperate Forest Ecosystems*, Reichle DE (ed), Springer-Verlag, New York, New York, USA, p 199–225, 1970.
39. Likens GE, Bormann FH, Pierce RS, Eaton JS, Johnson NM: *Biogeochemistry of a Forest Ecosystem*, New York, Springer, Verlag 1977.
40. Bormann FH, Likens FE, Melillo JM: Nitrogen budget for an aggrading northern hardwood forest ecosystem, *Science* 196: 981–983, 1977.
41. Silvester WB, Sollins P, Verhoeven T, Clive SP: Nitrogen fixation and acetylene reduction in decaying conifer boles. Effects of incubation time, aeration and moisture content. *Canad J For Sci* In Press, 1982.
42. Gessel SP, Cole DW, Steinbrenner EC: Nitrogen balances in forest ecosystems of the Pacific Northwest. *Soil Biol Biochem* 5: 19–34, 1973.
43. Youngberg CT, Wollum AG: Nitrogen accretion in developing *Ceanothus velutinus* stands: *Soil Sci Soc Amer J* 40: 109–112, 1976.
44. Youngberg CT, Wollum AG, Scott W: *Ceanothus* in Douglas-fir clearcuts: nitrogen accretion and impact on regeneration. In: *Symbiotic Nitrogen Fixation in the Management of Temperate Forests*, Gordon JC, Wheeler CT, Perry DA (eds) Corvallis, Oregon State University, p 224–233, 1979.
45. Tarrant RF, Miller RE: Accumulation of organic matter and soil nitrogen beneath a plantation of red alder and Douglas-fir. *Soil Sci Soc Amer Proc* 27: 231–234, 1963.
46. Van Cleve K, Viereck LA, Schlentner RL: Accumulation of nitrogen in alder (*Alnus*) ecosystems near Fairbanks, Alaska. *Arctic and Alpine Research* 3: 101–114, 1971.
47. Uemura S: Non-leguminous root nodules in Japan. *Plant and Soil*, Special Volume 1971, p 349–360.
48. Dommergues Y: Evaluation du taux de fixation de L'azote dans un sol dunare reboise en filao. *Agrochimica* 7: 335–340, 1963.
49. Wada E, Kadonoga T, Matsuo S: ^{15}N abundance in nitrogen of naturally occurring substances and assessment of denitrification from a global viewpoint. *Geochem J*. 9: p 139–148, 1975.
50. Bergerson FJ: Measurements of nitrogen fixation by direct means. In: *Methods for Evaluating Nitrogen Fixation*, Bergersen FJ (ed), Chichester, John Wiley and Sons, 1980, p 65–110.

51. Hauck RD, Bremner JM: Use of tracers for soil and fertiliser nitrogen research. *Advances in Agronomy* 28: 219–266, 1976.
52. Vose PB: Stable isotopes as tracers In: *Introduction to Nuclear Techniques in Agronomy and Plant Biology*, London, Pergamon Press, 1980, pp 151–176.
53. Burris RH: Methodology. In: *The Biology of Nitrogen Fixation*, Quispel A (ed), Amsterdam, North Holland 1974, pp 9–33.
54. Volk RJ, Jackson WA: Preparing nitrogen gas for nitrogen-15 analysis. *Analytical Chemistry* 51: 463, 1979.
55. Horstmann JL, Denison WC, Silvester WB: $^{15}N_2$ fixation and molybdenum enhancement of acetylene reduction by *Lobaria* species. *New Phytologist*. In press 1982.
56. Delwiche CC, Zinke PJ, Johnson CM, Virginia RA: Nitrogen isotope distribution as a presumptive indicator of nitrogen fixation. *Bot Gaz* 140, 1979.
57. Stewart WDP, Pearson MC: Nodulation and nitrogen fixation by *Hippophae rhamnoides* in the field. *Plant and Soil* 26: 348–360, 1967.
58. Fried M, Middleboe V: Measurement of nitrogen fixation by a legume crop. *Plant and Soil* 47; 713–715, 1977.
59. Edmeades DC, Goh KM: The use of the ^{15}N-dilution technique for field measurement of symbiotic nitrogen fixation. *Communications in Soil Science and Plant Analysis* 10: 513–520, 1979.
60. Legg JO, Sloger C: A tracer method for determining symbiotic nitrogen fixation in field studies. In: Proceedings 2nd International Conference on Stable Isotopes. Klein ER, Klein PD (eds), Illinois, Argonne National Laboratory, 1977.
61. Shearer GB, Kohl DH, Commorier B: The precision of determinations of the natural abundance of nitrogen-15 in soils, fertilisers and shelf chemicals. *Soil Science* 118: 308–316, 1974.
62. Shearer GB, Kohl DH, Chien S-H: The nitrogen-15 abundance in a wide variety of soils. *Soil Science Society America J.* 42: 899–902, 1978.
63. Broadbent FE, Rauschkolb RS, Lewis KA, Chang GY: Spatial variability of nitrogen-15 and total nitrogen in some virgin and cultivated soils. *Soil Science Society America J.* 44: 524–528, 1980.
64. Karamanos RE, Voroney RP, Rennie DA: Variation in natural N-15 abundance of central Saskatchewan soils. *Soil Science Society America J.* 45: 826–828, 1981.
65. Cheng HH, Bremner JM, Edwards AP: Variations of ^{15}N abundance in soils. *Science* 146: 1574–1575, 1964.
66. Virginia RA, Jarrell WM, Kohl DH, Shearer GB: Symbiotic nitrogen fixation in a *Prosopis* (leguminosae) dominated desert ecosystem. In: *Current Perspectives in Nitrogen Fixation*, Gibson AH, Newton WE (eds), Canberra, Australian Academy of Science, p 483, 1981.
67. Kohl DH, Shearer G, Harper JE: Estimation of N_2 fixation based on differences in natural abundance of ^{15}N in nodulating and non-nodulating isolines of soy beans. *Plant Physiology* 66: 61–65, 1980.
68. Steele KW: Natural abundance and ^{15}N enrichment techniques. In: *Current Perspectives in Nitrogen Fixation*, Gibson AH, Newton WE (eds), Canberra, Australian Academy of Science, 1981, p 336–337.
69. Dilworth MJ: Acetylene reduction by nitrogen fixing preparations from *Clostridium pasteurianum*. *Biochim Biophys Acta* 127: 285–294, 1966.
70. Hardy RWF, Burns RC, Holsten RD: Applications of the acetylene-ethylene assay for measurement of nitrogen fixation. *Soil Biol Biochem* 5: 47–81, 1973.
71. Postgate JR: The acetylene reduction test for nitrogen fixation. In: *Methods in Microbiology*, Norris JR, Robbins DW (eds), London Academic Press, 1972.
72. Turner GL, Gibson AH: Measurement of nitrogen fixation by indirect means. In: *Methods for Evaluating Nitrogen Fixation*. Bergersen FJ, (ed) London, John Siley and Sons, p 111–138, 1980.
73. Bergersen FJ: The quantitative relationship between nitrogen fixation and the acetylene reduction assay. *Aust J Biol Sci* 23: 1015–1025, 1970.
74. Tough HG, Crush JR: Effect of grade of acetylene on ethylene production by white clover (*Trifolium repens* L) during acetylene reduction assays of nitrogen fixation. *N Z Journal of Agric Res* 22: 581–583, 1979.

75. McNabb DH, Geist JM: Acetylene reduction assay of symbiotic N_2 fixation under field conditions. *Ecology* 60: 1070–1072, 1979.
76. Akkermans ADL: Nitrogen fixation and nodulation of *Alnus* and *Hippophae* under natural conditions. Leiden Doctoral Thesis, University of Leiden, 1971.
77. Sprent JI: Prolonged reduction of acetylene by detached nodules. *Planta* 88: 372–375, 1969.
78. Dixon RA, Cannon FC: Recent advances in the genetics of nitrogen fixation. In: *Symbiotic Nitrogen Fixation*, Nutman PS (ed), Cambridge, Cambridge University Press, 1976, pp 3–24.
79. Shanmugan KT, Valentine RC: Molecular biology of nitrogen fixation, *Science* 187: 919–931, 1975.
80. David KAV, Fay P: Effects of long-term treatment with acetylene on nitrogen fixing organisms. *Appl Environ Microbial* 34: 640–646, 1977.
81. Dobereiner J, Day JM, Dent PJ: Nitrogenase activity and oxygen sensitivity of *Paspalum notatum – Azotobacter paspali* association, *J Gen Microbiol* 71: 103–116, 1972.
82. Sloger C: Associative nitrogen fixation involving cereals and gasses. In: *Current Perspectives in Nitrogen Fixation*. Gibson AH, Newton WE, (eds), Canberra, Australian Academy of Science, p 321, 1981.
83. Gadgil RL: The nutritional role of *Lupinus arboreus* in coastal sand dune forestry III: Nitrogen Distribution in the Ecosystem before Tree Planting. *Plant and Soil* 34: 575–593, 1971.
84. Sprent JI, Silvester WB: Nitrogen fixation by *Lupinus arboreus* grown in the open and under different aged stands of *Pinus radiata*. *New Phytol* 72: 991–1003, 1973.
85. Sprent JI: Growth and nitrogen fixation in *Lupinus arboreus* as affected by shading and water supply. *New Phytologist* 72: 1005–1022, 1973.
86. Wheeler CT: The diurnal fluctuation in nitrogen fixation in nodules of *Alnus glutinosa* and *Myrica gale*. *New Phytol* 68: 675–682, 1969.
87. Minchin FR, Pate JS: Diurnal functioning of the legume root nodule. *J Exp Bot* 25: 295–308, 1974.
88. Trinick MF, Dilworth MF, Grounds M: Factors affecting the reduction of acetylene by root nodules of *Lupinus* species. *New Phytol* 77: 359–370, 1976.
89. Gordon JC, Wheeler CT, Perry DA (eds). Symbiotic Nitrogen Fixation in the Management of Temperate Forests, Corvallis, Oregon State University, 1979.
90. Trappe JM, Franklin JF, Tarrant RT, Hansen GM (ed). *Biology of Alder* Portland, U.S. Forest Service, 1968.
91. Tripp LN, Bezdirek DF, Heilman PE: Seasonal and diurnal patterns and rates of nitrogen fixation by young red alder. *Forest Science* 25: 371–380, 1979.
92. Silvester WB: Nitrogen fixation and mineralisation in kauri (*Agathus australis*) forest in New Zealand. In: *Microbial Ecology* (Proceedings in Life Sciences). Loutit MW, Miles JAR (eds), Berlin, Springer-Verlag, 1978.
93. O'Connell AM, Grove TS, Malajczuk N: Nitrogen fixation in the litter layer of eucalypt forest. *Soil Biol Biochem* 11: 681–682, 1979.
94. Larsen MJ, Jurgensen MF, Harvey AE: N_2 fixation associated with wood decayed by some common fungi in western Montana. *Canadian Journal of Forest Research* 8: 341–345, 1978.
95. Sharp RF, Millbank JW: Nitrogen fixation in deteriorating wood. In: *Experentia* 29: 895–896, 1973.
96. Jorgensen JR, Wells CG: Apparent nitrogen fixation in soil influenced by prescribed burning. *Soil Sci Soc Am Proc* 35: 806–810, 1971.
97. Todd RL, Meyer RD, Waide JB: Nitrogen fixation in a deciduous forest in the southeastern United States. In: *Environmental Roles* of Nitrogen-fixing Blue-Green Algae and Asymbiotic Bacteria. Ecol Bull (Stockholm) 26: 172–177, 1978.
98. Jones K: Nitrogen fixation in the phyllosphere of the Douglas fir; *Pseudotsuga douglasii*. *Ann Bot* 34: 239–244, 1970.
99. Richards BN: Fixation of atmospheric nitrogen in coniferous forests. *Aust For* 28: 68–74, 1964.
100. Richards BN, Voigt GK: Role of mycorrhizae in nitrogen fixation. *Nature* 201: 310–311, 1964.
101. Bergersen FJ, Costin AB: Root nodules on *Podocarpus lawrencei* and their ecological significance. *Aust J Biol Sci* 17: 44–48, 1964.

102. Baylis GTS: Mycorrhizal nodules and growth of *Podocarpus* in nitrogen poor soils. *Nature* 223: 1385–1386, 1969.
103. Bond G, Scott GD: An examination of some symbiotic systems for fixation of nitrogen. *Ann Bot* 19: 67–77, 1955.
104. Moore AW: Non symbiotic nitrogen fixation in soil and soil-plant systems. *Soils Fertil.* 29: 113–128, 1966.
105. Jurgersen MF, Davy CB: Non-symbiotic nitrogen fixing microorganisms in acid soils and the rhizosphere. *Soils Fertil* 33: 435–446, 1970.
106. Evans HF, Emerich DW, Maier RJ, Hanus FJ, Russell SA: Hydrogen cycling within the nodules of legumes and non-legumes and its role in nitrogen fixation. In: *Symbiotic Nitrogen Fixation in the Management of Temperate Forests*. Gordon JC, Wheeler CT, Perry DA (eds), pp 196–206. Corvallis, Oregon State University, 1979.
107. Minchin FR, Witty JF, Sheehy JE, Müller M: A major error in the acetylene reduction assay: decreases in nodular nitrogenase activity under assay conditions. *J Exp Bot* 34: 641–649, 1983.

7. Agricultural and horticultural systems: Implications for forestry

JANET I. SPRENT

Department of Biological Sciences, University of Dundee, Dundee DD1 4HN, Scotland

CROPS CAPABLE OF FIXING NITROGEN: LEGUMES

Taxonomy of legumes and associated rhizobia

Legumes. Two important publications on this subject appeared in 1981. The first (1) arose out of an international meeting on legume systematics held at Kew Gardens, England in 1978, but has been up-dated. The second (2) is the culmination of over forty years of work, principally on root nodules and their endosymbiotic rhizobia but including also many other fascinating attributes of legumes.

Both of these works use the major sub-divisions described by Hutchinson (3), namely the family Leguminosae being divided into three sub-families; Mimosoideae, Caesalpiniodeae and Papilionoideae. However, within these sub-families there are differences in usage. The classification adopted in this chapter follows the tribal arrangements of Polhill and Raven (1).

In general, the Caesalpiniodiae are confined to tropical regions, are mainly trees and only a small proportion of the species is nodulated. The Mimosoideae are principally trees and shrubs which have extended from the wet tropics into many dry areas, for example Australia: the majority of species are nodulated. The Papilionoideae contain few trees, being mainly shrubs and herbs: they have extended to almost all parts of the world and are almost universally capable of nodulation.

Many of the trees used directly for timber occur in the Caesalpiniodeae and Mimosoideae. Most of these have been exploited by felling of natural forests and no attention has been paid to their nitrogen fixing potential. However if there is to be re-afforestation in many of the traditional timber-source areas such as Africa and South America, this should be taken into consideration. To give one example, Allen and Allen (2) list 5 legume genera as being used for teak substitutes (table 1). *Afrormosia* is now usually incorporated into *Pericopsis:* it is unusual in being an arboreal member of the Papilionoideae and nodulation is, apparently, widespread. The other genera are all members of the Caesalpinioideae and are unlikely to be nodulated. Thus, on the grounds of its nitrogen fixing potential, *Pericopsis* is to be preferred.

Of the Caesalpinioid genera listed by Allen and Allen (2), the genera *Cassia*

Gordon, J.C. and Wheeler, C.T. (eds.). Biological nitrogen fixation in forest ecosystems: foundations and applications
© 1983, Martinus Nijhoff/Dr W. Junk Publishers, The Hague/Boston/London. ISBN 90-247-2849-5.
Printed in The Netherlands

Table 1. Legumes used as teak substitutes.[1]

Genus	Sub-family	Nodulation status
Afrormosia	Papilionoideae	nodulated
Baikiaea	Caesalpinioideae	not nodulated
Dicorynia	Caesalpinioideae	no information
Intsia	Caesalpinioideae	several negative reports, one positive
Pericopsis	Papilionoideae	nodulated

[1] Data taken from Allen and Allen (2).

and *Swartzia* stand out as containing most of the nodulated species. The latter is on the borderline of the Caesalpinioideae and for various reasons, including nodulation, has been included in the Papilionoideae (4). *Cassia* has now been sub-divided into 3 genera, *Cassia, Senna* and *Chamaecrista* (5). Most of the nodulated species are in the latter and they are mainly shrubs and herbs. Of the remaining Caesalpinioid genera with *confirmed* nodulation, only 5 or 6 are trees. This leads to the interesting conclusion that the comparatively rare nodulation in the Caesalpinioideae is largely confined to nonarboreal genera. Since Caesalpinioid trees mainly occur in rainforests where nitrogen is highly conserved, selection pressures would not favour N fixation (6). However, those tree genera that are nodulated are potentially very useful because the wood is of high quality. Projects are currently under way at various sites in Brazil by Dobereiner and co-workers to assess the growth and nitrogen fixation of some native species.

The Mimosoideae are in a different category, with tree-forms (albeit often small) predominant and with widespread nodulation. Many have adapted to exploit mainly dry and even desert areas in tropical regions. They may form canopies, but are more often sub-dominant plants. For example in Australian *Eucalyptus* forests species of *Acacia* are widespread as understory shrubs, with members of the Papilionoid tribe Mirbelieae taking their place in parts of Western Australia and the cycad *Macrozamia* (with a cyanobaterial N fixing symbiont) in areas of W. Australia and elsewhere. Where these plants grow, the soil is low in available nitrogen, and the leaf litter is subject to loss by fire: thus the ability to fix nitrogen is an important attribute.

Because a number of Papilionoid genera are important in agriculture, this sub-family has been the most widely studied. Agriculturally the species are divided into those harvested for seed (grain legumes) and those used for grazing, hay or silage (forage legumes). The general biology of these has been summarized by Sprent (7): see also Akkermans (this volume).

Host specificity of rhizobia. Most is known about those rhizobia which are of agricultural significance, principally nodulating the Papilionoid tribes Phaseoleae, Vicieae and Trifolieae. These rhizobia fall into two general categories, from tropical or from temperate regions. The major attributes of these groups are listed in table 2. Generally the tropical rhizobia show broader host specificity and this is one of the reasons why some consider them to be the more primitive. In addition the bacteroids of temperate rhizobia generally show a higher degree of increase in size and differentiation (compared with free living forms) than those of the tropical types: they also have lowered viability. The significance of differences in enzyme mobility is obscure, but it is interesting to note that superoxide dismutase may help to prevent inactivation of leghaemoglobin by the superoxide radical (13). Leghaemoglobin, the apoprotein of which is under host genetic control varies in structure from species to species. Recent work indicates that there is affinity amongst the temperate species (alfalfa, pea, clover and milk vetch) which differs from that of plants nodulated with slow growing rhizobia (soybean, lupin, jackbean and black locust). (14)

Now that more attention is being focused on legumes other than those used in Western agriculture, it is becoming clear that differences listed in table 2 are only 2 sets of possibilities. Many other geographical variants may yet be found. For example, species of *Mimosa*, *Acacia* and *Leucaena* (all Mimosoideae) and *Sesbania* (Papilionoideae) from tropical Papua and New Guinea are nodulated by fast growing strains of *Rhizobium*, having some affinity with *R. meliloti* (9). Other members of the Mimosoideae may be nodulated by slow-growing rhizobia, more typical of the tropical types described in Table 2. It is likely that in the next few years our concepts of the genus *Rhizobium* will change markedly. For

Table 2. Properties of rhizobia infecting some Papilionoid tribes.[1]

Character	Tribe	
Free-living cells	Vicieae/Trifolieae	Phaseoleae[2]/Desmodieae
growth rate on agar	fast	slow
tolerance of acid	poor	good
genetic manipulations achieved	frequently	rarely
nitrogenase activity	not substantiated	recorded many times
ability to take up and/or hydrolyse disaccharides	present	absent (11)
electrophoretic mobility of superoxide dismutase Rf at pH 7.5	0.55	0.15 (12)
Bacteroids		
DNA content	increased	little change
size	greatly increased	slightly increased
shape	pleomorphic	swollen rods
viability	little or none	some
detergent sensitivity	high	low

[1] For detailed references on most characters see Sprent (8, 10). The extent of data available varies greatly amongst characters.
[2] As exemplified chiefly by *R. japonicum* and the 'cowpea miscellany'. *R. phaseoli* is exceptional, being fast growing and nodulating *Phaseolus* spp. as well as having genetic affinity with rhizobia nodulating Vicieae/Trifolieae.

an excellent review of the general genetics of the genus readers should consult Beringer *et al.* (15).

Variations in nodules from different regions

Anatomy, morphology, physiology and biochemistry of legume nodules vary with species. In general, plants from related tribes and/or habitats have features in common. At present sufficient information for comparative purposes is only available for a few Papilionoid tribes (16). Some of the differences between nodules of the Vicieae/Trifolieae and the Phaseoleae are shown in table 3. It is likely that nodules from legumes adapted for other environments will show other features. For example in parts of Western Australia where rainfall is intermittent for three winter months and the rest of the year is hot and dry, nodules may have to be either ephemeral or perennial. The latter requires that they can tolerate hot dry conditions, almost certainly in an inactive state (17). Nodules from such areas are frequently found to have corky surfaces (2, 18). Preliminary observations on some genera in the Papilionoid tribe Mirbelieae suggest that nodules have a more-or-less continuous periderm which may give rise to aerating tissue (similar to the lenticels described for soybeans (19) during the active season and corky tissue during the 'dormant' season, Sprent and Pate, unpublished data). Thus surface features may be adapted to optimise gaseous exchange, with the inevitable concomitant of water loss, in the active period or water conservation in the dormant period (see also Dixon and Wheeler, this volume). In selecting legumes for use in open forest land, these features should be borne in mind.

Annual v. perennial growth habits

Annuals. Most annual legumes are in the Papilionoideae, a group which also

Table 3. Characteristics of the majority of nodules examined in some Papilionoid tribes.

Character	Tribe	
	Vicieae/Trifolieae	Phaseoleae/Desmodieae
growth period	several weeks to perennial	variable, but frequently only a few weeks
growth form	indeterminate	determinate
position of meristem	apical, often branching	$\pm$ spherical, not persistent
vascular system	branched, open ended	branched, fused at distal end
infected cells	vacuolate	non-vacuolate
vascular transfer cells	present	absent
surface regions for gaseous exchange	all over	confined to lenticels
principal regions of occurrence	temperate	tropical/sub-tropical

contains numerous perennials. Our comparisons will be largely confined to this sub-family.

Annual, and to some extent short-lived perennials, put most of their resources into seeds. These seeds compete with nodules for carbohydrates. This gives rise to the general seasonal pattern of activity shown in figure 1, with a peak just after flowering. The time of the peak may vary, for example with cultivar, as seen in soybean (*Glycine max*) (20). The magnitude of the peak also varies substantially with environmental conditions and with planting density. For example, widely spaced *Vicia faba* plants showed a large peak whereas closely spaced ones did not (21). The latter bore far fewer pods per plant, minimising internal competition for photosynthate.

Short-lived perennials such as *Lupinus arboreus* (tree lupin) with a shrubby habit may also have a high seed yield, but take a few years to achieve flowering size. During this time considerable amounts of nitrogen may be fixed (22).

Perennials. Because of the time and labour involved, relatively few studies of perennial N fixing plants have been made. Work on bush clover (*Lespedeza bicolor*) (23) and *Myrica gale* (an actinorrhizal plant) (24) and other species suggests that the following sequence obtains.

As their life span increases, the *total* nitrogen-fixed per unit area of ground increases. However the *rates* of fixation per unit of biomass drop steadily. This is because perennating organs (including trunks of trees) act as stores of N as well as C during adverse conditions (low temperature, drought, etc.). These stores

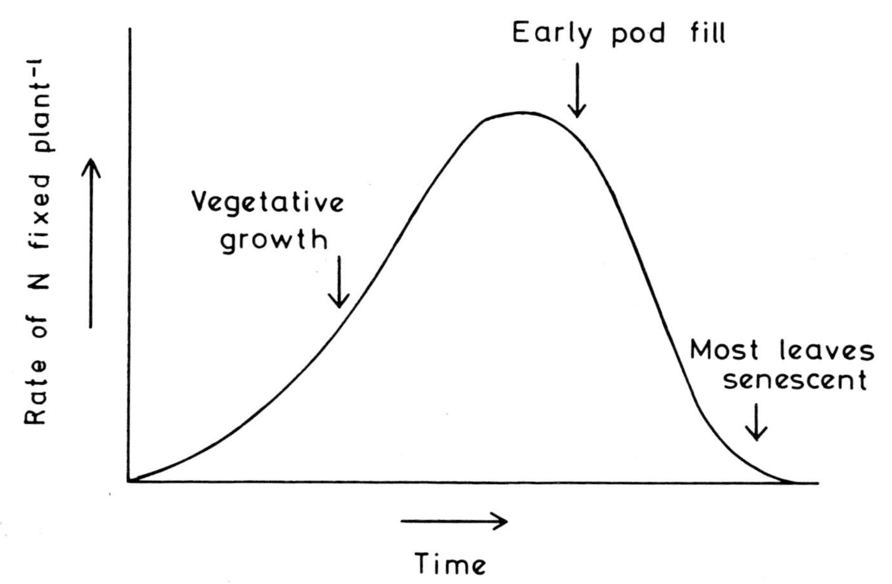

Fig. 1. Generalised pattern of nitrogen fixing activity for an annual legume. For variations see text.

may be recycled when growth is resumed. (Figure 2). Finally at the natural climax the nitrogen fixed balances that lost by fallen leaves and death of old tissues, so that there is no net gain of N after a season's growth. Plants grown on combined nitrogen will show the same pattern of N accretion and redistribution (in fact N fixing species will almost certainly use combined N as well). Net input again will decrease with increasing plant age as more N is held within the plant. For example teak (*Tectonia grandis* at 25 y has 70% of its N in the trunk and only 8% in the (renewable) leaves (25).

For an entire ecosystem at climax the net N gain equals net N loss (by run-off and denitrification). There is likely to be a small input contribution from some or all of several sources such as legumes, actinorrhizal plants, free-living bacteria, lichens and rainfall. The total N fixed by anyone of these is likely to be small.

Thus, as far as N fixation is concerned, there is a law of diminishing return as plants age. In using perennial plants these factors must be weighed against other benefits and management needs.

ATTRIBUTES OF LEGUMES OF POTENTIAL OR PROVEN USE IN FORESTRY

Total N fixed

In agricultural soils which are, apart from N, highly fertile, several hundred kg of N may be fixed per hectare each year. It is probable that in the humid tropics or other region where growth continues for the whole year, 1000 kg N ha^{-1} may be fixed. The limiting factors, apart from the lifespan and habit of the particular species are environmental, as discussed by Dixon and Wheeler (this volume).

Legumes are not altruistic. They fix N strictly for themselves. Thus, during the growth of a healthy young legume, normally little N is released into the soil. Some turnover of roots and nodules occurs and there may be limited leaf fall. Mineralization of such dead parts would result in N becoming generally available. This combined N may be absorbed by any plant *including the legume*. Transfer of N to other plants in this situation is thus likely to be limited. It is the ageing or stressed legume which acts as a significant N source for associated species. Essentially this means that when the general growth of a plant is slowed or stopped by internal or external factors, N fixing root nodules may become less essential or even a luxury which the plant cannot afford. They may then break down (or enter a 'dormant' state as proposed in 1.1.2.) Although their total mass is small (1–5% of total dry matter of young plants) their N content is high (about 5% of nodule dry weight). However total N input via nodule breakdown alone is relatively insignificant. When it is supplemented by root death and leaf fall appreciable ambounts of N become available to the ecosystem.

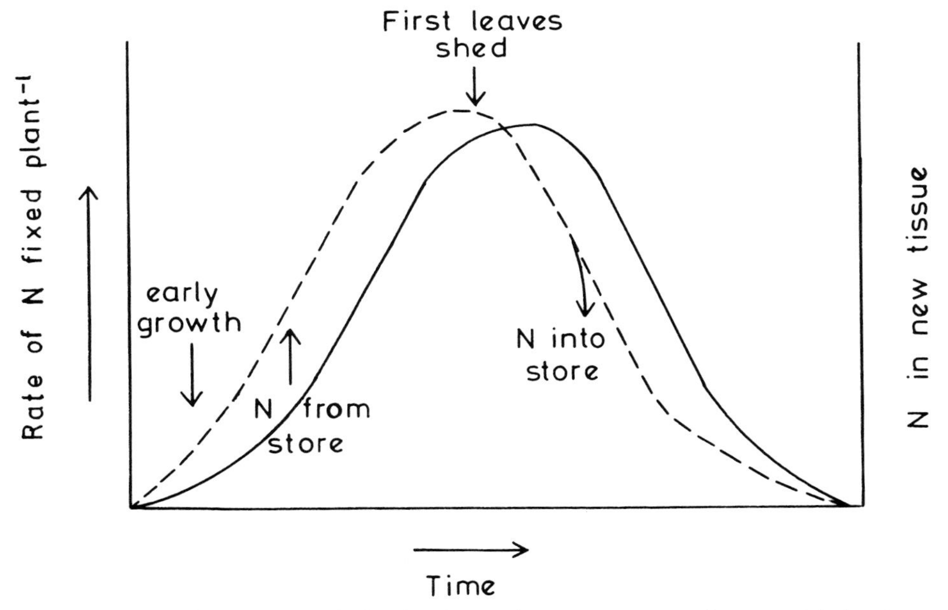

Fig. 2. Hypothetical pattern of N fixation during the growing season of a perennial plant in a climax system. The broken line indicates N passing into new tissue (expecially leaves). The areas under the two curves are the same.

Release of N following monoculture of legumes. The amount of N left *in situ* following growth of grain legumes varies widely with species and management practice. An understanding of the principal features may assist in the use of such plants as nurse crops or in agroforestry systems. If a crop is harvested for dry seed, most of the N will be removed: for example in *Vicia faba*, 80–90% of plant N was found in the dry seed (26). However, much of this may be transported to the seed relatively late in its development. With peas (*Pisum sativum*), which are widely grown for freezing, much of the N remains in the leaves and stems at harvest. Indeed Pate (27) argues that in terms of world food supplies we should harvest all seeds when they have achieved their maximum reserve potential, which would mean abandoning frozen peas in favor of dry peas. However, if the haulm is harvested for animal feed or ploughed back into the field, far more of the total N fixed is used profitably (see discussion in Sprent (28)).

Nodulated root systems remaining in soil after harvesting a grain legume contribute a variable amount of N to subsequent crops. Values of 50–80 kg ha^{-1} are reported for cowpea (*Vigna unguiculata*) and Bambara groundnut (*Vigna (Voandsia) subterranea* (29)). In a forestry context therefore maximum return in terms of N would accrue if plants were slashed prior to seed ripening. This would also minimise fire danger, a hazard with some nurse species such as broom (*Cytisus (Sarothamnus) scoparius)* (30)).

Intercropping. In an excellent publication on multiple cropping (31) the practice of growing two or more species together was defined as intercropping, being subdivided into mixed, row and strip intercropping according to whether the crops were grown with no special spatial arrangement or in rows or strips. Legume/grass pastures are a form of mixed intercropping which are considered under defoliation and will not be discussed here. All forms of intercropping are practiced with grain legumes and cereals, more particularly in peasant agriculture. The scientific bases of this practice are only now being assessed and they will have considerable interest for forestry. The species of legume used varies widely: *Phaseolus vulgaris* is predominant in South America, *Vicia faba* in the Middle East, *Vigna unguiculata* and other species in Africa. In all cases the grain is harvested dry. The potential for N return or transfer is thus relatively small, but could be significant when the legume crop is harvested ahead of the cereal, leaving time for mineralization of residues and uptake by the cereal. Such a possibility may be where corn (*Zea mays*) is grown with *Phaseolus vulgaris*. The latter is a rapidly growing crop, harvested before corn canopy closure. Overall yield in such situations often exceeds that of either species grown in monoculture (31). There may be definite benefits in saving of N fertiliser but, to the author's knowledge N transfer to the associated crop has not been fully quantified.

Defoliation in relation to N release

Defoliation is brought about principally by cutting and grazing, where the latter includes not only vertebrate herbivores (sheep, kangaroos, birds, etc.) but also invertebrate ones (slugs, snails, insect larvae). Seasonal defoliation of perennial species by drought or adverse temperatures normally involves a programmed sequence of events within the plant, with conservation of N and other resources in storage tissues as we have seen briefly above. Both cutting and grazing are important features of agriculture and forestry. Their effects upon plant growth and N fixation show distinct differences. Generally, if we exclude epidemics of locusts and similar phenomena, effects of grazing are spread over a longer period than those of cutting. Also, the mechanical effects of cutting equipment are smaller than those of continuous trampling by vertebrate grazers. The release of N following these methods of defoliation will be considered separately.

Cutting. In agriculture, this involves removal of the cut material for hay or silage. In forestry the slash may be left to break down (e.g. lupins in New Zealand, see Gadgil, this volume). The crop cut may be pure legume e.g. lucerne (*Medicago sativa*) or a grass-legume mixture. Simpson (32) for example grew three legumes, *Trifolium subterraneum*, *T. repens* and *Medicago sativa* in association with the grass *Dactylis glomerata*, over a three year period. Fifteen cutting treatments were made, where herbage was removed to a height of 3 cm

Table 4. Fate of some N fixed by forage legume species. Data from Simpson (32).

	Species *Trifolium subterraneum*	*Trifolium repens*	*Medicago sativa*
% N transfer to grass	20	6	3
% N transfer to soil	21	34	21
total legume N kg ha^{-1} over 3 y.	437	851	1162

above the soil. Yields of N varied considerably between species and harvest date, as did the amount of N passing to the (soil + grass). *M. sativa*, which was able to grow almost continuously yielded most N in herbage, but contributed least to the (grass + soil) (table 4). *T. subterraneum* is an annual and thus the whole plant, minus seed, can be broken down and mineralised. Under the conditions of the experiment, *T. repens* succumbed to heat and drought with consequent tissue death and N release. Again, therefore, we can relate legume death to ecosystem benefit. However, the extent of the benefit will depend upon the rate of mineralization and nitrification – a topic considered again in the next section. In table 4 it can be seen that the three legumes not only contribute different amounts of N to the associated grass, but also to the upper layer of soil (0–10 cm).

Working in a more tropical environment and with tropical grass and legumes, Johanson and Kerridge (33) found that 12–17% of the fixed N was transferred to the companion grass; soil N was also enriched, but the experimental procedures did not allow this to be quantified.

Grazing. Grazing animals such as sheep remove foliage very close to the soil surface. By virtue of their basal meristems, grass leaves can well survive this treatment, but only legumes with a suitable growth habit can survive severe, continuous grazing. The most widely studied of these is white clover, *T. repens*, which is stoloniferous. Grazing apparently exerts selection pressure on plant morphology (34) so that small leaved, more prostrate genotypes survive: these can more easily resume growth after grazing. These changes also lead to previously grazed white clover being better able to maintain N fixing activity after *cutting* than ungrazed clover (35). The ability of legumes to withstand grazing may be important for forestry. Seedling legumes regenerating after fire in *Eucalyptus* forest have been found to be heavily grazed (36).

When considering legumes for use in forests where grazing may be a problem, not only should the plants ability physically to survive grazing be considered, but also its palatability to the grazing animal. Agricultural legumes have been bred to maximise palatability, a considerable feat in some cases (e.g. lupins) because the Leguminosae are renowned for their toxic constituents (cyanogenic glycosides, alkaloids, fluoracetate, phytohaemagglutinins (lectins) etc.). Fauna often adapt to the toxins of plants of their own area, even to the extent of by-passing the Krebs cycle to avoid fluoracetate poisoning (37). When choosing plants for introduction, toxins may be put to advantage to avoid grazing problems.

Protection and shading

The use of legumes for protection has been exploited more for trees such as coffee and cocoa than for timber species, although the beneficial effect of leguminous 'nurses' have been described e.g. for sitka spruce (*Picea sitchensis*). The advantages and disadvantages of several nurse species have been discussed (30), particularly those of the common broom (*Sarothamnus scoparius*). Many species which have been used tend to smother young trees. This problem could be overcome by suitable management.

Shading, as well as protection, is a requirement for many coffee cultivars. One of the woody legume genera which has been used in Central and South America since pre-Columbian times is *Inga* (Mimosoideae) (2). In a comparison of two sites in Mexico, *I vera* was found not to be nodulated (although it nodulates elsewhere) and *I. jinicuil* to be profusely nodulated (38). A 37% higher yield of coffee was associated with the use of *I. jinicuil* as shade: nodules were active all year round. About 35 kg N ha^{-1} were estimated to be fixed each year, but, since the soil is regularly fertilized and contained 50 ppm NH_4^+, it was suggested that potential N fixation is much higher.

Shading of legumes by the associated forest crop has implications for plant survival as well as nitrogen fixation. This point has been considered in detail with respect to the use of *Lupinus arboreus* with *Pinus radiata* in New Zealand (39). In managing legumes (and probably actinorrhizas), it is important to realise how N fixation responds to shading. Dixon and Wheeler (this volume) have discussed the dependence of nodules on recent supplies of photosynthate. In most natural situations, plants are subjected to increasing amounts of shading, quite apart from seasonal and diurnal variations in the levels of incoming radiation. Basically, the shading is of two types: (a) mutual shading amongst members of the same species as the leaf area index exceeds one and (b) shading by competitor species. These effects are spread over a comparatively long time period, giving plants time to adapt to their changing environment. This is quite distinct from sudden changes in illuminance e.g. at a tropical sunset!

The response of legumes to shading is similar to that of many other species, for example an increased ratio of shoot to root growth. In all cases that have been studied plants only produce those nodules which they can maintain (See 40). Although some species, for example *Phaseolus vulgaris* need a very high level of irradiance to saturate photosynthesis (41), nevertheless they may produce very active nodules at extremely low irradiances (42). However legumes appear not to be able to survive well in the field under dense canopies and may die at canopy closure in some forests (40). However in more open forests, characteristic of relatively arid areas (e.g. large parts of Africa and Australia) associated legumes may receive a relatively large fraction of the incoming radiation. Perennial species such as *Daviesia horrida*, *Acacia* spp. may form a dense understory: their growth and N fixation is likely to be limited by factors other than light, for example water supply.

Soil binding and improvement

Apart from providing N, possible benefits of legumes are:
soil stabilization
increase in soil organic matter
suppression of weeds

All of these benefits also accrue from non-leguminous species. To what extent do legumes achieve these aims?

As far as soil stabilization is concerned, most work has been done on sand dunes and for once, in a forestry context. The shrubby legume *Lupinus arboreus* (40) and the tree *Acacia cyanophylla* (43) have both proved their effectiveness as soil stabilizers as well as improvers.

In certain citrus-growing areas such as Florida, cover crops are grown between trees at certain times of the year, for all the reasons listed above. Grasses have traditionally been used, but a recent report (44) discusses a number of legume alternatives. As a chapter of disasters as to what can go wrong if pests of legumes are not controlled the report is worth studying – nematodes, lesser corn stalk borers, grasshoppers, rabbits, rats and drought all took their toll.

In a more complex tropical situation, Sierra Leone, it has been suggested that legumes should be used in the fallow period between rice crops (45), as is done in parts of China (Li, personal communication). In Sierra Leone the areas on which rice is grown are changed every few years and land eventually reverts to trees. Scope for agro-forestry here is great. Ground cover plants are grown in association with rubber (*Hevea brasiliensis*) as a soil conservation measure. Creeping legumes such as *Pueraria phaseoloides*, *Centrosema pubescens* and *Calopogonium mucunoides* give better growth of *Hevea* than non-legumes (46).

Suppression of weeds is vital to agriculture. It is desirable to minimise weed leaf growth until the crop canopy closes. This can be done by cultivation, applying weedkillers or by growing companion crops which are competitive with weeds but not harmful to the crop species. Legumes have been used in this way, for example horse bean (*Vicia faba* ssp *equina*) is a good companion crop for grasses (47). In the case of perennial crops the companion species may yield an economic return during crop establishment.

Soil fertility

If an N fixing species is to be of maximum benefit to forest trees, release of fixed N must be related to the growth requirements of the tree species. This essentially means slow release and minimal loss by leaching. Slow release is clearly possible, as the data of Table 4 show, where not all the N lost by the legume was transferred to the companion species over a three year period. It is outside the scope of this book to discuss in detail the factors affecting mineralization and nitrification but in general the following features obtain:

1. Cultivation accelerates the processes
2. Decomposition is greatest at moderate temperatures (3)
3. Sufficient soil moisture is necessary: the effects here may be complex (48).
4. There must be a suitable C:N ratio. Leaves at fall should have a percentage nitrogen greater than 2 for rapid mineralization (49). In litter a C:N ratio in excess of 15:1 results in virtually no mineralization because all the N is transferred to the soil microflora.
5. A suitable pH is required, together with other soil factors such as suitable base exchange capacity, mineral content. For example low pH is generally inhibitory to mineralization.

In highly cultivated agricultural areas mineralization is rapid. In temperate regions a flush of mineralization often occurs in spring when the soil warms and sufficient moisture is available. Rapid nitrification is only of benefit if the nitrate can be immediately absorbed by plants, otherwise it is likely to be lost by leaching with all the concomitant pollution implications. Forest soils with their low pH and predominance of ammonium N rather than nitrate N are less prone to this leaching effect. However, when disturbed by felling and/or cultivation mineralization and nitrification may be rapid especially in the tropics (50).

How long nitrogen remains available following harvesting of a legume varies considerably. In some cases, for example white clover on peaty Scottish soils (51), most of the legume N may be immobilised for 2 years before slow release occurs. In Queensland (Australia) large areas of brigalow (*Acacia harpophylla*) have been cleared and put to agricultural use, chiefly for pasture. No nitrogen fertilizer is required for 10–12 years (Hart pers. comm.), the grass apparently being adequately nourished by mineralised N. Subsequently there is a rapid fall in productivity, which can be largely overcome by application of fertilizer N. This decline, which varies with soil type and weather (the area generally has hot wet summers and cool dry winters) is due more 'to reduced nitrogen availability than changes in N reserves' (52). This change in availability is attributed largely to the perennial grass cover, possibly associated with the semi-arid climate.

Legumes are sensitive to pH, in particular low pH. When actively fixing N they extrude protons and so contribute to the lowering of soil pH. However, there is great variability in pH tolerance amongst legumes and this should be borne in mind when selecting species for a particular area. As a very general rule, tropical species are more tolerant of low pH than temperate ones. Even so, tropical species vary considerably. Acid subsoils by virtue of their pH and mineral imbalances such as excess aluminium may restrict root growth and hence affect the ability of plants to withstand some adverse conditions, especially drought (53).

For legumes to be grown at minimum cost, they need soils low in N, otherwise their nodulation will be suppressed and they will assimilate mineral N. Because of the amount of N removed in harvesting, agricultural soils are frequently low in N prior to sowing. Forest soils, particularly in temperate areas often have considerable reserves of N much of which is unavailable for plant growth. As has been discussed above, N may become available following disturbance. A particular case where N deficiency may become especially acute is in forests, following fire, when litter is destroyed. Here, under natural conditions legumes frequently regenerate and replace lost N (17). Many are adapted so that seed germination is heat stimulated (36).

It is frequently observed that legumes nodulate better and fix more nitrogen if they are given a 'starter' dose of nitrogen (see Dixon and Wheeler, this volume, for corresponding data on *Alnus*). Thus it is common practice to apply some N to soils prior to sowing with legumes. This may not always be necessary, depending on the available soil N. By suitable management it should be possible to avoid the need for starter N. For example in areas where there is a spring flush of mineralization, concomitant sowing of legume seed would be advisable.

Temperate legumes have relatively high requirements for soil nutrients, notably P and K (e.g. 54). On the other hand, many tropical legumes, possibly because they have not been highly selected on fertile soils have lower nutrient requirements (55). The requirements for and availability of nutrients other than N should be borne in mind when choosing species for use in forestry.

Potential problems of nodule establishment and maintenance

Rhizobium inoculation. Inoculation procedures are standard in many countries such as Australia and the United States of America. In other areas, for example most of Europe, inoculation is less common. The main reason for this may be that the major legume crops have been grown for so long that large populations of compatible, effective rhizobia are present in soil. Introduction of new species may require inoculation. For example, *Rhizobium japonicum* which nodulates soybean *(Glycine max)* is absent from European and Middle Eastern soils. In other cases local rhizobia may be infective but not very effective. *Phaseolus vulgaris* is one species which seems to give nodulation problems almost every where it is grown. Thus it is always necessary to test local soil for compatible, effective rhizobia. Inoculation of newly introduced crops is comparatively straightforward, but where large numbers of poorly effective rhizobia are already in the soil, inoculated strains must be able to compete.

Nodule pests and diseases. Any pest or disease which weakens the host plant will tend to reduce nitrogen fixation. Fortunately there appear to be few diseases of nodules *per se* especially considering how highly nitritious they are! Eggs of some weevils (e.g. *Sitona*) are laid in nodules of, for example *Vicia faba*. The larvae develop at the expense of nodule tissue until they emerge, when only a shell remains. *Sitona* weevils can seriously reduce active nodule numbers on *Vicia faba* in some years (Sprent, unpublished observations). The nematode *Longidorus* sp., by invading nodules of the actinorhizal plant *Hippophae rhamnoides* enhances the decline of this plant during natural succession on sand dunes in the Netherlands (56).

Microbial interactions. Other soil microorganisms can interact positively or negatively with rhizobia. Survival of rhizobia in soil may be affected by pathogens such as *Bdellovibrio* (57) as well as protozoans especially ciliates (55). Treatment of legume seed with fungicides may have the unexpected benefit of reducing the numbers of such protozoans.

One of the most interesting microbial interactions described recently has been the possible role of 'helper' bacteria in the formation of actinorhiza nodules (59). It now appears that a similar phenomenon may obtain in legumes. The species *Glycine max*, *Vigna unguiculata* and *Trifolium repens* were found to develop more nodules in the presence of *Azotobacter vinelandii* than in its absence. This effect was not related to nitrogen fixation since it was manifest uning either nif^{+} or nif^{-} *Azotobacter* strains (60).

AGROFORESTRY SYSTEMS

This term is taken to cover all cases where trees are included in an agricultural system involving intercropping or crop rotations. 'Agrisilviculture' and 'agriforestry' are similarly used. The practice is most common in tropical areas and when one of the components (either tree or crop) is a legume a considerable benefit from nitrogen fixation may accrue. So much work is currently being undertaken to evaluate and exploit these long-established systems that caution against jumping to the 'agriforestry band wagon' has been proposed (61)! A number of the examples quoted in previous sections may be considered to belong to agroforestry.

Direct use of nitrogen fixing tree or shrub species

Proven examples of use of legumes are limited. In an agroforestry context the tree generally supplies more than timber – for example edible fruits or seeds, leaf meal for stock feed, drugs etc. In a project sponsored by the World Bank, tree farming at the smallholder level is being encouraged: *Albizia* a nodulated arboreal member of the Mimosideae is one of the trees being farmed for pulp (61).

In Nigeria and other parts of Africa and Asia agricultural crops are grown on areas which have previously been forest. Every few years the land is left to 'bush fallow', involving either natural regeneration or planting with small trees or woody shrubs to restore soil fertility. Some of these are legumes and may have a use as animal or human food, firewood or stakes. Of the species listed as being favoured by farmers (62) not all are legumes and not all the legumes are nodulated. How these restore soil fertility is an interesting question which should be investigated. One commonly used nodulated legume is *Leucaena leucocephala*. In an attempt to improve land utilization a system of 'alley cropping' has been proposed as alternative to the bush fallow system. Experiments have been conducted using rows of *Leucaena* interplanted each season (two per year) with maize (*Zea mays*). The competitive and shading effects of the *Leucaena* can be adjusted by varying planting distance and by pruning. If prunings are left on the ground they can provide soil nutrients. By this method quite good yields of maize can be obtained without addition of N fertilizer and the advantages of *Leucaena* for green manure, firewood and stakes obtained simultaneously. *Leucaena* is also acceptable to grazing animals. After pruning, it is likely that some nodule decay occurs, releasing N into the soil (similar to the defoliation effects described earlier) although there is no direct evidence for this.

It is evident that there is a lot to learn from 'traditional' farming, especially on marginal land. Recent work in the United States of America and in Australia has rediscovered plants used in arid regions in particular. A number of these have potential for both food and forest products (including fuel). Most extensive is the work of Felker and colleagues in Texas. Of various leguminous trees investigated, species of *Prosopis* (a nodulated member of the Mimosoideae) were thought to be close to the ideal (63). This obliging genus is salt, drought, cold and heat resistant, produces nutritious pods, good honey (with the aid of bees!) and for some species, excellent timber (2). There is considerable scope for improvement of its growth and nitrogen fixing potential (64, 65). If its potential can be realised, *Prosopis* would appear to hold great hope for many areas whose productivity is at present very low. Other tree legumes, including species of *Acacia* may be suitable for certain regions. The review of Felker (66) and the references cited therein should be consulted for further details.

The potential of tree legumes for harvest by coppicing has been neglected.

Acacia harpophylla is an example of a species which will produce dense thickets following felling of the main trunk (Sprent, personal observations). This could lead to release of nitrogen from existing nodules followed by formation of new nodules during regrowth – a situation parallel to the effects of cutting forage legumes.

Plants as fuel usually implies burning of wood. However there has been much interest recently in fermenting carbohydrates to produce ethanol as a substitute for mineral oil. With this possibility in mind Saxon (67) examined many tuberous legumes of Australia which have been used by the Aborigines as source of carbohydrate food. It was concluded that these could be a useful source of locally produced fuel. Most of the species used are herbaceous and could be grazed prior to harvest of tubers. Examination of the data on tuber yield and protein content suggests that in some species there could be about 40 kg N ha^{-1} in tubers alone. It would be interesting to know if, with suitable management, this N could be released into soil for use by other plants, including tree species.

Balance between nodules and mycorrhizas

This section will consider briefly nodules and mycorrhizas as alternatives for host plant nitrogen nutrition. Mycorrhizas do not fix nitrogen, in spite of a few early reports that they do. However, they are apparently excellent scavengers for soil nitrogen. Certain species with ectotrophic mycorrhizas ranked with nodulated legumes in their ability to colonise the nitrogen-poor anthracite wastes in Pennsylvania (68). In areas where nitrification is normally very low, for example, in certain climax communities, mycorrhizas may be especially important (69). Like nodulation, the formation of mycorrhizas may be depressed by the addition of combined nitrogen (70). It has been argued that at least some of the Caesalpinioideae use ectomycorrhizas as an alternative to nodules (71). Far more data of a physiological as well as descriptive nature are required before this generalization can be substantiated. Also the existence of genera such as *Inga* which may have both ecto- and endomycorrhizas as well as nodules suggests complementation of effects. There are certainly many situations where plants seem to manage remarkably well in soils low in combined N without fixing nitrogen. In the Sonora and Colorado deserts various leguminous trees predominate but one genus *Cercidium* (Caesalpinioidieae) is not known to nodulate. Felker and Clark (69) suggest that perhaps there are primitive, as yet undetected nodule on *Cercidium* because they find it hard to believe 'that it can dominate vast infertile regions without fixing nitrogen'. However, alternative strategies including mycorrhizal scavenging may obtain. These possibilities should be borne in mind when considering agroforestry practices involving, in particular, leguminous trees.

CONCLUSIONS

In a forest context legumes have been rather neglected. This in part relates to the fact that reafforestation has, until recently, largely been confined to temperate areas. With increased interest in conserving and replacing tropical rainforest and in the management of arid and semi arid areas, legumes may prove to be a very important group of plants. Agricultural and horticultural experience is also largely based on temperate areas. Useful pointers can be obtained from this experience, as described in the foregoing sections. However, we are to a large extent ignorant about those legumes which are likely to be the most useful on a world scale. Before attempting to exploit them we should ask many questions, including:

1. Do we need nodulated plants?
2. Is inoculation with rhizobia necessary?
3. Can they be used directly? If so how may nitrogen fixation be optimised? Short term rotations will probably be required.
4. If they are to be used indirectly (a) how may fixed N be released to maximise benefit to the associated crop (b) how can the competitive effects be controlled?
5. Can they grow and fix nitrogen adequately on the soils and under the conditions being considered?

REFERENCES

1. Polhill RM, Raven PH: *Advances in Legume Systematics*, Kew, Royal Botanic Gardens, 1981.
2. Allen ON, Allen EK: *The Leguminosae*, Madison, The University of Wisconsin Press, 1981.
3. Hutchinson J: *The Genera of Flowering Plants – Dicotyledons*, Oxford, Clarendon Press, 1964.
4. Cowan RS: Caesalpinioideae. In *Advances in Legume Systematics*, Polhill RM, Raven PH (eds) Kew, Royal Botanic Gardens, 1981, p 57–64.
5. Irwin HS, Baineby RC: Cassieae. In *Advances in Legume Systematics*, Polhill RM, Raven PH (eds), Kew, Royal Botanic Gardens, 1981, p 97–106.
6. Gutschick VP: Evolved strategies in nitrogen acquisition by plants. *Amer Nat* 118(5): 607–637, 1981.
7. Sprent JI: *The Biology of Nitrogen-fixing Organisms*, Maidenhead, McGraw-Hill Book Company (UK), 1979.
8. Sprent JI: Root nodule anatomy, type of export product and evolutionary origin in some Leguminosae. *Plant, Cell and Environment* 3: 35–43, 1980.
9. Trinick MJ: Relationships amongst the fast-growing rhizobia of *Lablab purpureus*, *Leucaena Leucocephala*, *Mimosa* spp., *Acacia farnesiana* and *Sesbania grandiflora* and their affinities with other rhizobial groups. *J appl Bact* 49: 39–53, 1980.
10. Sprent JI: Adaptive variation in legume nodule physiology resulting from host-rhizobial interactions. In *Nitrogen as an ecological factor*, McNeil S, Rorison IH, Lee JA, Gibson C: Oxford, Blackwells, 1983, 29–42.
11. Glenn AR, Dilworth MJ: The uptake and hydrolysis of disaccharides by fast and slow-growing species of *Rhizobium*, *Arch Microbiol* 129: 233–239, 1981.

12. Stowers MD, Elkan GH: An inducible iron-containing superoxide dismutase in *Rhizobium japonicum*, *Can J Microbiol* 27: 1202–1208, 1981.
13. Puppo A., Rigaud J., Job D: Role of superoxide anion in leghaemoglobin oxidation. Pl Sci Lett 22: 353–360, 1981.
14. Jing Y, Paau AS, Brill WJ: Leghaemoglobins from alfalfa (*Medicago sativa* L. vernal) root nodules I. Purification and in vitro synthesis of five leghaemoglobin components. Pl Sci Lett 25: 119–132, 1982.
15. Beringer JE, Brewin NJ, Johnston AWB: The genetic analysis of *Rhizobium* in relation to symbiotic nitrogen fixation. *Heredity* 45(2): 161–186, 1980.
16. Sprent JI: Functional evolution in some Papilionoid root nodules. In: *Advances in Legume Systematics*, Polhill RM, Raven PH (eds), Kew, Royal Botanic Gardens, 1981, p 671–676.
17. Hingston FJ, Malajczuk N, Grove TS: Acetylene reduction (N_2-fixation) by jarrah forest legumes following fire and phosphate application. *J appl Ecol* 19, 631–645, 1982.
18. Spratt ER: A comparative account of the root-nodules of the Leguminosae. *Ann Bot* 33: 189–199, 1919.
19. Pankhurst CE, Sprent JI: Surface features of soybean root nodules. *Protoplasma* 85: 85–98, 1975.
20. Ham GE, Lawn RJ, Brun WA: Influence of inoculation, nitrogen fertilizers and photosynthetic source-sink manipulations on field-grown soybeans. In *Symbiotic nitrogen fixation in plants*, Nutman PS (ed), Cambridge University Press, 1976, p 239–253.
21. Sprent JI: Nitrogen fixation by legumes subjected to water and light stresses. In *Symbiotic nitrogen fixation in plants*, Nutman PS (ed), Cambridge University Press, 1976, p 405–420.
22. Sprent JI, Silvester WB: Nitrogen fixation by *Lupinus arboreus* grown in the open and under different aged stands of *Pinus radiata*. *New Phytol* 72: 991–1003, 1973.
23. Song SD, Monsi M: Studies on the nitrogen and dry-matter economy of a *Lespedeza bicolor* var. *japonica* community. *J. Fac. Sci. Univ. Tokyo* section III, 9 (8–9): 283–332, 1974.
24. Sprent JI, Scott R, Perry KM: The nitrogen economy of *Myrica gale* in the field. *J Ecol* 66: 409–420, 1978.
25. Nwoboshi LC: Nitrogen cycling in a teak plantation ecosystem in Nigeria. In *Nitrogen Cycling in West African Ecosystems*, Rosswall T (ed) Stockholm, SCOPE/UNEP, 1980, p 353–361.
26. Sprent JI, Bradford AM: Nitrogen fixation in field beans (*Vicia faba*) as affected by population density, shading and its relation with soil moisture. *J agric Sci Camb* 88: 303–310, 1977.
27. Pate JS: Pea. In *Crop Physiology* Evans LT (ed), Cambridge University Press, 1975, p 191–224.
28. Sprent JI: Nitrogen fixation by grain legumes in the U.K. *Phil Trans Roy Soc Ser B* 296(1982): 387–295.
29. Nnadi LA: Nitrogen economy in selected farming systems of the savannah region. In *Nitrogen cycling in West African Ecosystems*, Rosswall T (ed) Stockholm, SCOPE/UNEP, 1980, 345–351.
30. Nimmo M, Weatherell J: Experiences with Leguminous nurses in forestry. *Rep on Forest Research*, 1961, 126–147.
31. Papendick RI, Sanchez PA, Triplett GB: (eds) *Multiple cropping*. Am Soc Agron Spec Pub. No 27. Madison, Wisconsin 1976.
32. Simpson JR: Transfer of nitrogen from three pasture legumes under periodic defoliation in a field environment. *Aust J exp Agric Anim Husb* 16: 863–869, 1981.
33. Johansen C, Kerridge PC: Nitrogen fixation and transfer in tropical legume-grass swamps in South Eastern Queensland. *Trop Grasslands* 13(3): 165–170, 1979.
34. King J: Ecotypic differentiation in *Trifolium repens*. *Pl Soil* 18(2): 221–224, 1963.
35. Haystead A, Sprent JI: Symbiotic nitrogen fixation. In *Physiological processes limiting plant productivity*. Johnson CB (ed) 1981, 345–364.
36. Shea SR, McCormick J, Portlock CC: The effect of fires on regeneration of leguminous species in the northern jarrah (*Eucalyptus marginata* Sm) Forest of Western Australia. *Aust J Ecol* 4: 195–205, 1979.
37. Mead RJ, Oliver AJ, King DR: Metabolism and defluorination of fluoroacetate in the brush-tailed possum (*Trichosurus vulpecula*). *Aust J Biol Sci* 32: 15–26, 1979.
38. Roskowski JP: Nodulation and N_2-fixation by *Inga jinicuil*, a woody legume in coffee planta-

tions. I. Measurements of nodule biomass and field C_2H_2 reduction rates. *Pl Soil* 59: 201–206, 1981.

39. Sprent JI, Silvester WB: Nitrogen fixation by *Lupinus arboreus* grown in the open and under different aged stands of *Pinus radiata*. *New Phytol* 72: 991–1003, 1973.
40. Sprent JI: Growth and nitrogen fixation in *Lupinus arboreus* as affected by shading and water supply. *New Phytol* 72: 1005–1022, 1973.
41. Sale PJM: Productivity of vegetable crops in a region of high solar impact. IV. Field chamber measurements on french beans (*Phaseolus vulgaris* L.) and cabbage (*Brassica oleracea* L.). *Aust J Pl Physiol* 2(4): 461–470, 1979.
42. Antoniw LD, Sprent JI: Growth and nitrogen fixation of *Phaseolus vulgaris* L. at two irradiances II Nitrogen fixation. *Ann Bot* 42: 399–410, 1978.
43. Nakos G: Acetylene reduction (N_2-fixation) by nodules of *Acacia cyanophylla*. *Soil Biol Biochem* 9: 131–133, 1979.
44. Anderson CA: Legume cover crop trials in citrus groves. *Soil Crop Sci Soc Fla Proc* 39: 80–82, 1980.
45. Nyoka GC: Weed control and nitrogen restoration with legumes in upland rice culture in Sierra Leone. In *Nitrogen cycling in West African Ecosystems*, Rosswall T (ed) Stockholm SCOPE/UNEP p 261–267, 1980.
46. Watson GA, Wong Phui Weng, Narayan R: Effect of cover plants on soil nutrient status and on growth of *Hevea* III. A comparison of leguminous creepers with grasses and with *Mikania cordata*. *J Rubb Res Inst* Malaya 18: 80–95, 1964.
47. Robinson RG: Pulse or grain legume crops for Minnesota. *Bull Agric Exp Sta Univ Minnesota* 513: 19 pp, 1975.
48. Robinson JBD: The critical relationship between soil moisture content in the region of wilting point and the mineralization of natural soil nitrogen. *J agric Sci Camb* 49: 100–105, 1957.
49. Silvester WB: Dinitrogen fixation by plant associations excluding legumes. In *A treatise on dinitrogen fixation* IV. Hardy RWF, Gibson AH (eds) New York Wiley Interscience 141–190, 1977.
50. Witousek PM: Nitrogen losses from disturbed ecosystems – ecological considerations. In *Nitrogen cycling in West African Ecosystems*, Rosswall T (ed) Stockholm SCOPE/UNEP 39–53, 1980.
51. Haystead A, Marriot C: Transfer of legume nitrogen to associated grass. *Soil Biol Biochem* 11: 99–104, 1979.
52. Graham TWG, Webb AA, Waring SA: Soil nitrogen status and pasture productivity after clearing of brigalow. (*Acacia harpophylla*). *Aust J exp Agric Anim Husb* 21: 109–118, 1981.
53. Pinkerton A, Simpson JR: Effects of sub-soil acidity on the shoot and root growth of some tropical and temperate forage legumes. *Aust J agric Res* 32: 453–463, 1981.
54. Duke SH, Collins M, Soberallske RM: Effects of potassium fertilization on nitrogen fixation and nodule enzymes of nitrogen fixation in alfalfa. *Crop Sci* 20(2): 213–219, 1980.
55. Henzell EF, 't Mannetje L: Grassland and forage research in tropical and subtropical climates. In *Perspectives in world agriculture*. Common wealth Agricultural Bureaux, Slough, 485–532, 1980.
56. Oremus PAI, Otten H: Factors affecting growth and nodulation of *Hippophae rhamnoides* ssp. rhamnoides in soils from two successional stages of dune formation. Pl Soil 63: 317–331, 1981.
57. Keyo SO, Alexander M: Factors affecting growth of *Bdellovibrio on Rhizobium*. *Arch Microbiol* 103: 37–43, 1975.
58. Lennox LB, Alexander M: Fungicide enhancement of nitrogen fixation and colonisation of *Phaseolus vulgaris* by *Rhizobium phaseoli*. *Appl Environ Microbiol* 41(2): 404–411, 1981.
59. Knowlton S, Berry A, Torrey JG: Evidence that associated soil bacteria may influence root hair infection of actinorrhizal plants by *Frankia*. *Can J Micro* 26(8): 971–977, 1980.
60. Burns TA Jr, Bishop PE, Israel DW: Enhanced nodulation of leguminous plant roots by mixed cultures of *Azotobacter vinelandii* and *Rhizobium*, *Pl Soil* 62: 399–412, 1981.
61. Spears JS: Can farming and forestry co-exist in the tropics? *Unasylva* 32: 2–12, 1980.
62. Kang BT, Wilson GF, Sipkino L: Alley cropping maize (*Zea mays* L.) and Leucaena (*Leucaena leucocephala* Lam) in Southern Nigeria. *Pl Soil* 63: 165–179, 1981.

63. Felker P, Bandurski RS: Uses and potential uses of leguminous trees for minimal energy input agriculture. *Econ Bot* 32(2): 172–184, 1979.
64. Felker P, Clark PR: Nitrogen fixation (acetylene reduction) and cross inoculation in 12 *Prosopis* (mesquite) species. *Pl Soil* 57: 177–186, 1980.
65. Felker P, Cannell GH, Clark PR: Variation in growth among 13 *Prosopis* (mesquite) species. *Expl. Agric.* 17(2): 209–218, 1981.
66. Felker P: Uses of tree legumes in semi arid areas. *Econ Bot* 35(2): 174–186, 1981.
67. Saxon EC: Tuberous legumes: preliminary evaluation of tropical Australian and introduced species as fuel crops. *Econ Bot* 35(2): 163–173, 1981.
68. Schramm JR: Plant colonization studies on black wastes from anthracite mining in Pennsylvania. *Trans Am phil Soc* 56: 1–194, 1966.
69. Raven JA, Smith SE, Smith FA: Ammonium assimilation and the role of mycorrhizas in climax communities in Scotland. *Trans Bot Soc Edinb* 43: 27–35, 1978.
70. Chambers CA, Smith SE, Smith FA: Effects of ammonium and nitrate ions on mycorrhizal infection, nodulation and growth of *Trifolium subterraneum*. *New Phytol* 85: 47–62, 1980.
71. Malloch DW, Pirozynski KA, Raven PH: Ecological and evolutionary significance of mycorrhizal symbioses in vascular plants (a review). *Proc Natl Acad Sci USA* 77(4): 2113–2118, 1980.
72. Felker P, Clark PR: Nodulation and nitrogen fixation (acetylene reduction) in desert ironwood (*Olneya tesota*). *Oecologia* (Berl.) 48: 292–293, 1981.

8. Nitrogen fixing plants in forest plantation management

NIGEL D, TURVEY*, PHILIP J, SMETHURST

A.P.M. Forests Pty. Ltd., P.O. Box 37, Morwell, Victoria 3840, Australia.

* Current address: School of Agriculture and Forestry, University of Melbourne, Creswick, Victoria 3363, Australia.

INTRODUCTION

There is a wide range of nitrogen fixing plants that have been used in forestry with the presumed objective of raising soil nitrogen levels and subsequently improving the growth of the non-nitrogen-fixing forest species. The nitrogen fixing plants that have been tried in managed forest systems range from trees to herbs and shrubs which fix nitrogen through root associations with either *Rhizo bium* strains in the case of legumes, or an actinomycete of the genus *Frankia* in the case of non legumes.

Apart from the objective of fixing nitrogen, there have been few objectives common to the management of nitrogen fixing plants in managed forests. In some cases the nitrogen fixing plant may have an intrinsic economic value as timber or fodder, and hence be managed and harvested as intensively as the forest crop either through rotational cropping or as plant mixtures. In other instances the nitrogen fixing plants may have no value other than for their ability to fix nitrogen, and hence may be ploughed into the soil or left as a mixture with the forest to fix nitrogen through the forest rotation.

Occasionally other objectives have attained primary importance in some forest situations e.g. the use of legumes in improving the quality and quantity of game for hunting (1), the use of fire-germinated legumes in the reduction of the soil borne *Phytophthora cinnamomi* plant pathogen (2) and the use of a legume cover in erosion control.

Common names of plants have been used in this chapter; botanical synonyms are listed in Table 1.

NITROGEN FIXING TREES

Alders

Whilst the value of alders as nitrogen fixing trees has been demonstrated world-wide (3), a great deal of work has been done on the use of red alder in the forests of the Pacific Northwest of the U.S.A. Here, DeBell and Radwan (4) assessed growth and nitrogen status in pure and mixed plantings of red alder and black

Gordon, J.C. and Wheeler, C.T. (eds.). Biological nitrogen fixation in forest ecosystems: foundations and applications

ISBN 90-247-2849-5.
Printed in The Netherlands

Table 1. Botanical names of species mentioned.

Common name	Botanical name
Alfalfa	*Medicago sativa*
Alaskan lupin	*Lupinus nootkatensis*
Arrowleaf clover	*Trifolium vesiculosum*
Bitter blue lupin	*Lupinus angustifolius*
Birdsfoot trefoil	*Lotus corniculatus*
Black alder	*Alnus glutinosa*
Black cherry	*Prunus serotina*
Black cottonwood	*Populus trichocarpa*
Black locust	*Robinia pseudoacacia*
Black walnut	*Juglans nigra.*
Bokhara clover	*Melilotus albus.*
Caribbean Pine	*Pinus caribaea*
Catalpa	*Catalpa spp.*
Caucasian clover	*Trifolium ambiguum*
Coriaria	*Coriaria arborea*
Crimson clover	*Trifolium incarnatum*
Crown vetch	*Cornilla varia*
Douglas-fir	*Pseudotsuga menziesii*
Eastern cottonwood	*Populus deltoides*
European gorse	*Ulex europaeus.*
Field pea	*Pisum arvense*
Flat pea	*Lathyrus sylvestris*
Golden prickly wattle	*Acacia brownii*
Hairy vetch	*Vicia villosa*
Honey locust	*Gleditsia triacanthos*
Hoop pine	*Araucaria cunninghamii*
Kauri pine	*Agathis robusta*
Kudzu	*Pueraria spp.*
Larch	*Larix spp.*
Lespedeza	*Lespedeza spp.*
Loblolly pine	*Pinus taeda*
Lotononis	*Lotononis bainesii*
Osage orange	*Maclura pomifera*
Pines	*Pinus spp.*
Poplars	*Populus spp.*
Radiata pine	*Pinus radiata*
Red alder	*Alnus rubra.*
Red oak	*Quercus rubra*
Rose clover	*Trifolium hirtum*
Scots pine	*Pinus sylvestris.*
Sericea lespedeza	*Lespedeza cuneata*
Silverleaf desmodium	*Desmodium unicatum*
Silver maple	*Acer saccharinum*
Siratro	*Phaseolus atropurpureus.*
Slash pine	*Pinus elliotii*

Table 1. (continued).

Common name	*Botanical name*
Snow brush	*Ceanothus velutinus.*
Spruce	*Picea spp.*
Subterranean clover	*Trifolium subterraneum*
Sweet gum	*Liquidambar syraciflua*
Sycamore	*Platanus occidentalis.*
Vetches	*Vicia spp.*
Washington blue lupin	*Lupinus polyphyllus.*
Wax myrtle	*Myrica cerifera*
Wetland deervetch	*Lotus uliginosus*
White clover	*Trifolium repens*
Yellow lupin	*Lupinus luteus*
Yellow poplar	*Liriodendron tulipifera*
Yellow tree lupin	*Lupinus arboreus.*

cottonwood. Planting was at 0.6 × 1.2 m spacing and coppicing was induced at 3 years after planting. Two years later coppice dry matter production and nitrogen content of black cottonwood was greater in mixed stands with alder rather than pure stands. Red alder also enhanced soil nitrogen levels. Mixed or pure plantings of red alder were considered to be beneficial in such short-rotation forest systems. Miller and Murray (5) reported the longterm effects of interplanting 4 year old Douglas fir with red alder in northwestern U.S.A. Red alder increased merchantable yields of the associated dominant Douglas-fir. Improved growth began at about 30 years from seed when the Douglas-fir emerged through the alder canopy. Control of stand density at an early age was necessary to maintain both Douglas-fir and alder in a dominant or codominant position. They considered from 50 to 100 uniformly distributed red alder per hectare should be retained to provide adequate nitrogen but not seriously reduce Douglas-fir growing stock.

Whilst Douglas-fir is a valuable tree crop, red alder has some economic value for both pulping and sawn timber. Management of such mixed stands then has a strong economic rationale based on stumpage values of both species. The management options involving crop rotations, lengths of rotations, crop mixtures, and inter crop competition and their economic considerations will be examined later in this chapter.

Haines *et al.* (6) reviewed the role of nitrogen fixing plants in forestry in the southeastern U.S.A. They reported several studies pertinent to plantation management of nitrogen fixing plants (7, 8, 9, 10, 11, 12). Plass (7) working on strip mine spoil in Kentucky, U.S.A. showed 10 year old growth of 5 hardwoods and 5 conifers to be improved when associated with a simultaneously planted nurse crop of black alder. Only loblolly pine and eastern cottonwood were overtopped by alders at the 2.1 m^2 spacing, resulting in decreased survival. Kellison and White (13) reported satisfactory performance of several provenances of black alder for growth in pure stands and mixtures with eastern cottonwood for wood production and soil amelioration.

Black locust

When conifers and the nitrogen fixer black locust were planted simultaneously in the central U.S.A. the latter overtopped and led to failure of the plantation (8); however, it was suggested that understorey plantings of black locust in native pine plantations or mixtures with rapidly growing hardwoods may be useful. Interplanting of black locust led to superior growth of several planted hardwood species in a study by Finn (9) where yellow poplar and black cherry benefited most through their association with black locust, but several other hardwoods also benefited from the improved soil nitrogen levels.

Ferguson (10) observed that catalpa growing with black locust had larger diameter and height growth than pure catalpa stands growing on the same site. Similarly McIntyre and Jefferies (11) found that catalpa growing within 2 m of black locust were 126% taller and 120% larger in diameter than those growing 15 m away. Ashby and Baker (14), working in a strip-mined area, Illinois, U.S.A., found that underplanted black locust improved height of black walnut, yellow poplar, silver maple and osage orange. Pure black locust stands have also benefited subsequent tree crops. Carmean *et al.* (12) reported that red oak, sweetgum, yellow poplar and black walnut planted on a previous 23 year old black locust site were larger at 16 years than those on previous pine or herbaceous vegetation sites. The growth benefits were attributed to better soil structure and increased foliar nitrogen levels. In the northeast U.S.A. Mellilo and Aber (15) reported studies that had been initiated on interplantings of black locust, black alder and honey locust with trees of high commercial value. As well as acting as a nurse crop they considered the nitrogen fixers would improve wood quality of associated species and provide fibre for paper production.

NITROGEN FIXING HERBS AND SHRUBS

The timing of establishment of a nitrogen fixing species in relation to the plantation tree species can be divided into three categories: prior plantings whereby the nitrogen-fixer is established greater than 1 year prior to plantation establishment; simultaneous plantings whereby the nitrogen fixer and plantation are established within 1 year of each other; and underplantings whereby the nitrogen-fixer is established greater than 1 year after the plantation.

Prior and simultaneous plantings

The use of yellow tree lupin has enabled the establishment of radiata pine on nitrogen-poor sand dunes in New Zealand. Lupins have been allowed to grow for about 3 years before being crushed, or sprayed with herbicides, to allow pine establishment. The lupins have regenerated and persisted until canopy closure, and have also regenerated at each thinning. Nitrogen fixed by the lupins is the major source of nitrogen for the site (16). A similar system of green manuring of yellow lupin is also recommended for some humus poor sandy soils in the USSR (17). The system is treated in greater detail in Chapter 12.

An important consideration in planting legumes prior to the tree crop is the opportunity cost of leaving the land vacant of commercial tree species while the nitrogen-fixer is grown. This cost, which has to be debited to the value of the nitrogen fixed, is avoided in the case of simultaneous or underplantings.

Miller and Zalunardo (18) reported the screening of eight legume cultivars for suitability in Douglas-fir forests of northwestern U.S.A. Legume seed was sown on fertilised and unfertilised logged and burnt clearfallen areas. Percent cover was greater for alfalfa, crown vetch and birdsfoot trefoil than for wetland deervetch, Washington blue lupin, bitter blue lupin, flat pea or hairy vetch. Fertilisation with nitrogen, phosphorus, potassium, and sulphur improved first year cover but neither subsequent plant cover nor height for the first three species mentioned. No effects on tree growth were measured, but the authors suggested that climbing species should be introduced at the pole stage of plantation development to avoid undesirable competition effects.

The use of subterranean clover and arrowleaf clover in loblolly pine establishment has been examined in southeastern U.S.A. (19). The plantation was fertilised (28 kg $N \cdot ha^{-1}$, 56 kg $P \cdot ha^{-1}$) and sown with inoculated clover seed at the time of establishment. After 4 years the clover x fertilisation treatments had greater dry weights of tree components and there was a greater proportion of the nitrogen in the ecosystem in pine plant tissue than in the controls, fertiliser only, or clover only treatments.

Jorgensen (20) screened over 70 legumes for their adaptablity, fertility requirements, influence on the site and growth of planted trees on a wide range of sites in the southeast U.S.A. Most were unsuitable but subterranean clovers, summer annual and perennial lespedezas were best adapted to the competition, low degree of site preparation, and adverse soil conditions of the typical forest planting. Legumes responded to phosphorus fertilisation, with the form of phosphorus being important on some sites. Legumes increased nitrogen in the site and the nitrogen content of sweetgum and loblolly pine foliage.

Youngberg *et al.* (21) reported that snowbrush stands that regenerated after logging and burning of Douglas-fir improved the total ecosystem nitrogen capital and improved survival, foliar nitrogen concentrations, and height growth of the Douglas-fir seedlings. Improved management techniques for these mixed plantings are now being considered. This includes establishment of snowbrush from seed in the absence of slash burning and the use of herbicides to release the trees from moisture and nutrient competition at age 7 to 10 years.

In northeastern U.S.A. white clover has been introduced into forest plantings via seed sown directly into soil in containerised tree seedlings (15). In Wisconsin, U.S.A. six legumes have been tested as nurse crops for poplar plantations (22). Crown vetch and birdsfoot trefoil were the best nurse crops, whereas hairy vetch and field pea outgrew the poplar and forced them to the ground. Longer poplar planting stock may overcome this problem.

The use of *Elaeagnus umbellata* in widely-spaced black walnut plantations in the midwest U.S.A., has lead to greater height and diameter crowth of the black walnut in mixed rather than pure stands (23). Haines *et al.* (6) reported several trials established to screen a number of herbaceous legumes for adaptablility to forest soils in southeastern U.S.A. In one trial four clovers were sown 5 months prior to planting of loblolly pine seedlings. After one growing season there were no clearly beneficial effects of the legumes on pine growth but competition for light on some fertilised clover plots lead to decreased pine survival.

Land (24) described the use of a grass-legume nurse crop for mixed forests in West Germany. In East Germany Thum (25, 26) reported the use of Washington blue lupin and bokhara clover as nurse crops to poplars. Tree heights at 5 years were; control 2.8 m, with lupins 3.3 m, with clover 3.9 m and with calcium ammonium nitrate fertilisation 4.5 m. Sowing of clover here also increases tree mortality and was recommended to be sown 1–2 years after tree planting.

In Iceland, Arnalds (27) reported excellent growth of larch when planted with Alaskan lupin.

When legume and tree seed were sown with phosphorus fertilisation in New Zealand on badly eroded mountain sites, survival and growth rates of spruce and pine were increased (28). The most useful legume species on these sites were white clover, lupin, lotus species, crown vetch, and caucasian clover. *Coriaria*, a native of New Zealand, has also been linked to improved growth of radiata pine in that country, but as yet the value of managing this species for the benefit of nitrogen fixation has not been assessed (29).

The use of subterranean clover pasture as an understorey to radiata pine plantations has been examined in Western Australia. Hatch (30) reported a rapid increase in the soil nitrogen status over 5 years when clover (with annual phosphorus, copper and zinc fertilisation) was established with pines at a wide spacing. This treatment provided the pines with adequate nitrogen for satisfactory growth.

In southeastern Australia a mixture containing ten varieties of clover sown at the time of establishment of radiata pine and fertilised with phosphorus produced greater height growth in pines at age 2 years than the nitrogen fertilisation treatment although there was growth depression of the pines by the clovers in the first growing season (31). Richards and Bevege (32) found that in northeastern Australia silverleaf desmodium, and siratro stimulated growth of the native hoop pine and Kauri pine, but depressed that of the exotic slash pine, loblolly pine and caribbean pine relative to tree growth with annual phosphorus and potassium fertilisation alone. It is known that exotic pines on the soils used in this study respond less to nitrogen fertilisation at establishment than do the native hoop and Kauri pines.

Nambiar and Nethercott (33) in South Australia, found that establishment of bitter blue lupin with radiata pine was possible on first and second rotation sites and that the foliar nitrogen concentration of pines 3 months after lupin maturity was greater in lupin treatment than in control and nitrogen fertilisation treatments. This lead to a corresponding increase in height 22 months after planting.

In southeastern Australia, we have investigated the use of a legume understorey in radiata pine plantations (34). We screened fifteen legumes (seven native and eight introduced) for their ability to grow and reproduce in a range of forest soils. Whilst almost all native legumes germinated poorly, the majority of introduced legumes germinated and grew on the cleared forest soil. Of these, subterranean clover, bitter blue lupin and yellow tree lupin have been selected for further trials with radiata pine.

Underplantings

Planting of a legume crop into an established forest plantation, or underplanting, has been carried out in many instances with Washington blue lupin to improve nitrogen nutrition and tree growth. In the USSR Zhilkin *et al.* (35) reported that 5 years of lupins in a spruce forest lead to improved soil fertility and nitrogen, calcium and potassium nutrition of the trees. Similarily Zhilkin and Rikhter (36) found the cultivation of Washington blue lupin in a 90 year old Scots pine stand resulted in more rapid decomposition of the litter and liberation of nitrogen and other elements into the biological cycle. Vorontsov (37) reported that underplanting of Washington blue lupin increased soil nitrogen status and resistance of pines to insect attack.

Lakhtanova and Beregova (38) showed that Washington blue lupin introduced into a 33 year old stand of Scots pine in the USSR increased profitability of forest production at 40 years by 19%. They also reported that lupins introduced to predominantly pine and pine/larch plantations 1–3 years after tree planting increased tree volume to 122–169% of stands with no lupins (39).

In West Germany, Rehfuess (40) reported the use of Washington blue lupin as an underplanting in Scots pine forests over the last 100 years. Lupin establishment had required extensive tilling and calcium and phophorus fertilisation. This procedure caused root damage to pines, premature shedding of needles, decreased volume increment of 5–10% during the first 3 or 4 growing seasons, and increased risk of windthrow. Eventually nitrogen nutrition and growth of pines has been improved substantially but only after a lag period of about 5–7 years. Melzer and Hertel (41) reported similar beneficial effects of this lupin on other tree species in East Germany.

In the U.S.A. Haines *et al.* (42) found tree growth in a 2 year old sycamore plantation increased over the following 4 years after three clovers and two vetches were grown with the trees. Of the legumes tested subterranean and crimson clovers were of greatest benefit.

Underplanting of a year old loblolly pine plantation in Southeastern U.S.A. with sericia lespedeza resulted in a superior growth rate and nitrogen nutrition status of the trees after six years, than trees that had received 112 kg of nitrogen per ha instead of the lespedeza (43).

Underplanting of a 13 year old radiata pine stand thinned to either 143 or 261 stems per ha with subterranean clover in Western Australia was reported by Batini and Anderson (44). An annual application of phosphorus, copper and zinc fertiliser was used for the clover. After 5 approximately years an increase in soil nitrogen status was observed and tree crowns had doubled in size. The trial indicated that the combination of agriculture and forestry was sustainable for six years under widely spaced 13–18 year old pine. After 5 years thinning was necessary to maintain pasture production. Woods (45) in South Australia was able to establish a subterranean clover sward with phosphorus, potassium and trace element fertilisation under 10 year old burnt radiata pine. Over the following 12 months the soil nitrogen status was markedly improved and he considered that nitrogen losses due to burning could be replaced in 4–5 years.

CRITERIA FOR USE OF NITROGEN FIXING PLANTS

With so much experimental evidence of the benefits of nitrogen fixing plants in forestry situations, the question must be asked as to why the use of such nitrogen fixing plants is not widespread throughout plantation forestry. Much of the work has been reported relatively recently and may yet lead to the establishment of successful operational systems. However, management inertia preventing introduction of such systems into forestry is caused by a variety of factors including; the cheapness and operational flexibility of applying inorganic nitrogen fertiliser, the economic uncertainty of nitrogen fixing systems, and an absence of transfer technology allowing easy dovetailing of the concurrent establishment of nitrogen fixing plants and tree seedlings.

Since the biological feasibility and benefits of nitrogen fixing plants in plantation forestry is well established, let us now examine the criteria for selecting a desirable nitrogen fixing plant, the management constraints on its introduction into the plantation, and the economic factors influencing implementation of biological nitrogen fixing systems.

To screen nitrogen fixing plants for use in pine plantations in southeastern Australia we defined several biological criteria concerning the plant's form and habit, nitrogen fixing rates and environmental conditions, and germination and reproduction ability (34). These criteria are comparable to criteria developed similarly by Jorgensen (20) for screening legumes for plantation forestry in the southeastern U.S.A.

Nitrogen fixation rate

Initial fixation rates should be between 50 and 100 kg nitrogen per ha per year. As shown elsewhere in this book (Chapter 7, 9) these appear to be realistic fixation rates for many legumes and non-legumes, and are high enough to ensure some growth response by most conifers during, and in the year immediately following, plantation establishment.

Soil type and fertilisation

Many agronomic species of legumes have been developed such that their economic limits to production are tied closely to specific soil pH, water and nutrient requirements. Many forest lands by definition are those unsuitable for agriculture and there is subsequent scepticism by agronomists about the ability of agronomic legumes to survive and fix nitrogen in harsh forest soils. It must be remembered, however, that the relatively high nitrogen fixation rates found in agriculture are not required or expected in the forest situation.

Rhizobium species developed for symbioses with agronomic legumes may suffer greatly in adverse soil conditions, but *Rhizobium* species tolerant of low pH levels have been found (46, 47) and may extend the range of such symbioses. Jorgensen (20) reported rose clover to be tolerant of acid soils but in very acid soils, such as peaty podzols, germination and growth of the host legume itself is doubtful. Similarily on acid soils productivity of the tree crop may be limited more by low pH than low nitrogen. On a haplohumod of pH 4.5 in south eastern Australia we observed limited germination and rapid mortality of germinants of lupins and clovers, despite fertilisation with phosphorus, copper and molybdenum. In their natural state these soils are covered with poorly productive eucalyptus trees and heath vegetation, amongst which are a number of native legumes. We do not know the rate of nitrogen fixation of these heath communities, but we have attempted unsuccessfully to screen several of the native nitrogen fixing species in our programme. Only one of the screened species, golden prickly wattle, germinated to eventually produce 100% ground cover after 2 years; all other native species failed to germinate or grew for only one season. In this instance the assumption of good adaptation to the soil type by both the native host and bacterium, if true, does not appear to help overcome other as yet unidentified factors influencing germination and growth.

Many forest plantations are fertilised at establishment. The element fertilised is that most deficient and limiting tree growth, and will probably be the major element limiting growth of an understorey. Several additional elements may be necessary for successful establishment of legumes (48).

In Australia and the southeastern U.S.A. phosphorus is most commonly supplied to plantations, whereas in Scandinavia and the north western U.S.A. nitrogen fertiliser is most often used. Whilst there are exceptions to these broad generalisations there is a clear progression of elemental addition in forest fertilisation. On the poorest soils phosphorus is most often the limiting element; once corrected, nitrogen, potassium, and trace element additions may be needed. On more fertile soils where phosphorus is adequate, the main limiting element may be nitrogen. There are isolated areas in the world (albeit regionally prominent) where forest plantations do not respond to fertilisation. The soils on which forests require little fertilisation are also those on which legumes are likely to be easily established. Conversely on poorer soils it will cost more, through supplementary fertilisation of legumes, to fix nitrogen than on more fertile soils where perhaps the requirement for nitrogen may be far lower. Legumes then are not a panacea for all soils and, paradoxically, both the poorest and most fertile soils will not be amenable to legume management in plantation forestry.

Seed availability

Whilst many plants screened for nitrogen fixation may be done from small seedlots collected at some expense, it is economically desirable that the plant can be farmed for seed production. As later discussion will show, the cost of seed is a large factor in determining the economic value of legume establishment.

Establishment and regeneration

Given good tolerance to adverse soils and adequate fertilisation as discussed above, the legume seed should germinate easily to produce vigorous seedlings. Many legume species have hard seed coats and are slow to germinate. Pretreatment such as boiling water, acid, hydrogen peroxide, or abrasion may be necessary to ensure adequate uniform and early germination.

If nitrogen fixation by the legume is dependant upon a specific *Rhizobium* strain, then inoculation via seed coating must be carried out. There is evidence to suggest that *Rhizobium* strains introduced into the soil microflora may survive for only a short period (49). This is of importance if indigenous strains reduce the rate of nitrogen fixation, if the symbiosis cannot survive long inactive periods between plantation thinning, or if the symbiosis does not rejuvenate after clear falling the forest plantation. An inhibition to the use of native nitrogen fixers is the current lack of specific *Rhizobium* inoculum for such species. These need to be developed if suitable strains are not endemic to the particular site or soil type.

To ensure longevity of the host plant it must be self seeding and the seed sufficiently hardy to survive and germinate with canopy openings during thinnings. A survival mechanism dependent more on perenniality than seed longevity is unlikely to survive the canopy closure for long periods between thinnings and between rotations.

Habit and weed control

Since climbing or tall plants are likely to interfere with seedling growth of the tree crop it is preferable that legumes have a low habit. Ground spreading plants are likely to be able to compete strongly with weed species. Similarly shrubby habits will have a weed control function.

Shade tolerance

If the legume is successful in its role of supplying nitrogen to the forest plantation and promoting tree growth its role will probably be eliminated by the closing canopy. The degree of shade tolerance is of value in extending the period of nitrogen fixation up to and through canopy closure, and initiating early and rapid nitrogen fixation following thinning. Jorgensen (20) reported that sericia lespedeza was somewhat shade tolerant. Knowles (50) identified about 40 species of forage legumes which show some degree of shade tolerance or occur naturally in woodlands of North America or Europe. These were screened for growth under exotic trees in New Zealand.

Growth and moisture balance

It is important, in areas with prolonged dry periods, that the legume grows most vigorously when the establishing forest has sufficient water. The problem of water use and competition is exacerbated in deep sands where the water table recedes rapidly, and on soils which have shallow surface horizons over clay such as natraqualfs where an ephemeral water table perched over the clay is soon exhausted. In these soils, shallow rooting by the legume would aid early senescence, but this must be offset by any reduction in growing period and nitrogen fixation. In the dry summers of southern Australia, researchers working with annual lupins and subterranean clover have found early senescence of the legume crop and little competition for moisture during critical periods (33). However, in our experience in southeastern Australia, yellow tree lupin, a perennial, overtopped the pine seedlings, and the dense lupin cover lasting into mid-summer may have depleted readily available soil water storage.

Effects on forest management activities

With the research worker's endeavour to maximise nitrogen fixation in the forest plantation it is easy to lose sight of forest management activities that may be adversely affected by the legume crop. Fire suppression and harvesting are the main activities affected. Recently senesced lupins and desiccated shrubby legumes in summer will increase the fire hazard in the plantation. Moreover, prickly or spiny legumes such as some acacias will make fire suppression and harvesting activities more arduous. Jorgensen (20) considered that nitrogen fixers such as wax myrtle, kudzu, some lespedezas, and European gorse, have some undesirable characteristics of flammability and antagonistic foliage.

In some parts of Australia a premium is paid on forest harvesting contracts carried out in blackberry-infested areas. A legume of similar habit would not be favoured and would likewise have a cost disadvantage.

Escapes

The search for dehiscent, hard seeded lupins has lead research workers back to strains rejected by agriculture. These strains suitable for forestry may have a higher alkaloid content than commercial varieties, and there is some concern that there may be some regression of the commercial crop if hard seeded varieties were used widely in forestry and cross fertilisation occurred.

The legume established in forests should not have the ability to spread rapidly and compete with other vegetation. There are numerous examples of escapes from agriculture, but two pertinent ones in this instance are the kudzu in south eastern U.S.A. (20) and the sand plain lupin which has become naturalised on the coastal sands of Western Australia.

A nitrogen fixing plant meeting all of the above criteria would be extremely valuable. However, the presentation of such criteria is greeted by agronomists and botanists with mirth and sceptisism; such a plant doesn't exist. The skill of the investigator must then be used to trade-off the least valuable criterion for each specific situation. For each criterion that is not met there is a penalty which will result in increased costs to silvicultural operations either as decreased nitrogen fixation or decreased tree growth.

ECONOMIC MANAGEMENT OF NITROGEN FIXING PLANTS IN PLANTATION FORESTY

Some basic concepts

Forest plantations are usually managed on the basis of a return in the form of wood on the capital invested in land and capital outlay in forest establishment and maintenance. The capital outlay and importance placed on the return on funds invested varies with the scale of perception from the national, through state and private industry, to the small woodlot owner, e.g. from the degree of self sufficiency deemed desirable at a time of international trade disruption, to secured finance to purchase a new farm tractor or an overseas trip on retirement. This serves to illustrate the range of tolerances about established economic criteria used to manage the forest plantation, and the degree of dependence upon the final product volume and piece size.

Much of the rationale behind legume research programmes has been the escalating cost of nitrogen fertilisers, and the hope that legume-fixed nitrogen will be much cheaper. Embodied in these incentives are two misconceptions. Firstly, as can be seen from figure 1 the cost of urea fertiliser in Australia (as elsewhere) has risen sharply since 1973. This price rise, however, has been linked closely to the rising cost of energy for production of nitrogen fertilisers, and consequently has also followed closely changes in the consumer price index (CPI). By contrast the cost of superphosphate has declined relative to the CPI. The second misconception is that nitrogen is fixed by legumes 'for free' and consequently must be cheaper than nitrogen fertiliser. Consistent with the rise in energy cost has been the cost of manufacture and running of agricultural machinery. Since legume establishment requires machinery and machine time additional to that required for plantation establishment, there is an increased energy and monetary cost to the plantation which has to be carried to the point of harvest.

Thus the economic prospects for wood grown with a legume understorey will improve over time only if the cost factors influencing its establishment do not increase with the cost of energy and the CPI, or if the cost of nitrogen fertiliser increases significantly more that the CPI. Otherwise the cost of legume establishment over time will change little relative to the cost of nitrogen fertilisation. If the nitrogen fixing crop has a monetary value as is the case of red alder, projected fluctuations in stumpage values (royalties) add a further dimension to the economics of managing nitrogen fixing plants in forestry.

Economics of wood rotations

The economic factors that influence decisions to grow red alder as a nitrogen fixing source for Douglas fir on the Pacific Northwest of the U.S.A. have been

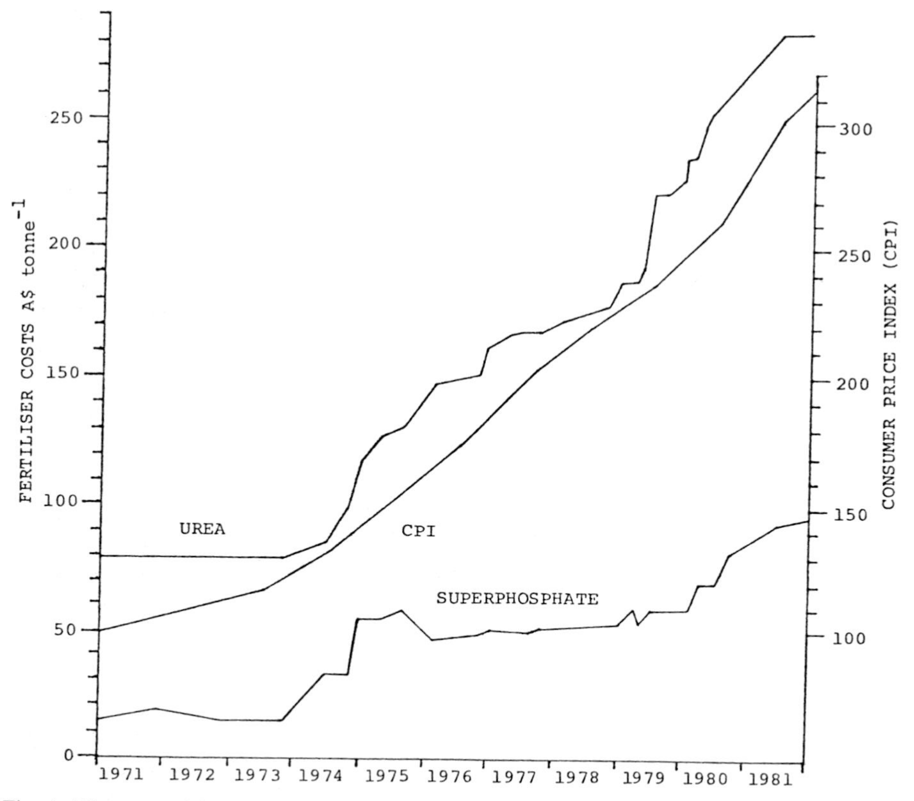

Fig. 1. The cost of Urea and Superphosphate fertilisers and the Consumer Price Index for Melbourne Australia, 1971–1982. Fertiliser costs include fluctuating Government subsidies and represent the true cost to the purchaser.

examined by Atkinson *et al.* (51) and by Miller and Murray (52). Both parties examined the various options of growing Douglas fir with nitrogen fertiliser, with alternating rotations of red alder, and as a mixture with red alder.

Atkinson *et al.* (51) costed the various options and compared the predicted returns on the basis of present net worth (discounted returns minus discounted costs at 7% annual interest rate). They found all options, including unfertilised Douglas fir to be profitable in that they all returned greater than 7% return on investment. All options involving alternate cropping of red alder and Douglas fir returned less than the option of continuous cropping of either fertilised or unfertilised Douglas fir. The least profitable option was that involving alternate cropping of red alder (13 years) for pulpwood and Douglas fir (45 years) for sawlogs. In this series of analyses it is apparent that the sum of the stumpage value of red alder together with the value of additional Douglas fir timber harvested due to alder inputs of nitrogen was less than that for continuous cropping of Douglas fir. In other terms the opportunity cost of taking land out of Douglas fir production was greater than the value of red alder wood and nitrogen fixation. Miller and Murray (52) examined discounted costs of supplying nitrogen through red alder. When expressed as \$ kg^{-1} Nha^{-1} the discounted costs reflected the expenditure in red alder site preparation and seeding, or planting additional to that for Douglas fir. As a result, fertilisation of Douglas fir with urea had one of the lowest discounted direct costs. They considered that insufficient data on wood volume existed to enable them to compute and compare the profitability of the various options, but some comparison was made by Atkinson *et al.* (51) which showed again that the large difference between stumpage values for the two species signified the low profitability of red alder stands. Miller and Murray (52) appended uncertainty rankings to the likelihood of obtaining projected additional nitrogen inputs. This risk factor was lowest for nitrogen fertilisation with low discounted costs.

It is clear that mixed or alternating wood crops are subject to the vagaries of more than one set of market fluctuations and stumpage values. By growing a nitrogen fixing understorey for one forest plantation crop, the value of the legume understorey is more simply equated with the cost of inorganic nitrogen fertilisation of the forest crop, and the value of the increased wood produced.

Economics of legume understoreys

By contrast with rotations of nitrogen fixing and non nitrogen fixing woody plants, management of a legume understorey involves harvesting and marketing of only one wood species. The wood crop will have presumably increased in volume growth through utilisation of nitrogen fixed by an herbaceous legume understorey.

In a previous economic analysis of the value of a legume understorey in a pine plantation in southeastern Australia (34) we assumed a nitrogen input of 90 kg Nha^{-1} for a relatively short duration from either a legume understorey or urea fertilisation. Assuming a similar tree growth response to both forms of nitrogen input, it was estimated that inorganic nitrogen fertilisation with urea grew wood that was some three times cheaper than that grown with legumes. However, it was shown that if the legume crop increased the mean annual increment of wood volume production by 5% greater than predicted, nitrogen fixation by a legume crop was economically comparable to inorganic nitrogen fertilisation.

This analysis did not take into account the return of the legume understorey following thinning, and the consequent continuing input of nitrogen through the rotation. In the ensuing pages we attempt to estimate the long term input of nitrogen from a legume understorey, and compare the cost of this input with that from urea fertilisation. We are not aware of any other economic analysis of a legume understorey in a planted forest, so in the following analysis we have made assumptions unique to our own situation.

Not only is there little information on the *in vivo* rate of nitrogen fixation by legumes in forest situations but, as Miller and Murray (52) pointed out, there is little or no information on the response in the form of conifer tree growth to nitrogen fixed by another co-existing species. Thus when legume management systems are postulated, often the only tree growth response data available are those for inorganically fertilised stands.

A managed legume scenario

In the discussions and examples which follow we will examine the use of legumes as an adjunct to forest plantation management by a forestry enterprise whose projected continuous wood output is committed to an integrated industry end process. This obviates problems associated with competitive marketing of wood associated with variations in piece size and volume cut according to fluctuations in wood markets. Whilst strong market forces affect these often large scale but short term fluctuations in wood cut, they seldom affect investment decisions at the plantation establishment phase.

In the following scenario we examine a legume understorey (e.g. lupin) under a crop of *Pinus radiata* grown for pulpwood over a 31 year rotation in south-eastern Australia. We examine the probable nitrogen fixation rates over the rotation, and compare the value of the predicted wood grown either with the legume crop or with inorganic nitrogen fertilisation.

Nitrogen fixation over the rotation

The progression of wood volume production due to increased levels of phosphorous and nitrogen fertilisation and weed control for this scenario are shown in figure 2. This illustrates the rate of canopy closure, with canopy production assumed for this exercise to be directly proportional to wood volume production. If the legume crop was sown and fixed no nitrogen, wood growth in the early years would be similar to the lowest curve in figure 2, and legume 'weed' cover would be only slowly excluded by the canopy. Conversely, if the legume crop was sown and fixed a maximum rate of 100 kg Nha^{-1} $year^{-1}$ which declined with shading, wood volume production would be similar to the top curve in figure 2, with canopy closure occurring near age 7.

Assuming that the legume crop was shade sensitive, nitrogen fixation rates may decline in direct proportion to shading and canopy production as shown in figure 3. Thus, up to canopy closure the initial crop of legumes would have subsequently regenerated annually to fix a total of 455 kg Nha^{-1} over 7.3 years (figure 4) or a mean annual nitrogen fixation over the period of 62 kg Nha^{-1}. This is for a legume crop able to supply 100 kg Nha^{-1} yr^{-1} under unshaded conditions. Experience with yellow tree lupin in New Zealand has

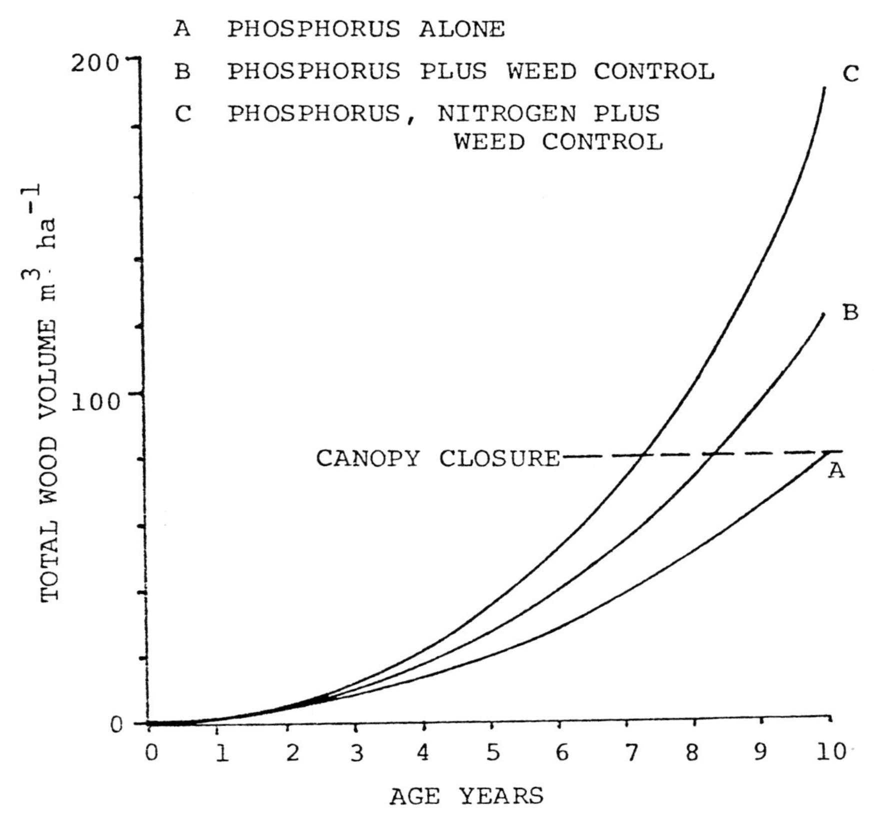

Fig. 2. The accumulation of wood volume and ocurrence of canopy closure in *Pinus radiata* grown in southeastern Australia under three fertiliser and weed control regimes.

shown that this legume will germinate after thinning and persist for four to five years (Gadgil, chapter 12). For this scenario maximum nitrogen fixation rate of 25 kg Nha^{-1} yr^{-1} was considered plausible under conditions following thinning, with the decline in production being less severe following the third thinning with a much reduced canopy (figure 5). Up to clear falling at age 31 years the legume crops were calculated as having fixed 711 kg Nha^{-1} for a wood volume of 741 m^3 ha^{-1} or a mean annual accumulation of 22.9 kg Nha^{-1} yr^{-1} with a mean annual wood increment of 23.9 m^3 ha^{-1} yr^{-1}. Some 64% of this nitrogen was fixed before canopy closure. Following clear falling the legume crop should germinate and may reach the maximum nitrogen fixation rates of the initial crop.

Fertiliser requirements

Data for *P. radiata* growth reponse to fertilisation at establishment on a range of soil types in south-eastern Australia is reasonably well founded (53, 54). How-

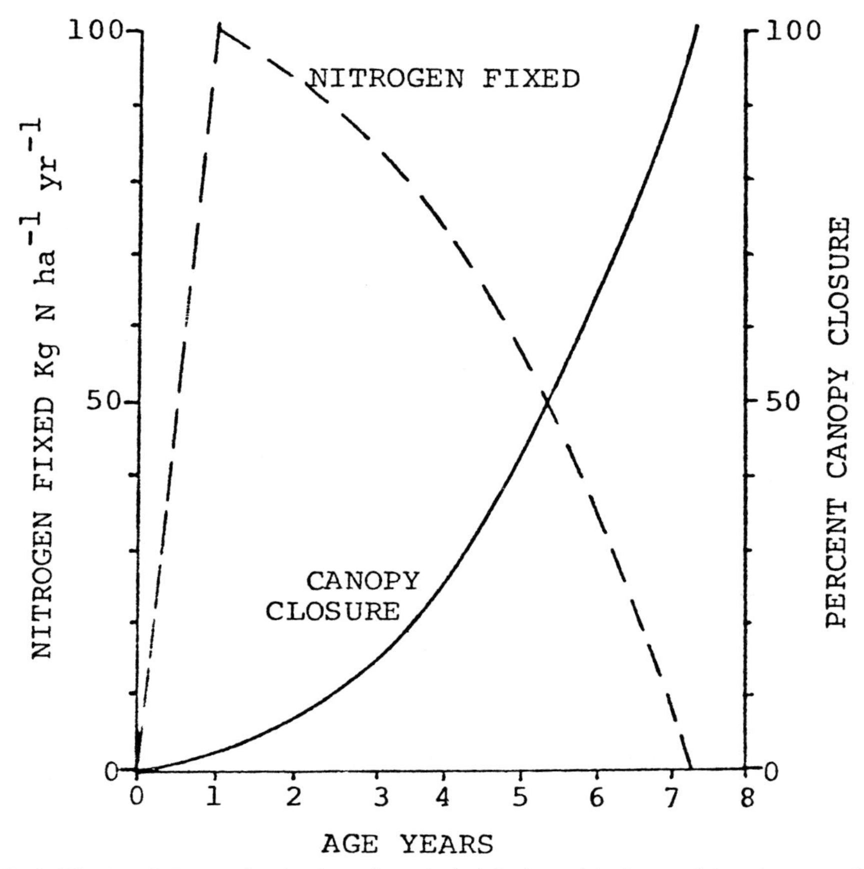

Fig. 3. The rate of nitrogen fixation for an hypothetical shade sensitive legume fixing nitrogen under a closing pine forest canopy.

ever there are few data available on the requirements for, and the response to, fertilisation later in the rotation. Crane (55) examined fertilisation of *P. radiata* following thinning of stands with low foliar nitrogen levels. The response of $6\,m^3\,ha^{-1}\,yr^{-1}$ for 4 years found by Crane (55) for *P. radiata* has been used as the likely response to both legume and urea fertilisation following thinning.

The fertiliser regimes used in this scenario are shown in table 2. Adequate phosphorus fertilisation is considered necessary to obtain a response to nitrogen fertilisation. Total nitrogen applied as urea is 1019 kg Nha^{-1}, whereas that assumed supplied by legumes is 711 kg Nha^{-1}. This assumes an efficiency of nitrogen supply by the legumes of 1.4 times that of urea. This allows for the relative inefficiency of urea as a nitrogen source (56).

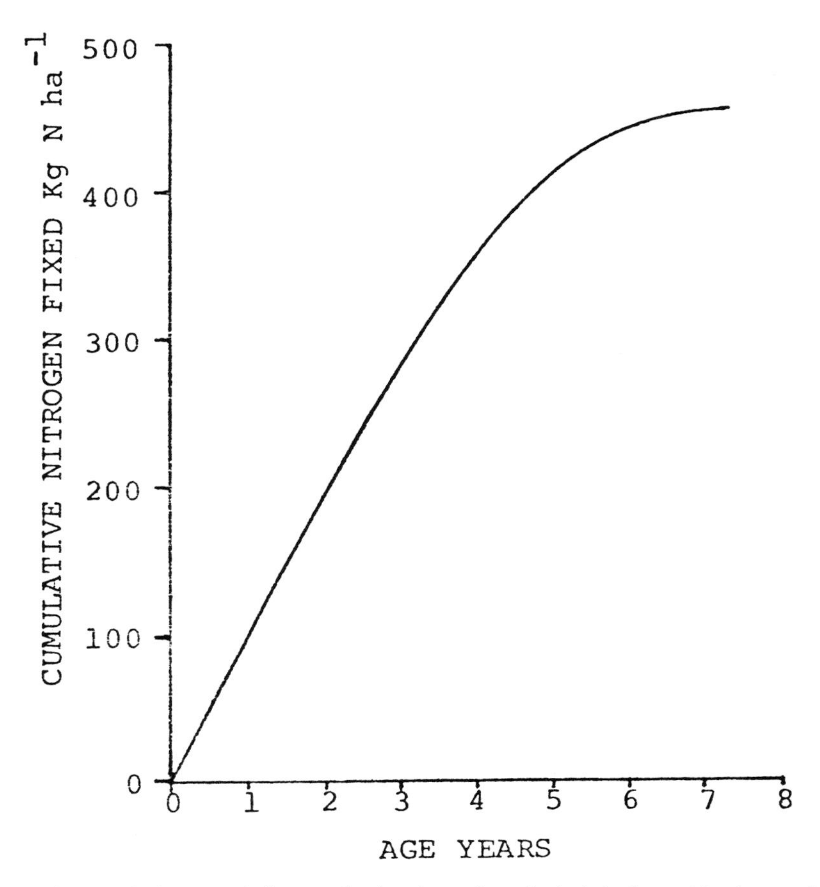

Fig. 4. The cumulative rate of nitrogen fixation for an hypothetical shade sensitive legume fixing nitrogen under a closing pine forest canopy.

Wood growing cost

In order to compare the economic value of wood grown with legumes and wood grown with urea fertilisation we have used the criterion of break-even wood growing cost. Since in this scenario the wood grown is committed to an end user, the criterion of net present value (discounted returns less discounted costs) is inappropriate because the returns on stumpage values are not determined by a free market (i.e. the returns are a function of dependence by the integrated end user). The break even growing cost is the discounted costs involved in growing the wood divided by the discounted wood yields, and represents the true cost over time to the investor in growing the wood. Thus we can assume equal volumes of wood grown either with legumes or with urea fertilisation using the regimes shown in table 2 and compare the costs of obtaining these wood vol-

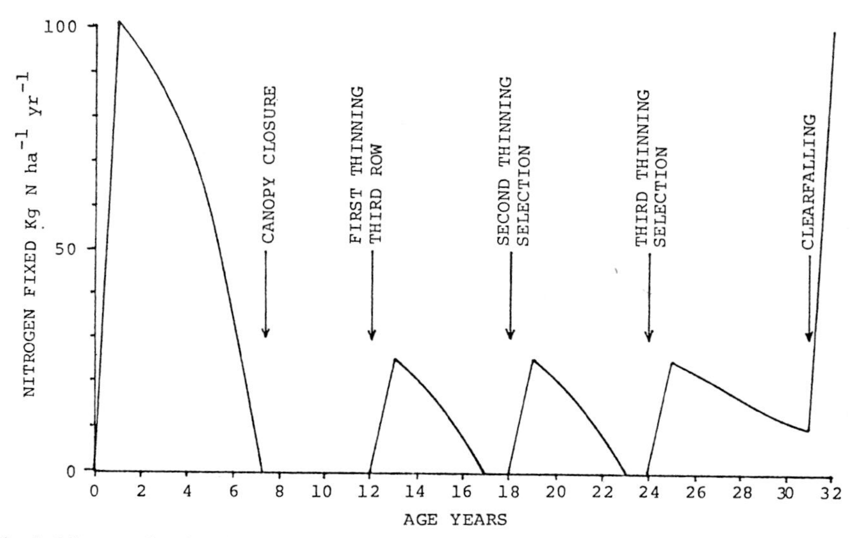

Fig. 5. Nitrogen fixation through a pine forest rotation by an hypothetical shade sensitive legume.

Table 2. Fertiliser requirements for biological nitrogen supply and inorganic nitrogen supply to produce equal wood volumes at clearfalling of *Pinus radiata*.

Year	Operation	Biological nitrogen Lupin + Fertiliser Element Kg ha^{-1}	Inorganic nitrogen Fertiliser alone Element Kg ha^{-1}
Establishment	Fertilise	60 P + 0.3 Mo	60 P
2	Fertilise	60 P + 0.3 Mo	44 N + 58 P
12	Thin Fertilise	130 P	325 N + 130 P
18	Thin Fertilise	130 P	325 N + 130 P
24	Thin Fertilise	130 P	325 N + 130 P
31	Clearfall		
	Total fertiliser applied	510 P + 0.6 Mo	1019 N + 508 P

umes. This is an extension of the type of comparison used by Miller and Murray (52) inasmuch as their discounted expenditure took no cognisance of wood volume produced.

From figure 6 it can be seen that the cost of legume establishment has a strong bearing on the growing cost of the wood. Since commercial availability of many suitable legume seeds is limited, current legume establishment costs can be relatively expensive (A \$ 500 ha^{-1}). Using 1981 fertiliser prices (Australia) it is evident from figure 6 that if legume establishment costs (in the form of seed and sowing costs) are A \$ 200 ha^{-1}, then the wood growing cost is equal to or less than the same volume of wood grown with the same base phosphorus fertilisation but including nitrogen fertilisation throughout the rotation. Thus if nitrogen fertiliser costs increase more rapidly than the cost of legume establishment, and

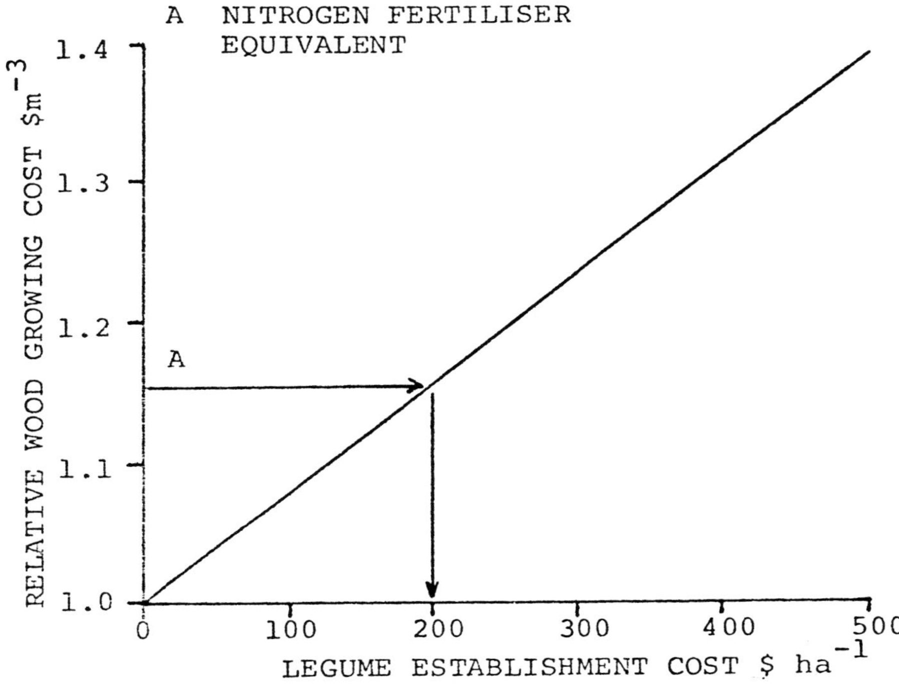

Fig. 6. The increase in relative wood growing cost with increases in legume establishment cost. Legume establishment cost is that cost incurred additional to normal plantation establishment costs. Wood growing cost is relative to zero legume establishment cost. The relative wood growing cost for nitrogen fertiliser required to grow an equivalent amount of wood is also shown.

with all other costs rising equally, either more expensive legume establishment treatments will become economically feasible, or the margin between wood grown with a legume understorey and wood grown with nitrogen fertiliser will change in favour of legume establishment.

Whilst the end result can be balanced in this way, the cash flow requirements for the two systems with equal growing costs are quite different. The total cash outlay in figure 7 shows that the total crude expenditure (not discounted) up to age 24 years for the legume system is 27% less than for the nitrogen fertiliser system. The establishment cost for the legumes system is higher than the nitrogen fertiliser system which itself demands a greater cash flow later in the rotation. When this cash flow is discounted at 9% p.a. (figure 8) it can be seen that the relatively high cost of legume establishment that has to be carried through time allows for heavy expenditure on nitrogen fertiliser in the parallel system. Thus the two quite different cash flow requirements of these parallel systems produce equivalent net discounted expenditure as shown in figure 8. Thus although the two investment decisions may result in the same discounted value of wood, the

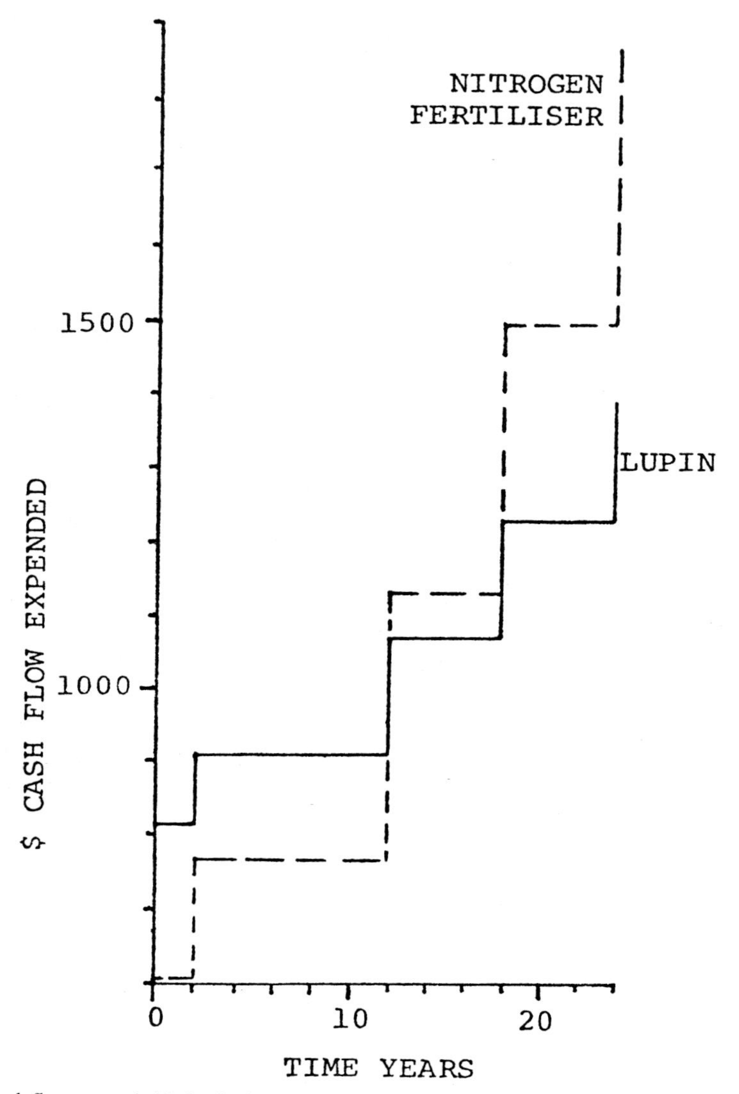

Fig. 7. Cash flow expended in both nitrogen fertilisation and legume growing to produce equivalent amounts of wood. Legume establishment cost \$ 200 ha^{-1}.

cash flow management is different for each system. The application of nitrogen fertiliser is a flexible system which allows an expenditure decision at mid rotation; this decision can be made by taking into account demands for wood and expected cash flow. The legume system requires higher capital investment over the first third of the rotation and, as Miller and Murray (52) point out, carries greater risks, both biological (crop failure) and monetary (changing interest rates), to the investment over time. These calculations were made without consideration of taxation which itself can dictate silvicultural strategies. Whilst these factors must be balanced by the manager, this analysis shows how economic factors mitigate against biological fixation of nitrogen in plantation forestry.

CONCLUSIONS

There is ample evidence of the benefits of biologically fixed nitrogen in plantation forestry through improvements in wood volume growth. However, the limited forays into economic analysis in plantation forestry either as tree mixtures in rotations (Douglas fir and red alder), or as a legume understorey (radiata pine and lupins), show that wood volume production can be achieved more cheaply through inorganic nitrogen fertilisation. It is also apparent that these situations will change only if red alder stumpage values rise in one case, and in the other if the costs of legume understorey establishment do not rise as rapidly as urea costs, energy costs, and the consumer price index. There is little likelihood of such changes occuring in the short term.

There is an aura of biological uncertainty surrounding the data used for economic analysis of nitrogen management in forestry. These uncertainties pertain to rates of nitrogen fixation, transfer into the crop trees, and the response in tree growth to nitrogen fixed. It seems pertinent then to conclude this chapter with some perceived research priorities.

The greatest priority in research into nitrogen fixing plants in forestry is to quantify the value of nitrogen fixed by the plants to the crop trees. There are clear

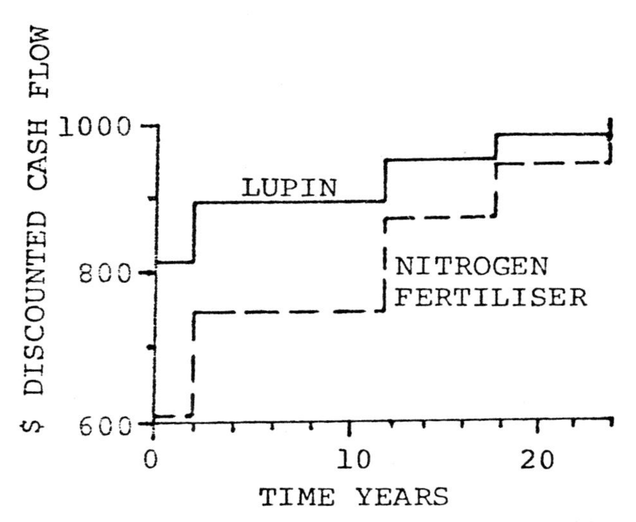

Fig. 8. Cash flow discounted at 9% p.a. for both nitrogen fertilisation and legume growing to produce equivalent amounts of wood. Legume establishment cost $ 200 ha^{-1}.

gaps in our knowledge of the efficiency and ancillary benefits of biologically fixed nitrogen. Many of these gaps are filled by 'common truths' which pervade the subject with notions that biologically fixed nitrogen must be 'a good thing'. However, for the calculation of economic efficiency of such systems in forestry it is important to know not only the rate of nitrogen fixed, but the efficiency of supply to the forest crop.

Clearly the risks surrounding the investment in establishment of nitrogen fixing plants are high. Such plants can only be accepted into forestry if the costs of establishing and maintaining them are reduced, thereby reducing the risk to the investment and predicted wood yields. Developmental research must focus on the establishment phase of the nitrogen fixing plants, and examine the efficiency of germination, factors of site preparation and fertilisation, the requirement for weedicide to reduce competition, and the efficiency of seed density and sowing.

The concept of perpetual nitrogen fixation in forest plantations has tantalised forest researchers and managers alike. Despite the existence of some very successful systems, the economic analyses available to us do not encourage utilisation of nitrogen fixing plants in forestry. Inorganic nitrogen fertilisation is currently both an economical and flexible means of supplying the forest's needs for nitrogen. It appears that use of nitrogen fixing plants may become more attractive economically if fertiliser prices increase at a faster rate than costs of establishment of biological nitrogen systems. In the lead time to such an eventuality there is ample scope for the improvement of predictions about the productivity of forest crops grown in association with nitrogen fixing plants.

REFERENCES

1. Cushwa CT, Martin RE, Hopkins ML: Management of bobwhite quail habitat in pine forests of the Atlantic piedmont: *Georgia Forest Research Paper* No. 65, 1971.
2. Shea SR: An ecological approach to the control of jarrah dieback. *Forest Focus* No. 21, 1979.
3. Tarrant RF, Trappe JM: The role of *Alnus* in improving the forest environment. *Plant and Soil*, Special Volume 1971: 335–348.
4. De Bell DS, Radwan MA: Growth and nitrogen relations of coppiced black cottonwood and red alder in pure and mixed plantings. *Bot Gaz* 140 (Suppl.): S97–S101, 1979.
5. Miller RE, Murray MD: The effects of red alder on growth of Douglas-fir. In: *Utilization and management of alder*, Briggs DG, De Bell DS, Atkinson WA (compilers) USDA For Serv Gen Tech Rep PHW-70, 1978, p 283–306.
6. Haines SG, Haines LW, White G: Nitrogen fixing plants in southeastern United States Forestry. In: *Symbiotic nitrogen fixation in the management of temperate forests*, Gordon JC, Wheeler CT, Perry DA (eds). Oregon State University Press, 1979, p 429–443.
7. Plass WT: Growth and survival of hardwoods and pine interplanted with European alder. *USDA For Serv Res Pap* NE-376, 1977.
8. Kellog LF: Failure of black locust-coniferous mixtures in the Central States. *USDA Forest Service Central States. For Exp Sta Note* No 15, 1934.
9. Finn RF: Foliar nitrogen and growth of certain mixed and pure forest plantings. *J For* 51: 31–33, 1953.

10. Ferguson JA: Influence of locust on the growth of catalpa. *J For* 20: 318–319, 1922.
11. McIntyre AC, Jeffries CD: The effect of black locust on soil nitrogen and growth of catalpa. *J For* 30: 22–28, 1932.
12. Carmean WH, Clark FB, Williams RD, Hannah PR: Hardwoods planted in old fields favored by prior tree cover. *USDA Forest Service Res. Paper* NC-134, 1976.
13. Kellison RC, White G: Black alder performance in the Southeast. In: *Symbiotic nitrogen fixation in the management of temperate forests* Gordon JC, Wheeler CT, Perry DA (eds). Oregon State University Press. 1979. p 345–355.
14. Ashby WC, Baker MB: Soil nutrients and tree growth under black locust and shortleaf pine overstories in strip-mine plantings. *J Forestry* 66: 67–71, 1968.
15. Mellilo JM, Aber JD: Symbiotic and nonsymbiotic nitrogen fixation in northern Rocky Mountain ecosystems. In: *Symbiotic nitrogen fixation in the management of temperate forests*, Gordon JC, Wheeler CT, Perry DA (eds). Oregon State University Press. 1979 p 309–317.
16. Gadgil RL: Alternatives to the use of fertilizers. In: *Use of fertilizers in New Zealand forestry*, Ballard R (ed), Forest Research Institute, Rotorua, New Zealand, 1977.
17. Mangalis I: (Role of bare following and green manuring in the raising of conifer seedlings and transplants in permanent forest nurseries.) Melnas un zalmeslojuma papuves nozime skuju koko sejenu un stadu izaudzesana pastavigajas meza kokaudzetavas. *Latvijas Lauksaimniecibas Akademijas Raksti* No. 93: 3–11, 1975.
18. Miller RE, Zalunardo R: Long-term growth of eight legumes introduced at three forest locations in southwest Oregon. *USDA For Serv Res Pap* PNW-225. 1979.
19. Vanderveer HL, Ballard R, Davey CB: Loblolly pine ecosystem response to clovers and fertilization (in press).
20. Jorgensen JR: Use of legumes in southeastern forest research. In: *Southern silvicultural research conference proceedings*, Atlanta, Georgia, U.S.A. Nov 6–7, 1980.
21. Youngberg CT, Wollum AG, Scott W: *Ceanothus* in Douglas-fir clearcuts: nitrogen accretion and impact on regeneration. In: *Symbiotic nitrogen fixation in the management of temperate forests*, Gordon JC, Wheeler CT, Perry DA (eds). Oregon State University Press. 1979. p 224–233.
22. Zavitkovski J, Hansen EA, McNeel HA: Nitrogen-fixing species in short rotation systems for fiber and energy production. In: *Symbiotic nitrogen fixation in the management of temperate forests*, Gordon JC, Wheeler CT, Perry DA (eds). Oregon State University Press. 1979. p 388–402.
23. Funk DT, Schlesinger RC, Ponder F: Autumn olive as a nurse crop for black walnut. *Bot Gaz* 140 (Suppl.): S110–S114, 1979.
24. Lang P: (The Pegnitz silvicultural system). Das Pegnitzer Waldbauverfahren. *Allgemeine Forstzeitschrift* 32, 23/24: 565–567, 1977.
25. Thum J: (Use of vegetation analysis to determine the success of cultivation measures for soil reclamation.) Vegetationskudliche Erfolgskontrolle bodenverbessernder Rekultivierungsmassnahmen. *Beiträge für die Forstwirtschaft* 12(3): 140–146, 1978.
26. Thum J: (Effect of growing *Lupinus polyphyllus* and *Melilotus albus* on tree growth and nutrition and on soil properties on a spoil mound.) Einfluss von Dauerlupine (*Lupinus polyphyllus Lindl.*) und Steinklee (*Melilotus albus Med.*) auf Baumwachstum, -ernährung und Bodeneigenschaften eines Kippenstandortes. Beiträge für die *Forstwirtschaft* 12(4): 167–172, 1978.
27. Arnalds A: (*Lupinus nootkatensis* (Donn ex Sims).) Rannsóknir á alaskalúpínu. *Ársrit Skograektarfélags* Islands, 1979, p 13–21.
28. Anon.: Legumes and protection forestry. *What's new in Forest Research* No. 33, Forest Research Institute, Rotorua, New Zealand, 1976.
29. Silvester WB, Carter DA, Sprent JI: Nitrogen input by *Lupinus* and *Coriaria* in *Pinus radiata* forest in New Zealand. In: *Symbiotic nitrogen fixation in the management of temperate forests*, Gordon JC, Wheeler CT, Perry DA (eds). Oregon State University Press. 1979. p 253–265.
30. Hatch AB: Pine legume mixtures in Western Australia. In: *Managing nitrogen economies of natural and man made forest ecosystems*, Rummery RA, Hingston FJ (eds). Perth, Western Australia. CSIRO Division of Land Resources Management, 1981, p 340–345.
31. Anon: Forestry and Timber Bureau Annual Report 1963, *Commonwealth of Australia*, 1964, p 17–18.

32. Richards BN, Bevege DI: The productivity and nitrogen economy of artificial ecosystems comprising various combinations of perennial legumes and coniferous tree species. *Aust J Bot* 15: 467–480, 1967.
33. Nambiar EKS, Nethercott KG: Nitrogen supply to radiata pine through legumes. In: *Australian forest nutrition workshop: productivity in perpetuity*. CSIRO Div For Res, Canberra Aust. 1981, p 330.
34. Turvey ND, Smethurst PJ: Biological and economic criteria for establishing a nitrogen fixing understorey in pine plantations. In: *Managing the nitrogen economies of natural and man made forest ecosystems*, Rummery RA, Hingston FJ (eds). Perth, Western Australia. CSIRO Division of Land Resources Management, 1981, p 124–145.
35. Zhilkin BD, Grigor'ev VP, Rozhkow LN: (Effect of perennial lupin on nitrogen nutrition of spruce.) *Agrokhimiya* No. 11: 14–20, 1970.
36. Zhilkin BD, Rikhter TA: Amount and properties of litter in mature *Pinetum vacciniosum. Lesovedenie i Les. Kh-vo*, No. 11: 35–43, 1976.
37. Vorontsov AI: Basic principles of intergrated protection of forest plantations. In: *Proceedings of VIII international plant protection congress*, Moscow, Vol. III, 1975.
38. Lakhtanova LI, Beregova TS: Influence of lupin on the increment of pine plantation phytocoenoses. *Botanika Issled* No. 18: 21–27, 1976.
39. Lakhtanova LI, Beregova TS: (Biological improvement of forest by the cultivation of perennial lupin.) *Lesnoe Khozyaïstvo* No. 5: 21–24, 1980.
40. Rehfuess KE: Underplanting of pines with legumes in Germany. In: *Symbiotic nitrogen fixation in the management of temperate forests*, Gordon JC, Wheeler CT, Perry DA (eds). Oregon State University Press. 1979. p 374–387.
41. Melzer EW, Hertel HJ: Lupine (*Lupinus polyphyllus* Lindl.) auf Buntsandstein und ihr Einfluss auf den Zuwachs der forstlich wichtigsten Koniferen. *Arch Acker- u Pflanzenbau u Bodenkd*, Berlin 25(5): 319–325, 1981.
42. Haines SG, Haines LW, White G: Leguminous plants increase sycamore growth in northern Alabama. *Soil Sci Soc Am J* 42: 130–132, 1978.
43. Bengston GW, Mays DA: Growth and nutrition of loblolly pine on coal mine spoil as affected by nitrogen and phosphorus fertilizer and cover crops. *Forest Sci* 24(3): 398–409, 1978.
44. Batini FE, Anderson GW: Agroforestry under 13 to 18 year old *Pinus radiata*, Wellbucket, Western Australia. In: *Managing the nitrogen economies of natural and man made forest ecosystems*, Rummery RA, Hingston FJ (eds). Perth, Western Australia. CSIRO Division of Land Resources Management, 1981. p 346–353.
45. Woods RV: Management of nitrogen in the *P. radiata* plantations of the south east of South Australia. In: *Managing nitrogen economies of natural and man made forest ecosystems*. Rummery RA, Hingston FJ (eds). Perth, Western Australia. CSIRO. Division of Land Resources Management, 1981. p 354–367.
46. Date RA, Halliday J: Selecting *rhizobium* for acid infertile soils of the tropics. *Nature* 277: 62–64, 1979.
47. Thornton FC, Davey CB: Response of the clover – *rhizobium* symbiosis to soil acidity and rhizobium strain. *Agron J* 1981 (in press).
48. Robson AD: Mineral nutrients limiting nitrogen fixation in legumes. In: *Mineral nutrition of legumes on tropical and subtropical soils*, Andrew CS, Kamprath EJ (eds). CSIRO. Cunningham Laboratory, Brisbane, 1978.
49. Brockwell J: Can inoculant strains ever compete successfully with established soil populations? In: *Current perspectives in nitrogen fixation*, Gibson AH, Newton WE (eds) Australian Academy of Science, Canberra, 1981, p 227.
50. Knowles B: Forage legumes suitable for growth under exotic trees in New Zealand: a literature revies. New Zealand Forest Service, Forest Research Institute, Report No. 140, 1980.
51. Atkinson WA, Bormann BT, De Bell DS: Crop rotation of Douglas-fir and red alder: a preliminary biological and economic assessment. *Bot. Gaz.* 140 (Suppl): S102–S107, 1979.
52. Miller RE, Murray MD: Fertilizers versus red alder for adding nitrogen to Douglas-fir forests of the Pacific Northwest. In: *Symbiotic nitrogen fixation in the management of temperate forests*, Gordon JC, Wheeler CT, Perry DA (eds). Oregon State University Press, 1979. p 356–373.

53. Turvey ND: A forest soil survey: II. The application of soil survey information to forestry operations. *Aust For* 43 (3) 172–177, 1980.
54. Cromer RN, Dargavel JB, Henderson VT, Nelson PF: More pulpwood from less land. *Appita* 31: 49–54, 1977.
55. Crane WJB: Growth following fertilisation of thinned *Pinus radiata* stands near Canberra in southeastern Australia. *Aust For* 44(1); 14–25, 1981.
56. Watkins SH, Strand RE, De Bell DS, Esch J: Factors influencing ammonia losses from urea applied to northwestern forest soils. *Proc Soil Sci Soc Amer* 36(2): 354–357, 1972.

9. Nitrogen fixation in North American forestry: research and application

ROBERT F. TARRANT

Forest Research Laboratory, Oregon State University, Corvallis, OR 97331 USA

INTRODUCTION

In most North American forest soils, nitrogen, in a form readily assimilable by plants, is not in sufficient supply to support best tree growth. Forest managers usually respond to this fact by accepting less than maximum productivity from their lands. Some apply synthetic fertilizers. But neither of these actions solves the critical problem of maintaining forest ecosystem productivity at a high level of the long term. New technology obviously is needed.

One approach to new technology is through utilizing biological nitrogen fixation, a product of symbiotic associations between certain plants and microorganisms. The idea is not new – N_2-fixing microorganisms and their plant partners in symbiosis have been employed to enrich agricultural systems for many decades. But only recently has the importance of nitrogen to tree growth been made evident.

During the past few years, research into biological nitrogen fixation in a forest management context has been accelerated with a number of microorganisms, both free-living and symbiotically-living types. Free-living N_2-fixers develop slowly because of relatively limited habitat and energy sources. In contrast, symbiotically-living microorganisms provide relatively large amounts of nitrogen, particularly those which form root nodule associations between vascular plants and either *Frankia* spp. or *Rhizobium* spp. Thus, symbiotically-living microorganisms and their plant hosts are of the greater interest for possible silvicultural use.

This review will briefly summarize the state of North American forest ecosystem practice and research involving endosymbionts and their host plants. A list of scientist or organizational contacts is included.

USE OF SYMBIOTIC NITROGEN FIXATION IN NORTH AMERICAN SILVICULTURE

In only one instance is a symbiotic nitrogen fixation system employed as a silvicultural tool on a relatively large scale. In the midwestern United States, a geographically replicated study of mixed plantations of black walnut (*Juglans nigra*) and autumn olive (*Eleagnus umbellata*), demonstrated that black walnut

Gordon, J.C. and Wheeler, C.T. (eds.). Biological nitrogen fixation in forest ecosystems: foundations and applications

ISBN 90-247-2849-5.
Printed in The Netherlands

growth was significantly improved as a result of association with the actinomycete-nodulated autumn olive (1). Because this research also offered impressive physical demonstration of long-term results, the mixed plantation silvicultural system has been widely adopted by growers of black walnut.

A number of other North American symbiotic systems have been shown to fix substantial amounts of N_2, transmit nitrogen to the ecosystem, and improve growth of associated plants. Yet, long-term demonstrations of the economic effectiveness of such management systems have not been offered. In the absence of hard proof of the value of symbiotic systems, forest owners are reluctant to manage their lands under such a departure from current practice. To ensure adaptation of knowledge from research, the value of new methods of forest management and their economic validity must be physically demonstrated.

Plants which develop N_2-fixing root nodule symbioses are used increasingly for so-called 'amenity' purposes (land reclamation and rehabilitation, wildlife and range habitat improvement, soil protection, etc.). At least 35 species of actinorhizal plants alone are planted for such purposes in Canada and the United States (2). A major impetus to this practice comes from the increasing area of lands requiring rehabilitation after surface mining for coal and oil shale. Perhaps successful demonstration of the use of symbiotic nitrogen-fixing systems for ecosystem repair will help speed expansion of similar practice to silvicultural applications.

CURRENT RESEARCH INITIATIVES

Approximately 40 North American universities, 10 industrial forest research organizations, and several regional forest research laboratories of Canadian and United States governments are involved in research with N_2-fixing symbiotic systems of potential value in silviculture. Subject-matter areas include: (1) Nature of the symbiosis; (2) Host plant improvement; (3) Nitrogen fixation and accretion; (4) Influence of symbiotic systems on the forest ecosystem; and (5) Management of symbiotic systems.

Forty eight genera of plants and two microsymbionts are the subjects of most research (Table 1). Thirteen of these plant genera associate in root-nodule symbiosis with *Frankia* spp. (actinomycete symbiosis), and 35 form symbioses with *Rhizobium* spp.

Nature of the symbiosis

Rhizobium symbioses have been observed and employed in agricultural systems for centuries, and in European forests for perhaps the past 100 years (3). In recent years, substantial research interest also has developed around the potential use of legumes in North America forests.

Actinomycete-nodulated woody plants have been recognized and studied for more than a century, but advances in knowledge have come slowly because pure cultures of the microsymbionts were not available (4). Recently, however, Callaham et al. (5) isolated an actinomycete that caused root nodulation of *Comptonia peregrina*, and Lalonde (6) independently confirmed the association. Soon afterward, Berry and Torrey (7) successfully isolated an actinomycetous endophyte from root nodules of *Alnus rubra*. As isolation techniques were improved, numerous pure cultures of *Frankia* spp. became available, and several genera of host plants (*Alnus*, *Eleagnus*, *Hippophae*, and *Shepherdia*) were shown routinely to become nodulated when inoculated with an appropriate host-specific *Frankia* (8).

Current research continues in three general directions:

I. Biology of *Frankia*
   A. Cultivation, maintenance, and preservation of strains of inoculum.
   B. Structure and function of the nodule
   C. Nature of the infection process
II. Processes Regulating N_2 Fixation by Actinorhizal Symbiosis
   A. Energetics of N_2-fixation
   B. Host-plant photoassimilate supply
   C. Allelopathic interference with nodulation
III. Adaptation and Improvement of Strains of *Frankia* and *Rhizobium* spp. for Effectiveness in Forest Ecosystems

Host plant improvement

A likely practical result of research into the nature of the symbiotic process is information that will facilitate genetic shaping of the microsymbiont for desired characteristics. Similar ability to regulate variations in host plant characteristics is necessary if optimum host-endophyte genotypes are to be developed for best adaptation to site, highly efficient nitrogen fixation, and desirable growth traits.

Genetic improvement of *Alnus glutinosa*, *A. rubra*, and *Robinia pseudoacacia* is underway. All three are tree-size plants of interest both for their role as symbionts in nitrogen fixation and fortheir potential value as wood fiber sources.

The probability of successful genetic improvement of *Alnus* appears to be high because of great species diversity and wide variation within species. Stettler (9) suggests that early sexual maturity, annual seed crops, and rapid juvenile growth combine to promise large genetic gains per unit of time and effort with *Alnus rubra*. Cloning, manipulation of chromosome number, and improvement of important symbiotic associates are considered important aspects of an *Alnus* improvement program (10).

Within about the past decade, a beginning has been made largely in the form of seed source screening, progeny testing, and establishment of provenance studies

Table 1. Symbiotic nitrogen-fixing systems of North America under study – microsymbiont and host plant genera with key* to contact with investigators.

System – genus of microsymbiont and host plant	Areas of research attention				
	Nature of the symbiosis	Host plant improvement	Nitrogen accretion – rate, amount	Influence of the symbiosis on the forest ecosystem	Host plant management
Frankia microsymbiont					
Alnus	1, 3, 5, 6, 7, 11, 13, 14, 17, 18, 21, 23, 25, 30, 31, 33, 34, 37, 38, 48, 49, 52, 55, 56, 58, 60, 65, 66	4, 9, 11, 18, 21, 23, 37, 38, 42, 51, 53	6, 9, 11, 14, 18, 20, 22 23, 26, 28, 30, 33, 35, 40, 41, 42, 44, 59, 62	6, 9, 11, 14, 16, 18, 20 22, 26, 28, 33, 35, 40, 41, 42, 44, 55, 66	6, 11, 12, 14, 16, 18, 20, 21, 22, 23, 26, 27, 28, 33, 35, 38, 40, 41, 42, 44, 47, 51, 52, 58, 63
Casuarina	1, 15, 56, 57, 66	15	15	15	15
Ceanothus	18, 48, 55, 60, 66		13, 18, 26, 27, 28, 29, 35, 42, 66	18, 26, 28, 35, 60, 66	18, 26, 28, 35
Cercocarpus	64, 66		27	27	27
Comptonia	1, 25, 30, 31, 55, 56, 66		30		
Coriaria	39, 66				
Cowania			27	27	27
Discaria	39				
Dryas	39, 66				
Eleagnus	1, 14, 23, 31, 33, 34, 39, 48	23, 33, 39	14, 33, 40, 41	14, 33, 40, 41, 54, 55	14, 23, 33, 40, 41, 54
Hippophae	31, 66				
Myrica	1, 3, 5, 16, 25, 31, 33, 34, 48, 54, 55, 56, 66				
Purshia	48, 64, 66		27, 67	27	27
Shepherdia	14, 31, 66		14, 35, 42, 66	14, 35	14, 35, 41
Rhizobium microsymbiont					
Acacia	36				
Albizia	36				
Amorpha	23		23		23
Amphicarpa	55				
Apios	55				
Astragalus	64		28	28	27, 28
Baptisia				54	54

Table 1 (continued)

System – genus of microsymbiont and host plant	Areas of research attention				
	Nature of the symbiosis	Host plant improvement	Nitrogen accretion – rate, amount	Influence of the symbiosis on the forest ecosystem	Host plant management
Cassia			19, 20, 50	19, 20, 50	19, 20, 50
Coronilla			8		41, 42
Cytisus	48		18	18, 54	18, 47, 54
Desmodium	36, 55				50
Discaria	39				
Eysenhardtia	36				
Glycine	32, 39				
Indigofera			50	50	50
Lathyrus	8, 24	24	8	54	24, 41, 42, 43, 54
Lespedeza	23		20, 23, 40, 41, 50	20, 23, 40, 41, 50	20, 23, 40, 41, 50
Leucaena	36, 64				
Lonicera				40	40
Lotus	2, 8		8, 41, 42, 43, 50	41, 42, 43, 50, 60	18, 27, 41, 42, 43, 50, 60
Lupinus	24, 36, 64	24	8, 20, 28, 29, 42, 43, 50	20, 28, 29, 43, 54, 61	20, 24, 27, 28, 41, 42, 43, 47, 50, 54, 61
Medicago	2, 24, 45	24			24, 41, 42
Melilotus	55				
Phaseolus					60
Pisum	32, 39		50	50	50
Pueraria	36				50
Robinia	21, 44, 55	21, 51, 53	21, 41, 44, 46, 51	21, 28, 40, 41, 44	21, 28, 40, 41, 44, 46, 50, 51
Thermopsis			42		47
Trifolium	2, 10, 24, 45, 55, 65	10, 24	8, 10, 19, 20, 40, 42, 50	10, 19, 20, 40, 42, 50, 61	10, 12, 19, 20, 24, 40, 41, 42, 43, 50, 61, 63
Ulex					50
Vicia	24, 55	24	20, 24, 40, 41, 42, 43, 50	20, 40, 41, 50	20, 24, 40, 41, 42, 43, 50, 63
Vigna	32, 39				50

* Numbers correspond to those before names of scientist contacts listed at end of chapter.

for *A. rubra* (11) and *A. glutinosa* (12). More recently, attention is being given to developing technology for vegetative cloning, a necessity for rapid maximization of genetic improvement.

Genetic improvement of *Robinia pseudoacacia* was begun only recently in Southeastern United States. As with *Alnus*, initial effort is directed toward selection and propagation of superior species.

Nitrogen fixation and accretion

The rate at which N_2 is fixed by plant host-endophyte symbiosis and the amount of nitrogen that accrues to the ecosystem depends on several factors: the nature of the symbiotic system, density of host plant occurrence, climate, photoassimilate supply, and undoubtedly, other influences yet to be discovered. The rate of N_2 fixation may be determined by the acetylene reduction assay, a measure of nitrogenase activity in the host plant nodule.

The amount of nitrogen accretion in the ecosystem is usually estimated from total nitrogen analyses of soil, forest floor, and plants of two adjacent ecosystems, one of which includes an N_2-fixing system ('treatment') and the other of which does not ('control'). A positive difference in total nitrogen content of the 'treatment' over that of the 'control' is presumed due to action of the N_2-fixing system.

Knowing the nitrogen-supplying effectiveness of a symbiotic system is necessary both in research-planning and in making management application decisions. Some such information is available for the major N_2-fixing host plant-endophyte systems (Table 2) but variability in physical parameters between different sites often makes it difficult to compare findings adequately.

Improved estimates of nitrogen accretion through symbiotic N_2 fixation are greatly needed. Soil, forest floor, and vegetative biomass should all be sampled to strengthen information on total ecosystem N capital and to increase understanding of how N capital is partitioned between various portions of the ecosystem.

Influence of symbiotic systems on the ecosystem

A major goal of intensive forest management is to produce a crop quickly in order to maximize return on investment. Management systems used to achieve this economic goal aim at maximizing crop tree growth rates commensurate with the potential biological productivity of the site.

The manager of forest lands dedicated to continuous production of commodity timber, must meet an additional consideration – that of maintaining or enhancing productivity of the land for future crops. Three factors are especially important in regard to this long-term biological goal:

1. Assuring a continuing, highly dynamic nutrient supply, especially that of nitrogen;
2. Maintaining a high level of soil carbon, essential to favorable soil structure, water-holding capacity, and nutrient ion exchange, and
3. Sustaining a vigorous, beneficial soil microflora and microfauna.

A number of North American nitrogen-fixing plants enhance ecosystem nitrogen supply and stimulate the growth of associated vegetation (Tables 2, 3). Addition of nitrogen, certainly a major cause of increased crop tree growth, probably is not the sole factor involved.

Soil carbon is increased under the influence of nitrogen-fixing plants including *Alnus rubra* (21) (44) (45) (46), *A. sinuata* (46), *Ceanothus velutinus* (29), and

Table 2. Nitrogen accretion in some North American ecosystems containing nodulated plants.

Plant species	Stand age	Nitrogen accretion	Reference no.
	years	kg ha^{-1} year^{-1}	
Alnus glutinosa	0–40	40	13
A. incana	0–5	362	14
	0–10	170	14
	0–20	156	14
A. sinuata	0–5	35	15
A. rubra	2–15	320	16
	0–30	208	17
	0–38	85	18
	0–14	163	19
	14–24	81	19
	24–65	40	19
	30–39	80–120	20
	0–40	100	21
A. rugosa	0–18	167	22
	0–16	85	23
Ceanothus spp.	—	60	24
	0–15	0–20	17
	0–10	72–108	25
	0–15	84	26
	0–17	80	27
	0–11	32	28
	0–12	100	29
Lespedeza spp.	0–4	100+	J. R. Jorgensen, pers. com.
Lotus scoparius	—	78	30
Lupinus excubis	—	55	30
Lupinus leucophyllus	—	7	J. M. Geist, pers. com.
Lupinus sericeus	—	10	J. M. Geist, pers. com.
Myrica cerifera	—	120	Fisher, pers. com.
Myrica pensylvanica	leaf litter	16–32	31
Purshia spp.	—	64	32
Robinia pseudoacacia	—	44	33
Trifolium subterraneum	0–4	100+	J. R. Jorgensen, pers. com., 1981

Table 3. Examples of crop tree growth stimulation when associated with a symbiotic nitrogen-fixing system.

Crop tree species	Symbiont host plant species	Reference no.
Acer saccharinum	*Robinia pseudoacacia*	34
Fraxinus americana	*Alnus glutinosa*	35
Fraxinus pennsylvanica	*Robinia pseudoacacia*	34
Juglans nigra	*Eleagnus umbellata*	1
	Robinia pseudoacacia	34, 36
Juniperus virginiana	*Robinia pseudoacacia*	34
Liquidambar stryaciflua	*Alnus glutinosa*	35
	Robinia pseudoacacia	34, 36
Liriodendron tulipifera	*Robinia pseudoacacia*	34, 36
Pinus echinata	*Alnus glutinosa*	35
Pinus ponderosa	*Ceanothus* spp.	37, 38
Pinus taeda	*Alnus glutinosa*	35
	Lespedeza cuneata	39
	Trifolium subterraneum	C. B. Davey 1980, pers. com.
	Trifolium vesiculosum	C. B. Davey 1980, pers. com.
Pinus virginiana	*Alnus glutinosa*	35
Platanus occidentalis	*Alnus glutinosa*	35
	Trifolium incarnatum	40
	Trifolium subterraneum	40
	Trifolium vesiculosum	40
	Vicia spp.	40
Populus spp.	*Alnus rubra*	41
Prunus serotina	*Robinia pseudoacacia*	34
Pseudotsuga menziesii	*Alnus rubra*	42
	Alnus rubra	43
	Ceanothus velutinus	26
Quercus alba	*Robinia pseudoacacia*	34

Robinia pseudoacacia (36). A significant increase in the rate at which soil carbon and mineral nutrients are cycled has been observed beneath *A. rubra* stands (18). Soil carbon improves physical soil properties to add a possibly strong increment to the potential biological productivity of ecosystems. Soil microflora and microfauna also may be beneficially affected, but little research has been done in this or other areas of interrelated effects of nitrogen-fixing systems on the ecosystem.

Potentially undesirable effects of introducing nitrogen-fixing plants into forest ecosystems include both physical and chemical impacts. Competition with crop trees for light, water, and nutrients is a factor that must be considered and evaluated. Increased damage to crop trees by insects or animals attracted by the nitrogen-fixing plant is also a possibility requiring better understanding. Chemical impacts include possible alteration of soil properties and allelopathic influences, neither of which are yet well understood. Increased soil acidity and decreased bases have been noted in the presence of *Alnus glutinosa* (13) (35), and

A. rubra (21) (45) (47). Allelopathic interference with crop trees by nitrogen-fixing plants is suggested as a possible area of concern by DeBell (48), although current research interest in this area conversely is centered about such effects by the crop tree on the *Frankia* microsymbiont (49) (50). We must understand better the interrelationships between symbiotic nitrogen-fixing systems and crop trees if we are to develop new management technologies for North American forests.

Management of symbiotic systems

Introducing nitrogen-fixing plant/microsymbiont mechanisms into managed forests implies the need for new silvicultural systems. The possible nature of such new systems has been discussed in some detail (48) (51 (52).

Two main systems appear feasible, one in which the nitrogen-fixing plant is a commercially valuable crop tree and the other, in which it is not. Each of the two main systems might have several variations (Table 4).

At the current state of knowledge, crop trees that fix nitrogen appear to be limited to *Alnus rubra*, *A. glutinosa*, *Casuarina*, *Eleagnus*, and *Robinia*. Silvicultural systems aimed at continuously-cropped, mediumlength (15–30 years) rotations of nitrogen-fixing plants grown for fiber or solid wood products would include *A. rubra* and *A. glutinosa* and, possibly, *Casuarina*. Shorter rotations

Table 4. Some silvicultural systems in which nitrogen-fixing plants might be employed.

Silvicultural system	N_2-fixing plants
I. N_2-fixing plants as crop trees	
A. Continuous cropping	
1. 25–30 year rotations for fiber and solid wood products	*Alnus rubra*, *A. glutinosa*
2. 5–15 year rotations for maximum biomass	*A. rubra*, *A. glutinosa*, *Casuarina* spp., *Eleagnus umbellata*, *Robinia pseudoacacia*
B. Alternate cropping with other crop tree species	All species above
C. Mixed plantations with other crop tree species	*A. rubra*, *A. glutinosa*
II. Non-commercially valuable N_2-fixing plants for ecosystem improvement	
A. Green manure crop, not harvested, followed by crop tree rotation	Herbaceous legumes, *Myrica*, *Dryas*, *Coriaria*
B. Understory with crop trees for part or all of rotation	*Alnus* spp. (other than *A. rubra* or *A. glutinosa*), *Cercocarpus*, *Comptonia*, *Cowania*, *Dryas*, *Eleagnus* spp., *Hippophae*, *Purshia*, *Myrica*, *Shepherdia*, Herbaceous legumes
C. N_2-fixing plant alone or in mixture with native plants for soil stabilization and amelioration in open-growing forests	*Ceanothus*, *Myrica*, *Purshia*, Herbaceous legumes

(5–15 years) to produce biomass from N_2-fixing trees alone would include the aforementioned species plus *Eleagnus* and *Robinia*. These same species might also be involved in systems of alternately cropping nitrogen-fixing woody plants and other commercial tree species. Mixed plantations, in which an N_2-fixing crop tree and one or more non-N_2-fixing crop trees would be grown simultaneously, might also include one of the five genera mentioned, according to geographical location.

The second major system is one in which the N_2-fixing plant component is not of commercial value, but is included to improve the ecosystem, particularly soil chemical and physical properties. Three variations of such a system appear to be potentially useful. One is to grow the N_2-fixer as a short-term green manure plant, after which a crop tree plantation would follow. Candidate plants for this option might include herbaceous legumes or low-growing actinorhizal plants such as *Coriaria*, *Dryas*, or *Myrica*.

In a second sub-system, the N_2-fixing plant would be grown as an understory with crop trees for part or all of a rotation. Nitrogen fixers of interest here include *Alnus* spp. (other than *A. rubra* or *A. glutinosa*), *Cercocarpus*, *Comptonia*, *Cowania*, *Dryas*, *Eleagnus*, *Hippophae*, *Purshia*, *Myrica*, and herbaceous legumes.

A third variation on the non-commercial N_2-fixer system is to use the N_2-fixing plant alone or in mixture with mature plants for soil stabilization and amelioration in open-grown forests. This system, applicable especially to interior western forests, would involve plant genera such as *Ceanothus*, *Myrica*, *Purshia*, or herbaceous legumes.

Contact with most North American centers of forestry research reveals that a considerable research effort on the management of nitrogen-fixing plants is underway in all major forested areas along four general lines:

I. Establishment and growth of N_2-fixing plants
   A. As crop trees
   B. In mixed stands
   C. On disturbed areas for site rehabilitation
II. Effect of N_2-fixing plants on growth of associates
III. Effect of traditional silvicultural practices (prescribed burning, soil scarification, herbicide use, soil fertilization, and tree spacing) on indigenous N_2-fixing plants
IV. Influence of site conditions on the microsymbiont

A high level of scientific effort is, of course, a necessary first step in developing a new technology. The amount of ongoing research with N_2-fixing plants adaptable to forest conditions is most encouraging. As biological and economic information is developed and displayed in pilotscale demonstrations, forest managers can be expected to adopt proven new systems that offer environmentally sound ways of increasing the productivity of North American forests.

RESEARCH WORKERS

The following list of investigator contacts is intended to provide information on location of centers of research – it is not a complete listing of North American scientists working in the subject matter area. Table 1 is keyed to this list by areas of investigation.

1. D. Baker, Kettering Research Labs. Yellow Springs, Ohio 45387.
2. D. K. Barnes, USDA SEA-CR, 1509 Gortner Ave., St. Paul, MN 55108.
3. D. R. Benson, The Biological Sciences Group, The University of Connecticut, Storrs, CT 06268.
4. C. L. Brown, School of Forest Resources, University of Georgia, Athens, GA 30601.
5. R. H. Burris, Department of Biochemistry, University of Wisconsin, Madison, WI 53706.
6. A. Carlisle, Petawawa National Forest Institute, Canadian Forestry Service, Chalk River, Ontario, Canada KOJ 1JO.
7. C. V. Carpenter, Weyerhaeuser Technical Center, Tacoma, WA 98477.
8. H. N. Chappell, International Paper Co., 34937 Tennessee Rd., Lebanon, Or 97355.
9. D. W. Cole, College of Forest Resources, University of Washington, Seattle, WA 98195.
10. C. B. Davey, School of Forest Resources, North Carolina State University, Raleigh, NC 27650.
11. J. O. Dawson, Department of Forestry, University of Illinois, Urbana, IL 61801.
12. P. M. Dougherty, Weyerhaeuser Co., 505 N. Pearl St., Centralia, WA 98531.
13. H. J. Evans, Laboratory for Nitrogen Fixation Research, Oregon State University, Corvallis, OR 97331.
14. R. J. Fessenden, Syncrude Canada Ltd., 10030 107 St., Edmonton, Alberta, Canada T5J 3E5.
15. R. F. Fisher, School of Forest Resources and Conservation, University of Florida, Gainesville, FL 32611.
16. G. Frisque, Laurentian Forest Research Center, 1080, Route du Vallon, P.O. Box 3800, Ste-Foy, Quebec, Canada G1V 4C7.
17. K. L. Giles, Life Sciences Department, Worcester Polytechnic Institute, Worcester, MA 01609.
18. J. C. Gordon, Forest Research Laboratory, Oregon State University, Corvallis, OR 97331.
19. C. Hagedorn, Department of Agronomy, Mississippi State University, Mississippi State, MS 39762.

20. S. G. Haines, International Paper Co., Rt. 1, Box 571, Bainbridge, GA 31717.
21. R. B. Hall, Department of Forestry, Iowa State University, Ames IA 50011.
22. P. Heilman, Western Washington Research Extension Center, Puyallup, WA 98371.
23. T. C. Hennessy, Department of Forestry, Oklahoma State Univ., Stillwater, OK 74074.
24. F. B. Holl, Department of Plant Science, University of British Columbia, Vancouver, B.C., Canada V6T 2A2.
25. J. G. Holt, Department of Bacteriology, 205 Science Bldg. I, Iowa State Univ., Ames, IA 50011.
26. L. Husted, Pacific Forest Products Ltd., 80607 E. Saanich Rd., Saanichton, B.C., Canada VO7 1MO.
27. Intermountain Forest and Range Experiment Station, 507 25th St., Ogden, UT 84401.
28. M. F. Jurgensen, Michigan Technological University, Houghton, MI 49931.
29. J. O. Klemmedson, Renewable Natural Resources Division, University of Arizona, Tucson, AZ 84721.
30. R. Knowles, Department of Microbiology, McGill University, Ste Anne de Bellevue, PQ, Canada H9X 1CO.
31. M. La Londe, Faculty de Forestiere et de Geodesie, Cite Universitaire, Université Laval, Quebec, Canada G1K 7P4.
32. T. A. La Rue, Boyce Thompson Institute, Tower Rd., Ithaca, NY 14853.
33. C. L. Lane, Clemson University, Clemson, SC 29631.
34. H. A. Lechevalier, Waksman Institue of Microbiology, Rutgers, The State University, P.O. Box 759, Piscataway, NJ 08854.
35. J. D. Lousier, MacMillan Bloedel Ltd., Nanaimo, British Columbia, Canada V9R 5H9.
36. C. J. Martinez, Nitragin Co., 3101 W. Custer Ave., Milwaukie, WI 53209.
37. C. A. Maynard, College of Environmental Science and Forestry, State University of New York, Syracuse, NY 13210.
38. L. H. McCormick, School of Forest Resources, The Pennsylvania State University, University Park, PA 16802.
39. W. Newcomb, Biology Department, Queen's University, Kingston, Ontario, Canada K7L 3N6.
40. North Central Forest Experiment Station, 1992 Folwell Ave., St. Paul, MN 55108.
41. Northeastern Forest Experiment Station, 370 Reed Road, Broomall, PA 19008.
42. Pacific Northwest Forest and Range Experiment Station, 809 N.E. 6th Ave., Portland, OR 97232.

43. Pacific Southwest Forest and Range Experiment Station, P.O. Box 245, Berkeley, CA 97401.
44. P. E. Pope, Department of Forestry and Natural Resources, Purdue University, West Lafayette, IN 47907.
45. W. A. Rice, Research Station, Agriculture Canada, Box 29, Beaverlodge, Alberta, Canada TOH OCO.
46. S. J. Riha, Department of Agronomy, Cornell University, Ithaca, NY 14853.
47. A. Rottink, Crown Zellerbach Corp., P.O. Box 368, Wilsonville, OR 97070.
48. K. R. Schubert, Department of Biochemistry, Michigan State University, East Lansin, MI 48824.
49. R. C. Schultz, Department of Forestry, 251 Bessey Hall, Iowa State University, Ames, IA 50011.
50. Southeastern Forest Experiment Station, P.O. Box 2570, Asheville, NC 28802.
51. K. Steinbeck, School of Forest Resources, Athens, GA 30602.
52. J. R. Thibault, Faculty de Forestiere et de Geodesie, Cite Universitaire, Université Laval, Quebec, Canada G1K 7P4.
53. E. Thor, Department of Forestry, The University of Tennessee, P.O. Box 1071, Knowville, TN 37091.
54. W. N. Tiffney, University of Massachusetts, Nantucket Field Station, P.O. Box 756, Nantucket, MA 02554.
55. J. D. Tjepkema, Harvard Forest, Harvard University, Petersham, MA 01366.
56. J. G. Torrey, Cabot Foundation for Botanical Research, Harvard University, Petersham, MA 01366.
57. J. H. Tyson, Department of Biology, University of South Florida, Tampa, FL 33620.
58. G. Vallee, Service de la Recherche-Forestiere, 2700 Einstein, Sainte Foy Quebec, Canada G1P 3W8.
59. K. Van Cleve, Forest Soils Laboratory, University of Alaska, Anchorage, AK 99701.
60. C. Van Raalte, School of Natural Science, Hampshire College, Amherst, MA 01002.
61. J. P. Vimmerstedt, Ohio Agricultural Research and Development Center, Wooster, OH 44961.
62. G. K. Voigt, School of Forestry and Environmental Studies, Yale University, New Haven, CT 06511.
63. G. White, Champion Timberlands, Box 250, Courtland, AL 35618.
64. S. E. Williams, Division of Plant Science, University of Wyoming, Laramie, WY 82071.

65. A. G. Wollum, Department of Soil Science, North Carolina State University, Raleigh, NC 27650.
66. C. T. Youngberg, Department of Soil Science, Oregon State University, Corvallis, OR 97331.
67. D. B. Zobel, Department of Botany and Plant Pathology, Oregon State University, Corvallis, OR 97331.

SUMMARY

Maintaining optimum nitrogen capital is a major silvicultural problem in most North American forest ecosystems. A rapidly growing body of information from research indicates that managed symbiotic nitrogen-fixing systems may offer a stable, long-term source of nitrogen. The most effective systems appear to be those in which root-nodule symbioses are formed between a variety of higher plants and the endophytes *Frankia* spp. or *Rhizobium* spp.

Several silvicultural systems have been suggested in which plant symbionts, either of intrinsic value as a crop or of interest solely for their contribution of nitrogen and organic matter, might be employed in silviculture. Such systems are not widely used in North America, however, despite encouraging knowledge gained from scientific study. The reluctance of forest managers to adopt new systems indicates the need for additional information. Before widespread adoption of such practices can be expected, economic feasibility must be demonstrated with physical examples.

Forest ecosystem-related research into symbiotic nitrogen fixation is proceeding at an increasing rate along five broad lines of investigation: (1) Nature of the symbiosis; (2) Host plant improvement; (3) Nitrogen fixation and accretion; (4) Influence of symbiotic systems on the ecosystem; and (5) Management of symbiotic systems.

REFERENCES

1. Funk DT, Schlesinger RC, Ponder F Jr.: Autumn-olive as a nurse crop for black walnut. *Botanical Gazette Supplement* 140: S110–114. 1979.
2. Fessenden RJ: Use of actinorhizal plants for land reclamation and amenity planting in the U.S.A. and Canada. In: *Symbiotic nitrogen fixation in the management of temperate forests.* Gordon JC, Wheeler CT, Perry DA (eds). Oregon State University, 1979, p 69–83.
3. Rehfuess KE: Underplanting of pines with legumes in Germany. In: *Symbiotic nitrogen fixation in the management of temperate forests.* Gordon JC, Wheeler CT, Perry DA (eds). Oregon State University, 1979, p 374–387.
4. Baker D, Torrey JG: The isolation and cultivation of actinomycetous noot nodule endophytes. In: *Symbiotic nitrogen fixation in the management of temperate forests.* Gordon JC, Wheeler CT, Perry DA (eds). Oregon State University, 1979, p 38–56.

5. Callaham D, Del Tredeci P, Torrey JG: Isolation and cultivation *in vitro* of the actinomycete causing root nodulation in *Comptonia*. *Science* 199: 899–902, 1978.
6. LaLonde M: Confirmation of the infectivity of a free-living actinomycete isolated from *Comptonia peregrina* root nodules by immunological and ultrastructural studies. *Canadian Journal of Botany* 56: 2621–2635, 1978.
7. Berry T, Torrey JG: Isolation and characterization *in vivo* and *in vitro* of an actinomycetous endophyte from *Alnus rubra* Bong. In: *Symbiotic nitrogen fixation in the management of temperate forests*. Gordon JC, Wheeler CT, Perry DA (eds). Oregon State University, 1979, p 69–83.
8. LaLonde M, Calvert HE, Pine S: Isolation and use of *Frankia* strains in actinorhizae formation. In: *Proceedings 4th Symposium on Nitrogen Fixation*. Canberra, Australia. p 296–299, 1980.
9. Stettler RF: Biological aspects of red alder pertinent to potential breeding pograms. In: *Utilization and management of alder*. Briggs DG, DeBell DS, Atkinson WA (eds). USDA Forest Service General Technical Report PNW-70, 1978, p 209–222.
10. Hall RB, Maynard CA: Considerations in the genetic improvement of alder. In: *Symbiotic nitrogen fixation in the management of temperate forests*. Gordon JC, Wheeler CT, Perry DA (eds). Oregon State University, 1979, p 322–344.
11. DeBell DS, Harrington CA: Mini-monograph on *Alnus rubra*. In: *Document presented to the FAO Technical Consultation of Fast Growing Plantations; Broadleaved Trees for Mediterranean and Temperate Zones*. Vol. 1. Invited Papers. Lisbon, Portugal, 1979, p 169–186.
12. Kellison RC, White G: Blak alder performance in the Southeast. In: *Symbiotic nitrogen fixation in the management of temperate forests*. Gordon JC, Wheeler CT, Perry DA (eds). Oregon State University, 1979, p 345–355.
13. Hansen EA, Dawson JO: Effects of *Alnus glutinosa* on hybrid *Populus* height growth in a short rotation, intensively cultured plantation. *Forest Science* 28: 49–59.
14. Van Cleve K, Viereck LA, Schlentner RL: Accumulation of nitrogen in alder (*Alnus*) ecosystems near Fairbanks, Alaska. *Arctic and Alpine Research* 3: 101–114, 1971.
15. Binkley D: Nitrogen fixation and net primary production in a young Sitka alder ecosystem. *Canadian Journal of Botany* 60: 281–284, 1982.
16. Newton M, El Hassan BZ, Zavitkovski J: Role of red alder in western Oregon forest succession. In: *Biology of alder*. Trappe JM, Franklin JF, Tarrant RF, Hansen GM (eds). Northwest Science Association Fortieth Annual Meeting Symposium Proceedings, 1968, p 73–84.
17. Zavitkovski J, Newton M: Effect of organic matter and combined nitrogen on nodulation and nitrogen fixation in red alder. In: *Biology of alder*. Trappe JM, Franklin JF, Tarrant RF, Hansen GM (eds). Northwest Scientific Association Fortieth Annual Meeting Symposium Proceedings, 1968, p 209–223.
18. Cole DW, Gessel SP, Turner J: Comparative mineral cycling in red alder and Douglas-fir. In: *Utilization and management of alder*. Briggs DG, DeBell DS, Atkinson WA (eds). USDA Forest Service General Technical Report PNW-70, 1978, p 327–336.
19. Lucken JO: *Biomass and nitrogen accretion in red alder communities along the Hoh River, Olympic National Park*. M.S. dissertation. Western Washington University, 1979.
20. Gessel SP, Turner J: Litter production by red alder in western Washington. *Forest Science* 20: 325–330, 1974.
21. Bormann BT, DeBell DS: Nitrogen content and other soil properties related to age of red alder stands. *Soil Science Society of America Journal* 45: 428–432, 1981.
22. Daly GT: Nitrogen fixation by nodulated *Alnus rugosa*. *Canadian Journal of Botany* 44: 1607–1621, 1966.
23. Voight GK, Steucek GL: Nitrogen distribution and accretion in an alder ecosystem. *Soil Science Society of America Proceedings* 33: 947–949, 1969.
24. Delwiche CC, Zinke PJ, Johnson CM: Nitrogen fixation by *Ceanothus*. *Plant Physiology* 40: 1045–1047, 1965.
25. Youngberg CT, Wollum AG: II. Nitrogen accretion in developing *Ceanothus velutinus* stands. *Soil Science Society of America Journal* 40: 109–112, 1976.
26. Youngberg CT, Wollum AG, Scott W: *Ceanothus* in Douglas-fir clearcuts: nitrogen accretion and impact on regeneration. In: *Symbiotic nitrogen fixation in the management of temperate forests*. Gorden JC, Wheeler CT, Perry DA (eds). Oregon State University, 1979, p 224–233.

27. Cromack K, Delwiche CC, McNabb DH: Prospects and problems of nitrogen management using symbiotic nitrogen fixers. In: *Symbiotic nitrogen fixation in the management of temperate forests*. Gordon JC, Wheeler CT, Perry DA (eds). Oregon State University, 1979, p 210–223.
28. McNabb DH, Geist JM, Youngberg CT: Nitrogen fixation by *Ceanothus velutinus* in northern Oregon. In: *Symbiotic nitrogen fixation in the management of temperate forests*. Gordon JC, Wheeler CT, Perry DA (eds). Oregon State University, 1979, p 481–482.
29. Binkley D, Cromack K, Fredriksen RL: Nitrogen accretion and availability in snowbrush ecosystems. *Forest Science* 28: 720–724, 1982.
30. Poth M: Biological dinitrogen fixation in chapparal. In: *Proceedings of the symposium on dynamics and management of Mediterranean-type ecosystems*, June 22–26, San Diego, California. Conrad CE, Oechel WC (eds). General Technical Report PSW-58, Forest Service, U.S. Department of Agriculture. Berkeley, California, 1982, p 285–290.
31. Benson DR: *Root nodules of Myrica pensylvanica (bayberry): structure, ultrastructure, and preparation of nitrogen-fixing homogenates*. Ph.D. dissertation. Rutgers University, 1978.
32. Jenny, M: The soil resource. Springer-Verlag New York, Inc. 1980.
33. Ike AF, Stone EL: Soil nitrogen accumulation under black locust. *Soil Science Society of America Proceedings* 22: 346–349, 1958.
34. Finn RF: Foliar nitrogen and growth of certain mixed and pure forest plantings. *Journal of Forestry* 51: 31–33, 1953.
35. Plass WT: Growth and survival of hardwoods and pine interplanted with European alder. *USDA Forest Service Research Paper* NE-376. 1977.
36. Carmean WH, Clark FB, Williams RD, Hanna PR: Hardwoods planted in old fields favored by prior tree cover. *USDA Forest Service Research Paper* NC-134. 1976.
37. Wallenberg WG: Effect of *Ceanothus* brush on western yellow pine plantations in the Northern Rocky Mountains. *Journal of Agricultural Research* 41: 601–612, 1930.
38. Maguire WP: Radiation, surface temperature and seedling survival. *Forest Science* 1: 277–284, 1955.
39. Jorgensen JR: Use of legumes in Southeastern forestry research. In: *Proceedings first biennial Southern silvicultural research conference*. Atlanta, GA. *USDA-FS General Technical Report* SO-34, 1980, p 205–211.
40. Haines SG, Haines LH, White G: Leguminous plants increase sycamore growth in northern Alabama. *Soil Science Society of America Journal* 42: 130–132, 1978.
41. DeBell DS, Radwan MA: Growth and nitrogen relations of coppiced black cottonwood and red alder in pure and mixed plantings. *Botanical Gazette Supplement* 140: S97–S101, 1979.
42. Tarrant RF: Stand development and soil fertility in a Douglas-fir red alder plantation. *Forest Science* 7: 238–246, 1961.
43. Miller RE, Murray MD: The effects of red alder on growth of Douglas-fir. In: *Utilization and management of Alder*. Briggs DG, DeBell DS, Atkinson WA (eds). *USDA Forest Service General Technical Report* PNW-70, 1978, p 283–306.
44. Tarrant RF, Miller RE: Accumulation of organic matter and soil nitrogen beneath a plantation of red alder and Douglas-fir. *Soil Service Society of America Proceedings* 27: 231–234, 1963.
45. Franklin JF, Dyrness CT, Moore DG, Tarrant RF: Chemical soil properties under coastal Oregon stands of alders and conifers. In: *Biology of Alder*. Trappe JM, Franklin JF, Tarrant RF, Hansen GM (eds). *Northwest Scientific Association Fortieth Annual Meeting Symposium Proceedings*, 1968, p 157–172.
46. Binkley D: Ecosystem production in Douglas-fir plantations: interaction of red alder and site fertility. *Forest Ecology and Management*. In press.
47. Bollen WB, Lu KC, Chen CS, Tarrant RF: Influence of red alder on fertility of a forest soil: microbial and chemical effects. *Oregon State University Forest Research Laboratory Research Bulletin* 12, 1967.
48. DeBell DS: Future potential for use of symbiotic nitrogen fixation in forest management. In: *Symbiotic nitrogen fixation in the management of temperate forests*. Gordon JC, Wheeler CT, Perry DA (eds). Oregon State University, 1979, p 451–466.
49. Dawson JO, Knowlton S, Sunn SH: Juglone inhibition of *Frankia* growth *in vitro*. *Plant Physiology Supplement* 67: 44 Abstract, 1981.

50. Thibault JR: *In vitro* allelopathic inhibition of nitrification by balsam poplar and balsam fir. *American Journal of Botany* 69: 676–679, 1982.
51. Haines SG, DeBell DS: Use of nitrogen-fixing plants to improve and maintain productivity of forest soils. In: *Impact of intensive harvesting on forest nutrient cycling. State University of New York (Syracuse)*, 1979, p 279–303.
52. Gordon JC, Dawson JO: Potential uses of nitrogen-fixing trees and shrubs in commercial forestry. *Botanical Gazette Supplement* 140: S88–S90, 1979.

10. Application of biological nitrogen fixation in European silviculture

PEITSA MIKOLA[a], PERTTI UOMALA[b] and EINO MÄLKÖNEN[c]

[a] *Department of Silviculture, Helsinki University, 00170 Helsinki 17, Finland*
[b] *Department of General Microbiology, Helsinki University, 00280 Helsinki 28, Finland*
[c] *Forest Research Institute, 00170 Helsinki 17, Finland*

INTRODUCTION: NITROGEN IN SILVICULTURE

Nitrogen cycling in forest ecosystems

Nitrogen is quite often a growth-limiting factor in both temperate and boreal forests, as is shown by the usually rapid response of trees to nitrogen fertilization. The total amount of nitrogen in most forest soils, however, is quite high. The limiting factor usually is not the amount of nitrogen but its availability to the trees.

Soil organic matter or humus is the main source of nitrogen to plants. Part of the nitrogen that trees take through their roots, returns to the soil in the litter and, thus, nitrogen is continuously circulating in the ecosystem. Numerous studies have been conducted to determine the amounts of nitrogen taken up by the trees, returned to the soil in litter, and retained by the growing biomass, as well as additions to the ecosystem from the atmosphere and leaching or other eventual losses. Depending on the climatic and soil conditions, and species and age of the tree stand, the annual nitrogen uptake varies from 34 to 123 kg ha^{-1}, of which 22–79 kg is returned to the soil in litter fall and 8–44 kg is retained in the biomass (1). The higher values refer to Central European deciduous forests and the lower ones to northern pine forests.

The total amount of nitrogen in soil is usually several thousands of kilograms per hectare, e.g. 4480–13760 kg ha^{-1} have been measured in deciduous forests in Belgium (2), 1905–15929 kg ha^{-1} in Bavarian forests (3), and 1017–1673 kg ha^{-1} in Finnish pine forests (4); the nitrogen content of soil organic matter usually varies from 1.5 to 3.0%.

The living biomass contains 100–400 kg ha^{-1} of nitrogen. Thus, the annual need of trees is relatively small, if compared to the total pool of the ecosystem. At any given time, however, only a small fraction, usually about 1%, of soil nitrogen is in available form. The bottle-neck is the slow rate of the mineralization of the organic-bound nitrogen in the soil.

Effects of silvicultural treatments on nitrogen cycling

European silviculture favors conifers. Conifers dominate naturally in the boreal

Gordon, J.C. and Wheeler, C.T. (eds.). Biological nitrogen fixation in forest ecosystems: foundations and applications
 ISBN 90-247-2849-5.
Printed in The Netherlands

zone of North Europe and in the Central European mountains, but they also have been extensively planted on Central European lowlands, where original forests had been deciduous.

In coniferous forests an unbalanced ratio generally prevails between the production and decomposition of organic matter. Production exceeds the rate of decomposition, which results in accumulation of raw humus, as has been thoroughly studied, for instance, by Wittich (5, 6) in Central European conditions and Sirén (7) in the far north. When raw humus accumulates, nitrogen also is blocked from circulation in non-available form in the humus layer and, thus, the shortage of available nitrogen increases along with thickening of the raw humus layer. Since cold climate further restricts the activity of soil microorganisms, the most acute shortage on nitrogen usually occurs in the north.

Clearcutting changes drastically ecological conditions and reverses humus development to the decomposition of raw humus and mineralization of its nitrogen reserves. In warmer climates mineralization of nitrogen can be so rapid that considerable amounts of nitrogen are lost through nitrification and leaching. In colder climates, however, as in Scandinavia, too rapid nitrification and leaching losses do not occur (8). Instead, in the Scandinavian countries the decomposition of raw humus and mobilization of nutrients has often been further promoted with prescribed burning. Nitrogen from the burnt organic matter is lost into the air; this loss, however, is compensated by the increased rate of mineralization of the remaining nitrogen (9). Soluble nitrogen probably is rapidly taken up by the ground vegetation and soil microorganisms which are strongly promoted by clearcutting and burning, thus preventing its leaching away. Soil cultivation, which is widely used in reforestation, accelerates nitrogen mineralization by improving temperature and moisture conditions for decomposing microorganisms.

Thinning also promotes growth of the ground vegetation and, to some extent, prevents the formation of raw humus. The remnants of herbaceous ground vegetation produce more readily decomposable humus than the needle litter of conifers.

Because the leaf litter of most deciduous trees is less acid, has a higher nitrogen content and decomposes faster than the coniferous needle litter, admixture of deciduous trees in coniferous forests has been repeatedly recommended, to prevent the accumulation of deleterious raw humus.

Improving nitrogen nutrition of trees

Two principal ways are available to improve the nitrogen nutrition of trees, viz. (a) activation of the decomposition of soil organic matter and accelerating nitrogen mineralization, and (b) adding nitrogen to the soil in easily available form, i.e. application of nitrogen fertilizers.

Prescribed burning has long been used in the Scandinavian countries, to reduce raw humus and to give an ash fertilization which, in turn, promotes the mineralization of nitrogen.

Several techniques have been developed in Central Europe to overcome the harmful effects of the raw humus and, if possible, convert it into a better type of humus with active nitrogen mineralization. These techniques, as has been described in details by Wittich (5, 10), include thorough soil cultivation, mixing humus into mineral soil, and heavy application of lime and often phosphorus and potassium also. Best results, however, have been obtained if biological nitrogen fixation has been combined with the above treatment, as will be described later.

The easiest way to overcome deficiency symptoms and to improve the nitrogen nutrition of trees is the application of mineral fertilizers. Therefore nitrogen fertilizers are widely used both on raw humus sites and on soils with a low humus and nitrogen content. In Finland, for instance, nitrogen is almost the only nutrient which is applied to mineral soils, whereas peat soils usually respond more strongly to phosphorus and potassium fertilization.

Nitrogen fertilization in forests, however, involves some drawbacks. Besides the relatively high price of the fertilizers, their effect lasts a rather short time and, to be efficient, the treatment must be repeated quite frequently. Baule and Fricker (11), for instance, recommend the application of 200–400 kg of fertilizer (20% N) per hectare at 2 or 3 year intervals throughout the whole rotation. In Finland nitrogen fertilization at 5–9 year intervals with 150 kg N/ha is recommended. Although leaching of nitrate, evaporation of ammonia, and denitrification are negligible in natural forests, after fertilization with mineral nitrogen they can attain alarming proportions. To avoid these drawbacks, the application of biological nitrogen fixation has been repeatedly suggested and experimented with in various ways in many European countries.

Biological nitrogen fixation

Nitrogen fixing plants in natural European forests are few. In the Finnish forests, for instance, only two leguminous species are relatively common, *Lathyrus vernus* and *Vicia silvatica*, and even they grow as solitary plants on the best sites and never form dense communities. Therefore legumes can not play any significant role in the boreal forests of Europe. In Western and Central Europe there are more native leguminous forest plants, e.g. bush-like species of *Cytisus*, *Genista*, and *Ulex*. In some localities they can grow in dense thickets, probably then fixing considerable amounts of nitrogen, but in normal forests they are mostly absent.

Alder (*Alnus*) species probably are the most important nitrogen fixing vascular plants in European forests. Grey alder (*Alnus incana*) is a common pioneering tree species and may have great importance in colonizing exposed soils and creating there the first humus and nitrogen reserves. *Alnus viridis* may have similar role in the tree-line zone of the Central European mountains, whereas the occurrence of *Alnus glutinosa* is more restricted to wet sites. The other European actinorrhizal species (*Myrica*, *Hippophae*, and *Elaeagnus*) do not ordinarily belong to forest vegetation.

The role of free-living nitrogen-fixing bacteria is small in most forest soils. Their activity is restricted by high acidity, low temperature and other unfavorable soil factors.

More important than free-living bacteria are symbiotic and semisymbiotic bacteria which are quite numerous in forest ecosystems. Symbiotic cyanobacteria are components in lichens, and many species are nitrogen-fixing (12). Other bacteria live in semisymbiotic association with mosses, and recently they have also been found as regular associates in the rhizosphere of many grasses and other plants of forest ground vegetation. Most probably these bacteria are the main cause of biological nitrogen fixation which has been detected in the northern forests. Granhall and Lindberg (13) measured an annual fixation rate of 0.2–0.3 kg ha^{-1} in a dry pine forest and 3.8 kg ha^{-1} in a moist spruce forest in central Sweden, and somewhat higher rates have been measured in temperate deciduous forests (14). Although the amounts are not large, apparently they are usually sufficient, together with the nitrogen brought down in the rain water, to compensate and even exceed eventual losses through leaching and evaporation. Continuous nitrogen fixation from the atmosphere makes possible the accumulation of organic nitrogen in the soil.

APPLICATION OF BIOLOGICAL NITROGEN FIXATION

Legumes in silviculture

Lupines have long been used for melioration of degraded forest soils in Germany. Experiments were started already nearly a hundred years ago and since then the growing of lupine has belonged to the regular silvicultural practice in some forest areas. Exact experimental data, however, on the effects of lupine on soil and tree growth are few.

Nitrogen deficiency is particularly severe in forests where litter gathering has been practised in former times. This practice of annual removal of all the litter and part of the humus layer for agriculture had continued in some parts of Germany for centuries and was abandoned as late as a few decades ago. Soils where litter has been repeatedly removed are low in humus and nitrogen and mineral nutrients. Their systematic melioration with lupine was first started in Ebnath Forest District in Bavaria during the first decade of this century (10). The technique was as follows: after clear-felling, the stumps and part of the raw humus were removed, 2000 kg ha^{-1} of lime was added and the whole area was cultivated. In the same autumn 2–3 kg of *Genista* seed was broadcast. Next spring spruce seedlings were planted and between them perennial lupine (*Lupinus polyphyllus*) was sown in rows.

The results were so encouraging, that the technique was rapidly applied to other parts of Germany, too (10, 15). Hassenkamp (16), in NW Germany, developed a still more intensive technique, in which the area after stump removal and complete cultivation received a full CaNPK fertilization and was then used two years for agricultural production (rye, oat or potato), and thereafter trees were planted and lupine and *Genista* was sown. Lidl (17) again, in Bavaria, did not cultivate the soil but only used heavy liming (1500–2500 kg/ha), and after a few years perennial lupine could be sown. Experiments and practical experiences on the use of lupine at afforestation have further been reported by Lent (18), Wehrmann (19), Seibt (20), and others.

On extensive studies Wittich (5, 10) has shown the drastic change of soil condition after the above melioration treatments. An example of these results is shown in Table 1. Lupine, together with soil cultivation and fertilization, changed the humus type completely, from an inert raw humus to active mull with rich animal life and rapid nitrogen mineralization.

Remarkable effects of lupine on the growth of young trees also have been reported. According to Wiedemann (21) and Seibt (20) the height and volume growth of conifers with lupine exceeded two- or threefold the growth of trees in other plots which had received the same melioration treatments but in which lupine had not been sown.

Besides their use at stand establishment, soil melioration with legumes has been practised in young stands to overcome growth stagnation, which commonly occurs at the pole stage in plantations on degraded soils (22–26). Although this technique often has given good results, there also are difficulties involved. Successful establishment of lupine necessarily needs soil cultivation, which usually causes severe root damage and, thus, may even cause growth depression for a few years and promote the root rot disease by *Fomes annosus*.

Table 1. Nitrogen content of soil organic matter in two experimental plots, 10 years after melioration with the Hassenkamp technique (after Wittich, 5).

	N in % of organic matter	
	Plot 143	Plot 177
Control	1.52	1.78
Melioration without N-fixing plants	1.73	2.11
Melioration with lupine	2.33	2.72
Melioration with alder		2.59

Therefore the technique is restricted mainly in pine stands. Soil cultivation also increases the risk of wind damage. The stand must be heavily thinned, to give sufficient light for the lupine; for this reason the technique is not suitable for the shade-tolerant tree species, whereas more light comes through pine canopy.

For the above reasons practising foresters prefer the simpler technique of using nitrogen fertilizers, and the growing of lupine in young and middle-aged stands is not as popular now as it was a few decades ago (26).

Baule and Fricker (11) give following recommendations for soil melioration with legumes:

1. The annual yellow lupine (*Lupinus luteus*) is suitable for dry sandy sites. For all the other sites the perennial blue lupine (*Lupinus polyphyllus*) is recommended.

2. Lupine is sown at a rate of 5–7 kg/ha. Inoculation of the seed with *Rhizobia* is advisable.

3. At stand establishment lupine is usually sown at the same time when trees are planted. Lupine, however, seems to favor the needle cast disease of pine (*Lophodermium pinastri*); therefore in pine plantation lupine should be sown a few years after tree planting.

4. For successful establishment of lupine, sufficient fertilization with Ca, K and P is necessary. A slight initial dose of N also is advisable, whereas too heavy nitrogen fertilization reduces the assimilation of atmospheric nitrogen.

5. A thorough soil cultivation is usually necessary.

Rehfuess (26) summarizes the German experiences on lupine growing as follows: 'Intermixing of legumes with conifers during the early phase of plantation establishment is practiced only on restricted areas, more in spruce and douglas fir than in pine cultures, under conditions where mechanical soil preparation and fertilization are necessary for other reasons, where rapid height growth enables the conifers to overcome the competition of lupines fast and where wide tree spacing allows lupines to grow for a long period.'

Lupine has also been used for soil melioration in Eastern Europe. First experiments to improve the growth of pine plantations on sandy soil were started in Lithuania in 1894 (27). According to the review by Ronkonen (28), lupines are today widely used in the Baltic countries, Belo-Russia, Ukraine, and in other parts of the Soviet Union. Statistical data, however, on the extent of this activity are not available.

Similar effects of lupine on soil properties have been reported from the Soviet Union as from Germany. Lupine has increased the humus and nitrogen content of the soil, improved the physical properties of humus and even promoted the potassium nutrition of pine. Through its deep root system lupine also is able to bring mineral nutrients from deeper soil layers into circulation (29, 30, 31).

On the basis of a 40-year old experiment Žilkin and Beregova (32) estimated that under Belo-Russian conditions soil melioration with lupine would increase

Table 2. Stemwood production of Scots pine, $m^3 ha^{-1}$, with and without lupine (32). (The figures up to 40 years are empirical, thereafter estimated).

	Stand age 20	30	40	50	60 $m^3 ha^{-1}$	70	80	90
Control (without lupine)								
Standing crop	27	89	144	202	251	290	327	352
Removed	6	22	69	109	154	196	233	268
Total production	33	111	213	311	405	486	560	620
With lupine								
Standing crop	52	115	197	274	341	393	438	473
Removed	11	29	77	125	174	215	256	289
Total production	63	144	274	399	515	608	694	762

the yield of a pine plantation during one rotation by 120–140 $m^3 ha^{-1}$ (Table 2).

The same species (*Lupinus polyphyllus*) is used in the Soviet Union as in Germany. Like in Germany, a thorough soil preparation is necessary for successful establishment of lupine. To reduce early mortality, planting of lupines is recommended, instead of direct seeding. 2-year old seedlings can be planted successfully even without soil preparation. On dry sandy soils lupines and pines can be planted simultaneously, whereas on other sites lupine is usually planted 2–4 years after pine. To maintain a vigorous growth of lupine for a long period, pine plantations must be thinned from time to time (33). According to Polish experience (34), possibilities for growing lupine on dry sites are restricted by lack of water.

Promising results of melioration of poor forest soils with lupine and the leguminous bushes broom (*Cytisus scoparius*) and gorse (*Ulex* spp.) have been obtained in Western Europe. Thus, for instance, in a Sitka spruce plantation, which was established after complete cultivation and liming on an old red sandstone soil in Ireland, the mean height at the age of 17 years was with legumes (lupine and broom) 61% and the basal area 144% greater than in the control plots without legumes (Dr. N. O'Carroll, personal communication).

The few sporadic experiments, which have been conducted earlier with lupines in North Europe, have largely failed, mainly because of poor wintering. Apparently the winter is too severe for lupine. Furthermore, it is true, in these experiments soil has usually not been so thoroughly cultivated and limed as is customary in the German techniques, for instance. According to more recent experiments, however, the Alaskan lupine (*Lupinus nootkatensis*) is a hardy species which thrives well even in North Scandinavia (35) and Iceland (36).

Actinorrhizal plants

Alders (*Alnus* spp.) are the only actinorrhizal plants which have been used to some extent in European forestry to fix atmospheric nitrogen; even these, however, mainly on an experimental scale.

The soil-improving capacity of alder has been known for a long time. The nitrogen content of alder leaf litter is usually 2–3‰, i.e. two to three times higher than in the litter of other European broad-leaved trees (37, 38, 39), and the nitrogen of alder leaf litter is easily mineralized (40). Viro (39) also showed that before the leaf-fall nitrogen is not translocated from alder leaves to other parts of the tree as much as is the case in other tree species.

The ancient Finns applied the nitrogen-fixing capacity of gray alder in their primitive farming system or 'swidden', although without knowing it. In this system trees were first cut and burned and then the area was used for growing grain. In a few years the soil fertility was exhausted and the area was abandoned. It was then rapidly colonized by gray alder; after 10 or 20 years the young alder stand was cut and burned, and again several good grain crops could be taken.

This system of shifting cultivation, which in Eastern Finland was practised until the early 1900's and in the Soviet Carelia somewhat longer, was possible because the alder rotation repeatedly built up a nitrogen reserve, which was sufficient for grain production for a few years.

A remarkably good growth of spruce plantations after gray alder also has been recorded (41) although the credit for that has not always been given to the previous alder generation.

In spite of the well-known favorable effects of alder on the soil, gray alder is usually considered by practising foresters as a weed which should be eradicated from the forests. The reason for this attitude is the strong competitive capacity and the low commercial value of gray alder. Conversion of gray alder forests into productive spruce stands in North and East Europe is considered as a silvicultural task of primary importance. Eventual utilization of the nitrogen-fixing capacity of alders is, therefore, only focused on the melioration of embryonic or degraded soils.

In some of the plots where Wittich (5) studied the effects of different melioration techniques, alder grew naturally and its favorable effect on soil appeared distinctly (Table 1). Gray alder also was included in one of the forest fertilization experiments, which Hausser (42) conducted on degraded soils in the Black Forest, where litter removal had been practiced for centuries. Some of the results are summarized in Table 3. To get well established, gray alder needed a substantial Ca and P fertilization, particularly P. Surprisingly, fertilization with mineral nitrogen (Plots CaN and CaPN) somewhat reduced the growth of both species from that without added nitrogen (Plots Ca and Ca P, respectively). Evers (43) explains this assuming that heavy application of mineral nitrogen reduced the nodulation of alder and the fixation of atmospheric nitrogen. In this experiment alder and spruce were planted in the same year. To reduce competition, alder had to be thinned heavily at the age of six years.

Table 3. The height of alder and spruce (planted in 1955) and the N content of spruce needles in a fertilization experiment on degraded soil in the Black Forest (after Hausser and Evers, 42, 43).

Treatment	Height of alder, in 1960, cm	Height of spruce, in 1963, cm	N % in spruce needles in 1963
Control	110	50	1.16
Ca	210	120	1.33
Ca N	180	110	1.24
Ca P	380	200	1.53
Ca PN	350	170	1.41

The soil improving properties of gray alder have also been reported from the Soviet Union (44, 45); growing gray alder is there particularly recommended together with pine on poor sandy soils (28).

The amazing results of the greenhouse experiment by Virtanen (46) gave an impetus to field experimentation in Finland, aiming primarily at melioration of poor sites. In the first experiment, which was started in 1950, gray alder and pine were planted on a coarse sandy soil. Stand measurement 14 years later (Table 4) showed that pine had grown considerably better with alder than alone, and the nitrogen and chlorophyll contents of pine needles also were higher in the mixed plantation than in the control plot (38). In this experiment the growth of alder was very poor on the dry sandy site and, therefore, pine did not suffer from competition.

Another experiment was conducted on the bottom of a peat-bog where peat had been removed for fuel (47). Pines were planted in 1964 and fertilization (20 g of NPK per seedling) was done next year; gray alder was planted in 1967. Assessment of the experiment in 1971 revealed a remarkable effect of alder (Table 5). At this stage the effect of the initial fertilization was apparently disappearing and the growth was declining (b), whereas with alder the-annual growth of pines was increasing even without fertilization (c).

Black alder also was tried in the above experiment. On the open lowland site, however, it suffered badly from spring frosts.

Table 4. The mean height and basal area of pines, and the nitrogen and chlorophyll content in top shoot needles in pure and alder-mixed pine plantations at 14 yrs (38).

	Pine alone	Pine with gray alder
Pine stems ha^{-1}	2900	3050
Mean height, m	3.0	3.9
Basal area, m^2 ha^{-1}	2.45	3.73
N in top shoot needles, %	1.20	1.41
Chlorophyll in top shoot needles, mg $100 g^{-1}$	534	634

Table 5. Development of pine seedlings up to the age of 8 years on a peat-bog bottom (47).

	Pine alone		Pine with alder	
	Without fertilization a	Fertilized b	Without fertilization c	Fertilized d
Mortality, %	55	22	9	24
Height, cm	36	125	113	161
Top shoot in 1970, cm	6	22	26	32
Top shoot in 1971, cm	6	21	31	36
Top shoot needles in 1971, mm	27	37	47	49

Other nitrogen fixing systems

Free-living aerobic bacteria probably are the most important nitrogen fixing organisms in mature temperate and boreal forests, even although their efficiency is very low (14, 48). The semisymbiotic bacteria in association with the roots of many herbs and grasses, again, may play principal role on clear-cut areas and seedling stands. Chances for silvicultural manipulation of these systems to promote their nitrogen fixation are minimal, however.

The role of lichens in nitrogen fixation has received attention particularly in arctic and subarctic ecosystems (e.g. 48, 49). A few species only, such as *Stereocaulon pascale*, *Peltigera aphthosa*, and *Nephroma arcticum* (Table 6) are able to fix appreciable amounts of nitrogen, whereas the *Cladonia* spp., the main winter diet of reindeer, are inefficient. Few prospects exist for silvicultural application.

Among mosses the highest rate of nitrogen fixation has been detected in cyanobacteria in association with some *Sphagnum* species (51). Since Sphagnum fens, because of excessive water, are unsuitable for tree growth, this nitrogen fixing system cannot be utilized in forestry.

FUTURE DEVELOPMENT

Comparison of different alternatives

There are three principal alternatives to increase the source of available nitrogen in the soil, (a) promoting the mineralization of soil organic-bound nitrogen, (b) biological nitrogen fixation, and (c) mineral fertilization. All of them have their advantages and drawbacks. A sound policy would be, therefore, the use of all the three methods, with preference depending on several biological and economic factors.

Table 6. Annual nitrogen fixation by some lichens in North Scandinavia.

Species	kg N ha^{-1}	Reference
Stereocaulon pascale	1.0–7.7	50
Nephroma arcticum	2.1	49
Peltigera aphthosa	0.1–0.2	49

Fertilization with mineral nitrogen is an easy technique and, hence, attractive to foresters. Trees usually respond rapidly to fertilization and its economic profitability is therefore easily demonstrated. A serious drawback, however, is the inefficient uptake of fertilizer nitrogen by the trees. Less than 10% of the added nitrogen may find its way in the first growing season directly to trees, whereas more than 90% is lost through leaching, denitrification and volatilization, or bound in ground vegetation and soil humus (52, 53, 54, 55). Even a young stand at its fastest growth phase can take little more than 25% of the added nitrogen (56). Because the available nitrogen is rapidly taken by the trees or lost fertilization must be repeated frequently, usually at 5–7 year intervals if maximum response is to be achieved. The possibility of harmful side effects of nitrogen fertilization, such as eventual rise of nitrate content in ground water, also has created concern, particularly among environmentalists.

Theoretically, biological nitrogen fixation would be an ideal method which avoids all the disadvantages of the mineral nitrogen fertilizers. Its practical application in European forestry, however, is hampered by several technical difficulties. The most serious obstacles are (a) competition and (b) the low commercial value of the nitrogen fixing species. Gray alder, in particular, is a strong competitor on sites suitable for it, and therefore alder mixture or overstorey often suppresses more than promotes the main tree crop. If alder is used as nitrogen fixer, it must be carefully watched, thinned and removed on time to eliminate harmful competition. Since it is a non-commercial species, gray alder is disliked by foresters; planting gray alder would be regarded as poor silviculture and may be against present forest law in some countries.

The superiority of biological nitrogen fixation to the use of mineral fertilizers is hard to demonstrate. Nitrogen fertilization affects forest growth immediately but for a short time only, whereas the influence of nitrogen-fixing plants appears later but lasts longer. These effects are, however, difficult to compare quantitatively and still more difficult to estimate in money. Considering the long time scale of experiments and the variable inflation rate and other unknown economic factors, monetary calculations may have only an academic value.

Because of the above reasons, biological nitrogen fixation in forestry practice has mainly remained at experimental stage and even conclusive results of the experiments are scanty.

Apparently, possibilities for the application of nitrogen-fixing plants in forestry are limited in the future, too. Primarily the three alternative systems, as discussed by DeBell (57) for American forestry, also come into consideration in Europe, viz. (a) nitrogen-fixing plants as understorey during early stand growth, (b) mixed stands, and (c) alternate rotations of nitrogen-fixing plants and more valuable commercial trees.

The first alternative appears most promising. The use of biological nitrogen fixation will most probably remain mainly as a method of reclamation of waste areas, such as spoil banks of mines, old gravel pits, peat-bog bottoms after industrial peat exploitation, and other denuded soils with low nitrogen content. On such sites new vegetation is usually established most successfully by planting alder, lupine, or other nitrogen-fixing plants (e.g. *Cytisus* or *Ulex*) together with the main tree species. Alder also can be first grown as a pioneer crop and a shade tolerant species, e.g. spruce, is planted later under the alder canopy. On the other hand, experience of planting alder or lupine afterwards under the crown canopy of existing tree stands has been disappointing (26).

Growing mixed stands of nitrogen-fixing and non-fixing trees is restricted by the different growth pattern of the species. Alder, however, could perhaps be grown in mixture with other short-rotation trees, such as willow. As a consequence of the energy crisis, there has recently appeared a great interest in short-rotation trees for fuel. The 'energy forests', dense plantations of fast growing trees, use large amounts of nutrients and therefore need heavy nitrogen fertilization, which perhaps could be at least partly replaced by the biological nitrogen fixation of alder. Growing pure alder stands for fuel with short rotation also has been taken into account.

Alternating rotations of nitrogen-fixing trees and valuable timber trees, pine or spruce, seem less attractive. Even a short rotation of alder is too long if the site is suitable for valuable longrotation species. Growing more valuable deciduous hardwoods, either as admixed trees or in alternating rotations with conifers, may be more advisable because their litter promotes both the activity of free-living nitrogen-fixing bacteria and other biological soil processes (5, 14).

Biological nitrogen fixation can never totally replace nitrogen fertilizers or other methods of site melioration. Sound silviculture, however, combines the benefits of all alternative techniques. Nitrogen-fixing plants usually increase the humus content of the soil and improve humus properties. In moist boreal coniferous forests, for instance, the total amount of nitrogen can be very large, up to 6000 kg ha^{-1} (58), of which, however, less than 1% usually is in available form. Thus, there is no need to increase the total nitrogen, but alder or lupine can greatly improve the humus quality and accelerate the mineralization of organic nitrogen present (5). On sandy barrens and denuded soils, again, the contents of both humus and nitrogen are low and both of them can be corrected by nitrogen fixing plants. To be successful, however, in both cases melioration may also need mechanical soil preparation and perhaps some mineral fertilizers, lime and phosphorus and even a small dose of nitrogen.

Biological nitrogen fixation belongs to the techniques of the most intensive forest management. One can anticipate, therefore, that along with increasing intensity of forest management, nitrogen-fixing plants also will find more applications in European silviculture.

Current research activities

Generally speaking, interest in biological nitrogen fixation has gone up and down in parallel with the price of nitrogen fertilizers. When biological nitrogen fixation was first discovered and rapidly applied to agriculture, possibilities for its use in forestry also were foreseen and preliminary research was started in various countries, resulting in field application as has been described above. When, after the discovery of the Haber-Bosch techniques, cheap nitrogen fertilizers came to the market, foresters turned their interest from biological nitrogen fixation to mineral fertilization. Even existing field experiments were often neglected and forgotten. Intensive research on forest fertilization has been in progress for a couple of decades and, thus, both the biology and economy of nitrogen fertilization in forests are relatively well known today. The energy crisis of 1970's and consequent rise of the prices of fertilizers again revived the interest in biological nitrogen fixation. At the present time, however, research in this field is still a small side branch of extensive forest fertilization research.

The main emphasis of recent research has been in basic studies. The total nitrogen budget of forest ecosystems and the role of biological nitrogen fixation in it are still poorly known. Therefore both national and international research programs have been organized to cover all the various phases of this complicated area. Thus, nitrogen relations were intensively studied in the forest projects of the International Biological Program (59) and the Man and Biosphere program (60), for instance. These studies have produced much valuable basic information for both the fertilization of forests and the application of biological nitrogen fixation in forestry.

Field experiments and other practically oriented studies also have recently been started in many forest research organizations. The main interest is directed in different nitrogen-fixing plants, depending on climate and former tradition; alder in Northern Europe (35, 61), lupine in Central Europe, and broom and gorse in the British Isles. Some new approaches also have been taken, e.g. the Alaskan lupine in Northern Sweden (35).

Since field experiments with forest trees require long time and produce results slowly, the most urgent task at present would probably be an inventory and remeasurement of existing old experiments.

REFERENCES

1. Duvigneaud P, Denaeyer-De Smet S: Biological cycling of minerals in temperate deciduous forests. In: *Analysis of temperate forest ecosystems*, Reichle DE (ed). Berlin, Springer-Verlag, 1970, p 199–225.
2. Duvigneaud P, Froment A: Recherches sur l'ecosystème forêt. Série E: Forêts de Haute Belgique. *Bull Inst Roy Sci Nat Belg* 45: 1–48, 1969.
3. Emberger S: Die Stickstoffvorräte bayrischer Waldböden. *Forstw Cbl* 84: 156–193, 1965.
4. Mälkönen E: Annual primary production and nutrient cycling in some Scots pine stands. *Comm Inst Forest Fenn* 84(5): 1–87, 1975.
5. Wittich W: Der heutige Stand unseres Wissens vom Humus and neue Wege zur Lösung des Rohhumusproblems im Walde. *Schriftenreihe d Forstl Fak d Univ Göttingen* B 4, 1952.
6. Wittich W: Die Grundlagen der Stickstoffernährung des Waldes und Möglichkeiten für ihre Verbesserung. In: *Der Stickstoff – Seine Bedeutung für die Landwirtschaft und die Ernährung der Welt*. Stalling AG, Oldenburg, 1961.
7. Sirén G: The development of spruce forest on raw humus sites in Northern Finland and its ecology. *Acta Forest Fenn* 62(4): 1–408, 1955.
8. Nykvist N: Changes in the amounts of inorganic nutrients in the soil after clear-felling. Nutrient cycle in tree stand – Nordic symposium. *Silvia Fennica* 11(3): 224–229, 1977.
9. Viro PJ: Prescribed burning in forestry. *Comm Inst Forest Fenn* 67(7): 1–49, 1969.
10. Wittich W: 50 Jahre Ebnath (Weitere Untersuchungen über die Melioration extrem ungünstiger Rohhumusböden). *Forstw Cbl* 75: 407–422, 1956.
11. Baule H, Fricker C: *Die Düngung von Waldbäumen*. Bayer Landwirtschaftsverl, Augsburg, 1967.
12. Henriksson E: Kvävefixering hos lavar. In: *Processer i kvävets kretslopp*, Rosswall T (ed), Stat Naturv Rap 1213, Solna, Sweden, 1980, p 74–77.
13. Granhall U, Lindberg T: Nitrogen input through biological nitrogen fixation. In: *Structure and function of northern coniferous forests – An ecosystem study*, Persson T (ed), *Ecol Bull* (Stockholm) 32: 333–340, 1980.
14. Eregova SV, Kalininskaya TA: Nitrogen fixation in a native Norway spruce forest and derivative forest types of Moscow district. *Lesovedenie* 5: 31–37, 1981.
15. Wiedemann E: Die Leguminosendünging in Ebnath. *Forstw Cbl* 49: 13–15, 1972.
16. Hassenkamp W: Die Umwandlung von Rohhumusboden in Mullboden durch Waldfeldbau und Leguminosenanbau. *Forstarchiv* 12: 41–57, 1941.
17. Lidl O: Erfahrungen mit Kalkdüngungsversuchen im bayerischen Voralpengebiet. In: *Der Wald braucht Kalk*, Landw Verl, München, 1948.
18. Lent J: Die Dauerlupine in der deutschen Waldwirtschaft. *Mitt d Deutsch Landw Ges* 49, 1943.
19. Wehrmann J: Die Nährelementversorgung der Dauerlupine auf Sandböden der Oberpfalz. *Forstw Cbl* 75: 357–366, 1956.
20. Seibt G: Forstliche¨ngung in der Lüneburger Heide. In: *Der Wald braucht Kalk*, Kölner Univ Verl 1959.
21. Wiedemann E: *Die Fichte* 1936. Hannover, Verl M u H Schaper, 1937.
22. Krauss HH: Die Methodik der Meliorationsmassnahmen im norddeutschen Diluvialgebiet. *Forst und Jagd, Sonderheft Waldbodenmelioration* II, p 24–34, 1958.
23. Bredow-Stechow W von: Lupinen-Unterbau in älteren Beständen. *Allg Forstzeitschr* 16: 669–671, 1961.
24. Bredow-Stechow W von: Unterbau mit Dauerlupine. *Allg Forstzeitschr* 18: 510, 1963.
25. Rehfuess K: Ernährung und Wachstum der Dauerlupine (*Lupinus polyphyllus* L.) unter Kiefernschirm auf Oberpfälzer Standorten. *Forstw Cbl* 84: 265–292, 1965.
26. Rehfuess KE: Underplanting of pines with legumes in Germany. In: *Symbiotic nitrogen fixation in the management of temperate forests*, Gordon JC, Wheeler CT, Perry DA (eds). Corvallis, Oregon State Univ, 1979, p 374–387.
27. Žilkin BD: Povyšenie produktivnosti lesov kulturoj ljupina. *Izd-vo škola*, Minsk, 1965.
28. Ronkonen NI: Vvedenie počvoulučšajuščih rastenij v kultury sosny na vyrubkah. *Povyšenic effektionosti lesovosstanovitelnyh meroprijatij na severe*, Karelskij filial AN SSSR, Institut lesa. Petrozavodsk, 1977, p 78–87.

29. Žilkin BD, Lahtanova LJ: Osobennosti kornevogo pitanija sosny i ljupina pri sovmestnom proisrastanij. *Fiziol-biohim osnovy vzaimodejstvija rast v fitocenozah* 3: 83–87, 1972.
30. Žilkin BD, Rihter TA: Vlijanie biologičeskoj melioracii jelovyh molognjakov kulturoj mnogoletnego ljupina na biologčeskij krugovorot veščestov. *Lesnoj Žurnal* 5: 8–12, 1973.
31. Rihter TA, Rihter IE: Vlijanie mnogoletnego ljupina na himičeskie svojstva lesnoj podstilki molodnjakov sosny i jeli. *Lesovedenie i les h-vo* 8: 47–52, 1974.
32. Žilkin BD, Beregova TS: Opyt opredelenija ekonomiceskoj melioracii sosnjaka vereskovogo kuluroj mnogoletnego ljupina. *Lesnoj Žurnal* 4: 136–140, 1973.
33. Žilkin BD, Lahtanova LI, Rihter IE: Vlijanie gustoty posadki, rubok uhoda i mnogoletnego ljupina mnogolistnogo na pokazateli rosta sosny obiknovenmoj. *Lesovedenie i les h-vo* 6: 56–62, 1972.
34. Krolikowski L, Janiszewski B, Matusz S, Strzelec Z, Tyszka-Roth: Studies on lupine application and mineral fertilization of young plantations of pine on poor sandy soils. *Prace Inst Bad Lesn* 418: 49–96, 1973.
35. Lundmark J-E, Huss-Danell K: Odlingsförsök med gråal och lupin på tallhedar i Norrbotten. *Sv Skogsv Tidskr* 3/81: 17–26, 1981.
36. Bjarnason H: Um fridun lands og frjosemi jardvegs. *Arsrit Skoraekterfelags Islands*, p 4–19, 1971.
37. Mikola P: Experiments on the rate of decomposition of forest litter. *Comm Inst Forest Fenn* 43(1): 1–50, 1954.
38. Mikola P: The value of alder in adding nitrogen in forest soils. *Final report of research conducted under grant authorized by US public law 480.* Univ Helsinki, Dept Silviculture, 1966.
39. Viro PJ: Investigations on forest litter. *Common Inst Forest Fen* 45(6): 1–65, 1955.
40. Mikola P: Liberation of nitrogen from alder leaf litter. *Acta Forest Fenn* 67(1): 1–10, 1958.
41. Cajander EK: Untersuchungen über die Entwicklung der Kulturfichtenbestände in Süd-Finnland. *Comm Inst Forest Fenn* 19(3): 1–101, 1933.
42. Hausser K: Ein Düngungsversuch zu Fichten bei der Streuflächen-Aufforstung der Gemeinde Klosterreichenbauch auf oberem Buntsandstein. *Die Phosphorsäure* 24: 227–241, 1964.
43. Evers FH: Boden- und nadelanalytische Auswertung eines Düngungsversuches zu Fichte auf einer ehemaligen Streunutzungsfläche. *Die Phosphorsäure* 24: 242–256, 1964.
44. Grozdov BV: Bystrorastuščie i hozjajstenno-cennie porogy v lesnom hozjajstve. *Puti povyšenija produktivnosti lesn h-vo*, p 60–68. Brjansk, 1961.
45. Milto NN: Ispolzovanie pocvoulučšajuščih svojstv olhi seroj. *Lesovedenie i lesn h-vo* 3: 37–42. Vyš škola, Minsk, 1970.
46. Virtanen AI: Investigations on nitrogen fixation by the alder without combined nitrogen. *Phys Plant* 10: 164–169, 1957.
47. Mikola P: Afforestation of bogs after industrial exploitation of pet. *Silva Fennica* 9(2): 101–115, 1975.
48. Granhall U, Lindberg T: Nitrogen fixation in some coniferous ecosystems. In: *Environmental role of nitrogen-fixing blue-green algae and asymbiotic bacteria*, Granhall U (ed). *Ecol Bull* (Stockholm) 26: 178–192, 1978.
49. Kallio S, Kallio P: Nitrogen fixation in lichens at Kevo, North Finland. In: *Fennoscandian tundra ecosystems, Part I*, Wielgoleski FE (ed). *Ecol Studies* 16: 292–304, 1975.
50. Huss-Danell K: Nitrogen fixation by *Stereocaulon pascale* under field conditions. *Can J Bot* 55: 585–592, 1977.
51. Basilier K: *Nitrogen fixation associated with Sphagnum and some other musci.* Ph.D. Thesis, Dept of Physiol Botany, Univ Uppsala, 1979.
52. Björkman E, Lundberg G, Nömmik H: Distribution and balance of ^{15}N labelled fertilizer nitrogen applied to young pine trees (*Pinus silvestris* L.). *Stud For Suec* 48: 1–23, 1967.
53. Nömmik H, Popović B: Recovery and vertical distribution of ^{15}N labelled fertilizer nitrogen in forest soil. *Stud For Suec* 92: 1–20, 1971.
54. Paavilainen E: Studies on the uptake of fertilizer nitrogen by Scots pine using ^{15}N labelled urea. *Comm Inst Forest Fenn* 79(2): 1–47, 1973.
55. Paavilainen E: Effect of fertilization on plant biomass and nutrient cycle on a drained dwarf shrub pine swamp. *Comm Inst Forest Fenn* 98(5): 1–71, 1980.
56. Viro PJ: Estimation of the effect of forest fertilization. *Comm Inst Forest Fenn* 59(3): 1–42, 1965.

57. DeBell DS: Future potential for use of symbiotic nitrogen fixation in forest management. In: *Symbiotic nitrogen fixation in the management of temperate forests*, Gordon JC, Wheeler CT, Perry DA (eds). Corvallis, Oregon State Univ, 1979, p 451–466.
58. Staaf H: Kvävebudget för mellansvensk tallhed. In: *Processer i kvävets kretslopp*, Rosswall T (ed). *Stat Naturv Rap* 1214: 17–24. Solna, Sweden, 1980.
59. Nutman PS: Symbiotic nitrogen fixation in plants. *Intern Biol Progr* 7, Cambridge Univ Press, 1976.
60. Lohm U: Soil processes studied within the Swedish Coniferous Forest Project – an introduction. In: *Structure and function of northern coniferous forests – An ecosystem study, Persson T (ed). Ecol Bull* (Stockholm) 32: 329–332.
61. Huss-Danell K, Lundmark J-E: Can *Alnus incana* be used for soil improvement in coniferous forests in northern Sweden? In: *Symbiotic nitrogen fixation in the management of temperate forests*, Gordon JC, Wheeler CT, Perry DA (eds). Corvallis, Oregon State Univ. 1979, p 479.

11. Nitrogen fixation in Southeast Asian forestry: research and practice

IRENEO L. DOMINGO

Paper Industries Corporation of the Philippines, Bislig, Surigao del Sur, Philippines

INTRODUCTION

The natural forests of Southeast Asia have been decreasing, and the denuded areas rapidly turned into vast grasslands and brush that are mostly useless. With the everincreasing population, pressure on the forest will continue. This means much reduced wood production.

Moreover, parts of the natural forests are increasingly called upon to provide non-timber goods and services to the exploding population. The continuing increase in population will increase the need for converting forest lands for living space, expanded agriculture, watershed protection, and recreation. All these will reduce further the forest land yielding wood for various wood products.

Yet, the continuing increase in population will also continue to increase the need for wood. The increasing demand for wood, not only within the region but also in the international market, will require the production of more wood from a smaller area. This is a tall order for Southeast Asia, considering that the region has produced wood from extensive areas of virgin forest.

The natural forests in the region are composed mostly of valuable timber, mostly rainforest dipterocarp species in the Philippines, Indonesia, and Malaysia and teak in Thailand and Burma, species which grow too slowly to lend themselves to intensive timber management designed to produce more wood on less land at a shorter time. With reduced land for forest production, the present natural forest species will never be able to yield wood sufficient to meet the increasing demand for wood products.

Fortunately for the region, there are many fast-growing species that could be grown intensively to produce more wood than can be produced under traditional, extensive forestry. However, it is not easy because tropical soils are usually poor in nutrients especially nitrogen. Tropical soils vary so much in texture and nutrient status that generalizations are difficult, but a few facts stand out. Because of the pattern and intensity of rainfall and consequent erosion, many of the available soil nutrients are leached out or washed away through surface runoff. Since most of the nitrogen is contained in the top layer of tropical soils, which is easily washed out in steep, denuded slopes, the soil is usually deficient in nitrogen.

Additions of inorganic fertilizer to increase or maintain soil fertility to support

Gordon, J.C. and Wheeler, C.T. (eds.). Biological nitrogen fixation in forest ecosystems: foundations and applications

ISBN 90-247-2849-5.
Printed in The Netherlands

growth is almost out of the question. The cost is too high and because of the high rainfall, not much of the added fertilizer will be retained and used anyway.

The removal of nutrients tied in the upper biomass after harvesting plantations will further reduce elements and will almost certainly reduce yields in second and subsequent generations of plantations.

The use of nitrogen-fixing trees or cover crops is therefore one alternative that offers great promise in establishing forests in denuded lands and in producing more wood in industrial plantations.

This chapter presents a survey of the use of nitrogen-fixing systems in Southeast Asian silviculture, reviews research initiatives, and presents the silvicultural systems of the few nitrogen-fixing species that are being grown extensively in plantations.

Southeast Asia is a region that lies within the equatorial belt from China and the Pacific Ocean in the north to Australia and the Indian Ocean in south and from Burma in the west to west Iran in the east.

Temperatures in Southeast Asia are relatively uniform throughout the year but rainfall varies greatly, from about 1,000 mm per year in interior areas of Burma and Thailand that are sheltered by mountains to over 4,200 mm in eastern Mindanao in the Philippines. In many areas in the north, the total annual rainfall may come down within only 4 to 5 months; in other areas such as those near the equator, rainfall is fairly evenly distributed throughout the year.

SURVEY OF NITROGEN-FIXING SYSTEMS IN SOUTHEAST ASIA

The removal of the forest cover in tropical Southeast Asia usually does not result in bare soil. The area is taken over quickly by plant communities dominated by species of trees of intermediate successional stages. Given time and freedom from human interference, the area undergoes secondary plant succession, progressively towards the original climax conditions. What usually happens, however, is that after removal of the forest cover, or even after legitimate selective cutting, the area is taken over by shifting cultivators who burn the remaining green cover, cultivate the land and plant agricultural crops. In a few years, after soil fertility is lowered, the area is abandoned and grass or brush takes over. Very often the forest thus is reduced to coarse grassland of little, if any, economic value. The grass, usually the tenacious *Imperata cylindrica*, variously known as cogon, alang-alang, and by other names, grows a dense network of roots and underground rhisomes, and is very inflammable in the dry season. It grows even better after it is burned.

The difficulty of reforesting the vast grass – and brush lands in Southeast Asia has encouraged the use of nitrogen-fixing tree species in reforestation, either as a permanent tree crop or as a nurse tree. On the other hand, the need to produce more wood on less land and over a shorter time is an overriding consideration in the use of fast growing species, such as the legumes, for industrial tree plantations in the wood industry.

Nitrogen-fixing tree species

At present, there are several nitrogen-fixing species being planted for various end-uses, such as fuelwood, furniturewood, pulpwood, sawtimber, or simply to reclaim denuded lands. Most of these species are being planted in government reforestation and other tree planting projects. Only a few are planted in large scale industrial tree plantations.

Pterocarpus indicus Willd. (Papilionoideae). This species is used in government reforestation projects. This legume is excellent material for furniture because of the beauty of its grain and color, workability, and durability. Its roots have nodules although there has been no effort to study its nitrogen fixation rate. The use of this species seems to be mainly due to its economic uses and its relatively fast growth compared to other high-premium wood species. The fact that it is a legume apparently helps very much in its establishment and relatively fast growth even on very poor sites.

The use of *Pterocarpus indicus* in reforestation, so far, has been limited to government reforestation projects and in planting parks and highways. It has not been planted yet into an industrial tree plantation, although the furniture industry is a big business in Southeast Asia.

Leucaena leucocephala (Lam.) de Wit (Mimosoideae). Leucaena has been used extensively in Southeast Asia, as in other tropical regions, for several decades.

In small, and even in adult plants, lateral roots are few and they usually grow downward at a sharp angle. But small lateral roots occur near the soil surface and carry the nitrogen-fixing *Rhizobium* nodules which are usually 2.15–15 mm in diameter and are frequently multilobed. The functioning nodules are bright pink inside. The leucaena *Rhizobium* partnership is capable of annually fixing more than 500 kg of nitrogen per hectare. This is equivalent to 2500 kg ammonium sulphate per hectare per year.

However, nitrogen fixation occurs only if the correct *Rhizobium* strain is present in the soil. Unlike many other tropical legumes, leucaena is highly specific in its inoculation requirements (1). Leucaena plants that are not nodulated are usually stunted, unproductive, frequently have pale green or yellow foliage which is low in protein. Where the Leucaena is naturalized the bacteria are normally widespread (e.g. in Southeast Asia). However, in areas where leucaena has never been grown before, the seed must be inoculated with an appropriate *Rhizobium* strain just before it is sown. Alternatively, soil from beneath nodulating trees may be used to inoculate new plantings (2).

Among the various known uses of leucaena are firewood, industrial fuelwood, wood products such as pulp, lumber, and fence posts, props for ripening banana bunches to prevent them from breaking off or the whole plant from toppling over, nurse crop for coffee, cacao, tea, cinchona, citrus, rubber, coconut, and oil palm, and the leaves are good for animal feeds.

In Indonesia extensive reforestation is already being carried out using leucaena. More than 30,000 ha on the island of Flores are being planted in contour hedgerows of leucaena for the purpose of rejuvenating unstable slopes (2). In the Philippines three large corporations are planting the species in thousands of hectares for fuelwood; two will use it for electricity generation, and the other to produce charcoal for the production of calcium carbide and vinyl plastics, and another corporation is planting it for pulpwood.

Acacia auriculiformis A. Cunn. ex Senth. (Mimosoideae). This species is native to Papua New Guinea and Northern Australia but has been introduced into Indonesia, Malaysia and the Philippines for reforestation, mostly in denuded, difficult areas, and as ornamental and shade tree along city streets.

This tree is already established on a large scale private fuelwood plantation and in government forest areas in Indonesia. It grows in a wide range of deep and shallow soils including sand dunes, clay, limestone, and lateritic soils. The plant produces profuse bundles of nodules and can survive on land very low in nitrogen and organic matter. In Malaysia, it has grown well on spoil heaps after tin mining. In Indonesia and the Philippines the species has been successfully planted on steep, unstable slopes for erosion control (3).

Acacia mangium Willd. (Mimosoideae). This nitrogen-fixer is native to Queensland, Australia. It was introduced into Sabah about a decade ago. Unexpectedly, it grew exceedingly well, comparable to *Gmelina arborea* and *Albizia falcataria.* In untended stands of 9-year old trees yield was as high as 415 m^3 per hectare. Sabah foresters have now converted 1,200 hectares of degraded *Imperata* grassland into productive forest of *A. mangium* (4).

From its performance in Sabah, the species became well-known and is now being tried in several places in Southeast Asia, especially on poor sites where it out performs most fast-growing hardwood species.

The tree can be used for pulp and paper and large diameter logs can be peeled or sawn into lumber. It makes excellent particle board and could be useful for furniture and as cabinetwood.

Calliandra calothyrsus Meissn. (Mimosoideae). This legume is native to Central America but was introduced into Java in 1936 for erosion control and firewood. Its performance as a plantation crop for fuelwood has been so successful that over the last 25 years plantations of this species have been steadily expanding and now cover more than 30,000 hectares in Java.

By its nitrogen fixation and heavy litter production, this species improves soil productivity so rapidly that farmers often rotate or intercrop agricultural crops with calliandra plantations.

Sesbania (Papilionoideae). *Sesbania bispinosa* (Jacq.) Wight and *S. grandiflora* (L.) Pers. are two legume species that have varied uses in Southeast Asia. *S. bispinosa* is used mainly in Vietnam as a firewood crop and at the same time to fertilize the soil in preparation for the planting of food crops. It nodulates vigorously and its foliage is a choice green manure. *S. grandiflora* is more widely distributed in Southeast Asia, used mainly in agroforestry systems. It has long been used as firewood in Southeast Asia although the specific gravity of the wood, 0.42, is not very high. In several areas in Indonesia it has been planted to provide fuel and other products in 'turinization' projects. The tree is often planted for beautification along roadside, fencelines and other boundaries. The leaves, flowers and tender pods are favorite Asian vegetables. In Java the tree is extensively used as a pulp source.

The two species are able to grow in a wide range of soils, even poor ones, including black, poorly structured clays. Their extraordinary nodulation undoubtedly helps restore fertility to these soils. In Timor, *S. grandiflora* is often found in abandoned swidden land. This coupled with its rapid growth, suggests that its soil improvement qualities, though unmeasured, may be exceptional (4).

Albizia falcataria (L.) Fosberg (Mimosoideae). This is one of the fastest, if not the fastest, growing tree species in the world. It is native to the Moluccas, the island of New Guinea and the Solomon islands. It was introduced to Java in the 1870s, and thereafter, it was spread westward, northward and eastward. It is now found all over Southeast Asia. The Paper Industries Corporation of the Philippines (PICOP) manufactures good quality newsprint from 100% *Albizia falcataria* pulp. PICOP and other companies in Eastern Mindanao also use it for lumber, blockboard, and plywood. It is estimated that in Eastern Mindanao alone, companies now have close to 30,000 hectares of industrial plantations of the species. Moreover, in the same region, about 4,000 farmers have established also about 30,000 hectares of tree farms planted to *Albizia falcataria*.

Its extremely fast growth on most sites is undoubtedly due partly to its nitrogen-fixing ability and its heavy leaf litter production which is high in nitrogen.

Casuarina equisetifolia L. (Casuarinaceae). *Casuarina equisetifolia* is the more common and widely used species of the genus *Casuarina*. It is indigenous to Indonesia, Malaysia, the Pacific islands and other countries, but it is also found everywhere now in Southeast Asia. It has been called the best firewood in the world, used both for domestic and industrial fuel, although its wood has a number of other uses (3).

Although not a legume it has root nodules containing nitrogen-fixing actinomycete microorganisms. It is salt tolerant, wind resistant and adaptable to moderately poor soils except heavy soils such as clays.

Other species. The following other species which are used also in plantation forestry are also known to be nitrogen-fixers (3):

Legumes

Gliricidia sepium (Jacq.) Steud.
Albizia lebbek (L.) Benth.

Non-legumes

Alnus japonica (Thumb.) Steud
A. nepalensis D. Don
Parasponia andersonii
P. rugosa
P. rigida
P. parviflora

Cropping systems

Haines and Debell (5) discussed a number of possible cropping systems by which nitrogen-fixing plants are used in forest management programs. In Southeast Asia, the following systems are practiced.

Continuous use of nitrogen-fixer as the principal crop. This system is the most common practice, both in government reforestation projects and in industrial plantations. The fast-growing, nitrogen-fixing, tree species is planted in plantations and managed as a monoculture for single use such as fuelwood or for multiple uses such as pulpwood, sawtimber, and fuelwood. *Albizia falcataria* is the best example of a nitrogen-fixing tree species being grown on a large scale by this system. In the industrial plantations in Eastern Mindanao, the species is grown as a monoculture, ranging in spacing from 1.5 m × 1.5 m to 5 m × 5 m. Depending on the spacing, a non-commercial thinning is conducted one to two years after planting, then one or two thinnings are conducted for pulpwood, then the stand is clearcut at ages 12 to 16 years for both pulpwood, hardboardwood, and sawtimber for blockboard, plywood, or lumber for cabinet interiors, ceilings, or other interior uses.

After clearcutting, the plantation is regenerated by coppice or natural seeding.

Mixed species systems. This system is usually practiced in the denuded, grass-dominated reforestation lands. There are a number of variations. One common system involves first planting a fast-growing, nitrogen-fixing tree species to build up the soil and eliminate the grass, after which a more commercially needed, higher-value species is underplanted. The former species serves as a nurse tree to the latter. The mixed plantation is managed to produce different products from the two species. An example of a nurse tree is *Leucaena leucocephala*, a fuelwood species, and the high value species may be a *Vitex parviflora* which grows slowly but is highly prized for railroad ties and termite proof house posts. Eventually, the *Vitex* will overtop the *Leucaena* but the latter will continue to grow underneath the *Vitex* canopy.

Another system is enrichment planting. *Albizia falcataria* is planted after site preparation in selectively logged, but inadequately stocked, dipterocarp forest. The canopy of the residual dipterocarps is open enough the allow development of the *Albizia falcataria*. The volume of the nitrogen-fixer will add to the volume of dipterocarps to be harvested at the end of the cutting cycle. At the same time, the growth of the dipterocarp residuals is enhanced and their reproduction encouraged.

Another system is the growing together of the nitrogen-fixing tree species as the principal tree crop and agricultural food crops. An example is *Albizia falcataria* mixed with coffee, cocoa, or pineapple. The *Albizia* is either planted at the same time with the food crops or the food crops are underplanted in an older *Albizia* stand. In either case the food crops benefit from the additional nitrogen provided by the *Albizia* through nitrogen-fixation and leaf-litter.

Non-commercial nitrogen fixers such as species of *Stylosanthes*, *Calopogonium*, and *Centrosema*, have been planted at the same time or after establishment of the main tree crop. These legumes add nitrogen to the soil while the tree crop is being established or before the tree canopy closes. Eventually the nitrogen-fixer will be eliminated when light becomes limiting.

The same legumes have been similarly planted for several years in rubber (*Hevea braziliensis* Meul Arg.) plantations in Malaysia. It was estimated that during a 5-year period nitrogen-fixation by the legumes added nearly 880 kg of nitrogen per hectare. Despite the fact that the legumes died out after the sixth year, increase in rubber yield persisted for about 20 years (6).

Crop rotation. Opportunities for the practice of this system in forestry is limited because of the number of years involved. In Papua New Guinea, however, shifting cultivators have already integrated nitrogen-fixation and soil building into their shifting cultivation rotation. Before moving on, they plant *Leucaena leucocephala* or *Casuarina equisetifolia*. Both species have nitrogen fixing root nodules and as they grow the soil nitrogen increases and a deep leaf litter forms. Instead of waiting 10 years or more, the farmers can return to the *Leucaena*- or *Casuarina*-planted patches after only about two years, because during this time the nitrogen in the soil has increased adequately (6).

CURRENT RESEARCH INITIATIVES

Research on biological nitrogen fixation in Southeast Asia is just starting, mainly at the Forest Research Institute and the University of the Philippines at Los Baños in the Philippines and at the Forest Research Institute in Malaysia. However, silvicultural research on certain promising nitrogen-fixing tree species has been going on for quite some time in all of the region. Moreover, promising nitrogen-fixing tree species have been used and some cropping systems taking advantage of the potentials of nitrogen-fixers have been practiced for some time.

Survey and identification of symbionts

Research on biological nitrogen-fixation is starting with the survey of nitrogen-fixing tree species and the isolation and identification of the micro-organisms responsible for their nodulation. In the Philippines, at the Forest Research Institute and the University of the Philippines at Los Baños, several species, mostly legumes, have been observed to contain nodules in the roots. However, actual isolation and identification of the microorganisms have been done so far on only a few species (Table 1).

A survey is for nitrogen fixing organisms is also being conducted on grasses found in reforestation areas and in food crop plants. So far, isolates have been obtained from some species of grass and crop plants. Tests are underway for effectiveness and cross inoculation on tree species used for reforestation. Initial positive responses have been observed and attributed to both hormonal effects and to nitrogen fixation (Garcia, M.U., personal communication, see Research Contacts.)

Use of ground cover legumes

Another direction of research is the test of certain ground cover legumes in increasing survival and growth of planted tree seedlings. At the Forest Research Institute in Malaysia *Calopogonium mucunoides*, *C. caeruleum*, *Centrosema pubescens*, *Pueraria phaseoloides* and *Stylosanthes gracilis* are being test-planted in agro-forestry systems (Ahmad, N, personal communication, see Research Contacts).

At the Paper Industries Corporation of the Philippines, in Eastern Mindanao, *Stylosanthes* is also being tested for effectiveness in increasing growth and survival of planted seedlings of *Eucalyptus deglupta* and *Albizia falcataria*. The *Stylosanthes* seeds are directly sown at the same time and place that the tree seedling is planted. It is hoped that the early establishment of *Stylosanthes* would discourage growth of competing weeds and at the same time it would add nitrogen to the soil around the roots of the planted tree seedling.

Table 1. List of tree species where *Rhizobia* were isolated (Unpublished data of Dr. S.S. Quiñones, Forest Research Institue, College, Laguna, Philippines).

Tree species	Sources of specimens
Albizia falcataria (L.) Fosb.	Los Baños, Laguna
	Carranglan, N.E.
	Mambusao, Capiz
	Malaybalay, Bukidnon
	Puerto Princesa, Palawan
	Bislig, Surigao del Sur
Albizia julibrisin Durazz.	Puerto Princesa
Amherstia nobilis Wall.	Los Baños
Bauhinia purpurea L.	Puerto Princesa
	Malaybalay
Cassia fistula L.	Buhisan, Cebu
Cassia nodosa Buch. Ham. ex Roxb.	Buhisan
Derris cummingii Benth.	Baguio City
Gliricidia sepium (Jacq.) Steud.	Mambusao
Intsia bijuga (Colabr.) O. Ktze.	Lamacan, Cebu
Leucaena leucocephala (Lam.) de Wit (giant)	Los Baños
	Carranglan
	Mambusao
	Malaybalay
	Puerto Princesa
	Bangued, Abra
	Baguio City
	Minglanilla, Cebu
	Buhisan
	Lamacan
Leucaena leucocephala (Lam.) de Wit (native)	Los Baños
	Mambusao
Piliostigma malabaricum (Roth)	Carranglan
Benth. var. *acidum* (Korth.)	Malaybalay
de Wit	Buhisan
Pithecellobium dulce (Roxb.) Benth	Los Baños
	Carranglan
Samanea saman (Jacq.) Benth.	Los Baños
	Carranglan
	Mambusao
	Buhisan
	Lamacan
Schizolobium exelsum Vog.	Malaybalay
	Buhisan
Pterocarpus indicus Willd.	Los Baños
	Carranglan
	Mambusao
	Malaybalay
	Bangued
	Minglanilla
	Lamacan
	Batac, Ilocos Norte
	Ballesteros, Cagayan
Tamarindus indica L.	Mambusao
	Malaybalay
	Buhisan
	Lamacan

Silviculture of nitrogen-fixing species

Extensive research on the silviculture of certain fast-growing tropical hardwoods that happen also to be nitrogen-fixing species has been underway for some time in the Southeast Asian countries.

Albizia falcataria has been the subject of extensive trial plantings and evaluation as early as the 1950s in Malaysia and even earlier in Indonesia. In the Philippines, it has been the subject of silvicultural research since after the war under the initiative of the wood industry. Being the pulpwood species for newsprint manufacture at the Paper Industries Corporation of the Philippines, this Company's research arm has been concentrating on the silviculture as well as other aspects of wood production of the species. An integrated research program, which includes silviculture, on *Leucaena leucocephala* is being implemented by government research institutions in the Philippines with support from the International Development and Research Center of Canada.

Research in silviculture of other nitrogen-fixing tree species is not concentrated and integrated but several projects are underway.

Interplanting

At the University of the Philippines at Los Baños, Zabala (7) interplanted *Albizia falcataria* with *Anthocephalus chinensis*. Likewise, at the Paper Industries Corporation of the Philippines *A. falcataria* was interplanted with *Eucalyptus deglupta*. In both studies the expectation was for *A. chinensis* and *E. deglupta* to benefit from the mixture because of the nitrogen added to the soil from *A. falcataria* through its nitrogen-fixation and heavy leaf litter production. The results in both studies, however, showed that *A. chinensis* and *E. deglupta* both suffered from the mixture (Table 2). Apparently, light was a more important factor than nitrogen. *A. falcataria* grows much faster in height than either *A. chinensis* or *E. deglupta*, shading the latter two species in 2 to 3 years. Since both species are shade intolerant, their survival and growth were adversely affected.

Table 2. Height, diameter, volume and survival of *Albizia falcataria* (Af) and *Anthocephalus chinensis* (Ac) in pure and mixed (1:1) stands $5\frac{1}{2}$ years after planting (7).

Attributes	Pure stand		Mixed stand	
	Af	Ac	Af	Ac
Height, m	17.7	8.6	18.8	6.8
DBH, cm	19.4	12.0	21.3	8.0
Volume per tree, m^3	0.70	0.19	0.84	0.06
Survival, %	40.3	36.6	44.5	25.4

Research directions

With the increasing cost of nitrogen fertilizer, the active research that has just been started in the Philippines and in Malaysia is likely to accelerate over the next few years. The present survey and identification of symbionts will continue but new initiatives are indicated on studies of the nature of the symbiosis, field tests and production of inoculants for specific tree crops.

At the same time more nitrogen-fixing tree species will be tested and silviculture research will continue to focus on *Albizia*, *Leucaena* and other promising fast growing nitrogen-fixers.

SILVICULTURE SYSTEMS OF MAJOR SPECIES

A silvicultural system is a program of silvicultural prescriptions underway or planned, for a particular timber stand. It consists of a number of logical steps, such as establishment, intermediate treatments and regeneration (8). In Southeast Asia, only two species have so far been used on such a scale to demand a full implementation of a normal silvicultural system, namely *Albizia falcataria* and *Leucaena leucocephala*.

Albizia falcataria (L.) Fosberg

The species is known as falcata and Moluccan sau in the Philippines, batai in Malaysia, and djeungjing in Indonesia. In this chapter it shall be called falcata. It is being planted in the Philippines in industrial plantations and agro-forestry farms for newsprint pulpwood, blockboard, plywood and lumber. It grows naturally in Indonesia where it is utilized for light house construction. In Malaysia it is still confined to small, scattered trial plantations.

Propagation. Falcata is propagated by seeds. The seeds are hard-coated and they germinate irregularly and unevenly when untreated. Treatment of the seed may be done in a number of ways but the best is to soak the seeds wrapped with a piece of cloth in hot water. The seeds are then allowed to stay in the wet cloth for two days, after which the radicles come out and the germinated seeds are planted directly in veneer or plastic tubes or other containers. The seedlings are grown in the nursery for 45 days and then set out in the plantation area.

Cropping system. Being a nitrogen-fixer falcata is grown continuously in *pure industrial plantations* for pulpwood and sawtimber established on government forest land, in Eastern Mindanao.

Mixing falcata with *Eucalyptus deglupta* was tried but trials resulted in the thinning out of the latter, due mainly to inadequate light and space conditions underneath the canopy of the falcata. The falcata and *Eucalyptus deglupta* were planted in alternate rows four meters apart and four meters along the row. It was hoped that the *E. deglupta* would benefit from the mixture, falcata being able to add nutrients to the soil through its root nodules and leaf litter. In 5 to 6 years, the falcata had a higher canopy and occupied more space, causing over-crowding in the canopy and resulting in more mortality and reduced growth of *E. deglupta*. More trials will have to be done to determine the compatible ratio of mixture. However, it is now considered better to grow the species separately in pure plantations.

In agro-forestry farms falcata is planted as the main crop with food crops planted between the falcata rows such as cocoa, coffee, pineapple, vegetables, rice, and corn. The last 3 crops are grown during the first two years when the falcata canopy has not closed yet. The cocoa, coffee and pineapple are grown underneath the falcata canopy at any age, even extending to a number of rotations of the falcata. The food crops, which are expected to benefit from the falcata, are intended to provide income to the tree farmer while waiting for 8 years when the falcata is ready for harvest.

An innovative cropping system being practiced in Eastern Mindanao is *enrichment planting* with falcata in logged-over dipterocarp forest. This is conducted in areas selectively logged but which, for some reason, contain only a few immature trees considered inadequate to yield sufficient volume in the next cutting cycle.

The second-growth dipterocarp forest is site-prepared for planting by cutting down trees of miscellaneous, unwanted species, the defective and overmature trees that were not removed during the previous harvest cutting, vines and the shrubby vegetation on the forest floor. Utilization of the miscellaneous species, defective and overmature trees for pulpwood and fuelwood pays for the operation. At the same time, the operation opens up the canopy to allow sufficient light to reach the forest floor. Falcata seedlings are then planted. While falcata requires full sunlight for maximum growth, the partial shade is sufficient to allow development of falcata to merchantable size in time for the next cutting cycle of the dipterocarps in 30–35 years.

At the end of the next cutting cycle of the second-growth dipterocarp stand, the total volume that can be harvested will be much greater than had been there no enrichment planting. The additional volume would come from the enrichment-planted falcata. Moreover, the residual dipterocarps would grow faster because of the greater space created when the miscellaneous species and defective and overmature trees were cut during site preparation. The biological nitrogen fixation by and the leaf fall from the falcata would enrich the soil and contribute to the greater growth of the residual dipterocarp trees. The generally clear floor underneath the falcata and the leaf litter make a better seedbed and promotes development of dipterocarp reproduction. Thus, natural regeneration of the dipterocarp species is encouraged.

Establishment. The establishment of falcata plantations or tree farms is accomplished either by planting nursery-grown seedlings or by direct seeding. By either method, a thorough site preparation is necessary. It is accomplished by cutting down all standing vegetation on the plantation site and slashing the downed material with a chainsaw into shorter pieces. For the planting of nursery-grown seedlings, the specific spots where the seedlings are to be planted are staked and thoroughly ring-weeded resulting in a completely bare soil surface. One seedling is planted on every staked and ring-weeded spot. Spacing varies from 1.5 m × 1.5 m to 4 m × 4 m. Every seedling is ring-weeded three to four times in the first year, and also in the second year for plantations established on wide spacing.

For direct seeding, it is advisable to burn the plantation area after the vegetation has been cut and dried to make sure that the seeds come directly in contact with the soil and that there is no obstruction to the development of the young seedlings. From one-half to one liter of seeds may be broadcast by hand to a hectare depending on the expected germination capacity of the seeds used.

Intermediate treatments. The plantation is usually considered established at the end of the second year when it can grow by itself without weeding except for occassional vine cutting.

For a plantation established at close spacing, a noncommercial thinning may be needed sometime within the second to the fourth year after planting. Likewise a non-commercial thinning is needed during the same period for a plantation established by direct seeding.

In the 6th to 8th year a commercial thinning is conducted for pulpwood and the remaining trees are harvested after the 12th to 14th year for pulpwood and sawtimber.

Growth and yield. Falcata is one of the fastest, if not the fastest, growing trees in the world. It has been called the 'miracle tree' because of its amazingly fast growth (Fig. 1). Measurements made in Indonesia (Java and Kalimantan), Malaysia, Hawaii and in numerous Philippine locations indicate that on the good sites falcata may reach 7 m in height in about a year, 13–18 m in 3 years, 21 m in 4 years. Diameter may increase at about 5–7 cm per year (7, 8, 9, 10, 11, 12, 13). In 10 years it may attain a height of up to 50 m and a diameter of up to 60 cm. Undoubtedly, its amazingly fast growth is due partly to nitrogen additions through its root nodules (Fig. 2) and its heavy leaf litter production. Although it tolerates a wide range of site conditions, its growth varies with site quality, from just over 100 m^3 to more than 500 m^3 per hectare in 10 years in Eastern Mindanao in the Philippines (Table 3).

Fig. 1. Seven year old *Albizia falcataria* industrial tree plantation of the Paper Industries Corporation of the Philippines in Eastern Mindanao. *Albizia falcataria*, a nitrogen-fixer, is one of the fastest, if not the fastest, growing tree species in the world.

Table 3. Volume yield in m^3 per hactare of *Albizia falcataria* (L.) Fosberg up to 10 cm top diameter inside bark in Northeastern Mindanao, Philippines (13).

Age,	Site index in meters, based at age 10 years										
years	20	22	24	26	28	30	32	34	36	38	40
2	27.7	33.8	40.6	48.0	56.1	64.8	74.2	84.3	95.0	106.4	118.5
3	38.8	47.4	56.9	67.3	78.6	90.9	104.0	118.1	132.2	149.2	166.1
4	49.3	60.3	72.3	85.6	99.9	115.5	132.2	150.1	169.3	189.6	211.2
5	59.4	72.6	87.1	103.0	120.4	139.1	159.3	180.8	203.9	228.4	254.2
6	69.3	84.3	101.4	119.9	140.1	161.9	185.4	210.5	237.3	265.8	296.0
7	78.7	96.1	115.3	136.4	159.3	184.1	210.8	239.4	270.0	302.3	336.6
8	87.9	107.3	129.0	152.4	178.1	205.8	235.6	267.6	301.6	337.7	376.2
9	97.0	118.4	142.2	168.2	196.4	227.0	259.9	295.2	332.7	372.7	415.0
10	105.9	129.3	155.2	183.6	214.5	247.8	283.7	322.2	363.3	406.9	453.1
11	114.6	140.0	168.1	198.8	232.2	268.3	307.2	348.9	393.3	440.5	490.5
12	123.3	150.6	180.7	213.7	249.6	288.5	330.3	375.1	422.9	473.6	527.4
13	131.8	160.9	193.1	228.5	266.9	308.4	353.1	401.0	452.0	506.3	563.8
14	140.2	171.2	205.0	243.0	283.9	328.1	375.6	426.5	480.8	538.6	599.7
15	148.5	181.3	217.7	257.4	300.7	347.5	397.9	451.8	509.3	570.4	635.2

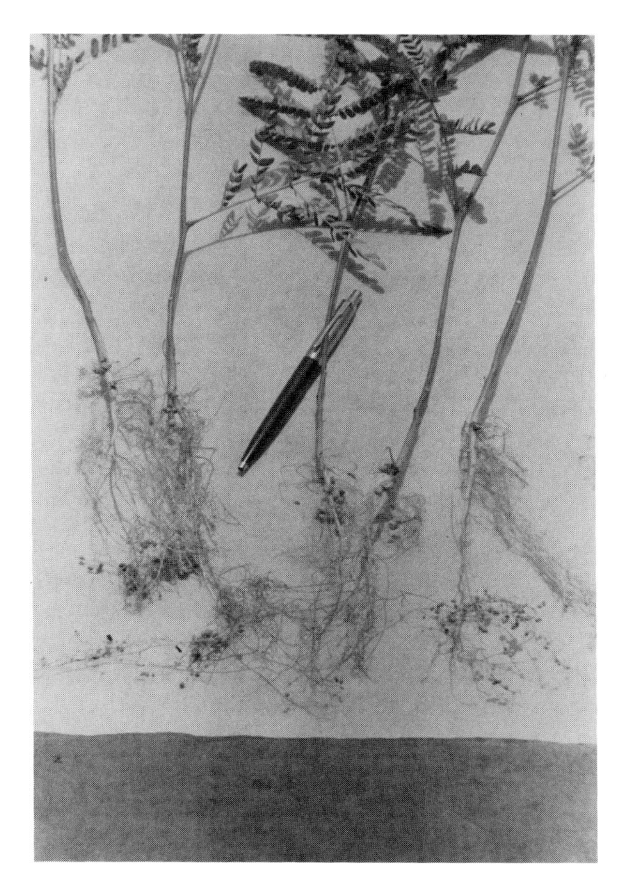

Fig. 2. *Albizia falcataria* seedlings showing *Rhizobium*-filled nodules in the roots.

Regeneration. Falcata is a prolific seeder during seed years which are frequent. The seeds are hard-coated and viability may last for several years. Seeds stored in the soil or in the litter may germinate if given favorable conditions, even after several years.

The stumps after harvest also produce many sprouts.

Regeneration, therefore, of harvested falcata plantations may be done either by coppice or natural seeding or both. After harvesting the area should be cleared thoroughly, exposing the soil. Individual stumps have to be cleaned every two to three months to allow the sprouts to come out and develop properly. After six months, the best one or two sprouts per stump may be selected and the rest are removed. At the same time, some seedlings will emerge from the seeds left on the ground before harvest. If coppice and natural seedlings are too thick, the stand should have a non-commercial thinning after a year or two.

Leucaena leucocephala (Lam.) de Wit

This species, which is called ipil-ipil in the Philippines, lamtoro in Indonesia and Koa haole in Hawaii, is a legume which has gained popularity not only in Southeast Asia but throughout the tropical and sub-tropical world. Originating from Central America, the shrubby type was introduced into the Philippines in the early 1600s, from where it was introduced to the rest of Southeast Asia, Hawaii and the rest of the Pacific islands (14).

The aboreal type was discovered in Central America in 1910 and brought to Hawaii in 1960. Since then, some 341 varieties/strains of the species have been collected worldwide and several were developed at the University of Hawaii, including the 'giants'. In 1976, the Philippine Council for Agriculture and Resources Research and the US National Academy of Sciences sponsored an international consultation on the species at the University of the Philippines at Los Baños where the giant varieties were introduced to foresters, agronomists, animal scientists, and animal and wood industries. Since then the giants have been distributed throughout Southeast Asia.

Ipil-ipil has been planted both as an industrial plantation for various industry uses and in agro-forestry systems.

Propagation. Ipil-ipil is propagated by seeds. Like the falcata, the seedcoat is hard and impervious and in addition, waxy, making it difficult to germinate without treatment. The easiest and most readily adopted method of seed treatment is by dipping the seed wrapped with a piece of cloth in boiling water for 3 to 5 seconds, then allowing it to cool before planting.

In areas where ipil-ipil has been grown, the proper Rhizobia may be present in the soil and inoculation may not be necessary. In areas where ipil-ipil has not grown previously, inoculation of the treated seeds before planting is required for good growth and survival. The University of the Philippines at Los Baños has prepared packets of 'legumin' good for inoculating 4 kg of ipil-ipil seeds with the right strain of *Rhizobium* bacteria. For seedbeds, the inoculant may be dissolved in water and watered in after the seeds have germinated. But seed pelleting is preferred. Pelleting involves mixing the inoculant with a sticker solution, coating the seeds with the sticker/inoculant mixture, and then adding a dry coating material such as powdered lime and/or superphosphate (15).

When nursery-grown seedlings are desired as planting materials, the treated seeds are allowed to sprout and then planted in soil-filled plastic bags or veneer tubes when available. The inoculant may then be watered into the pots.

Establishment. An ipil-ipil plantation may be established either by planting nursery-grown seedlings or by broadcasting the treated seed on the plantation area, depending on the specific situation and purpose.

For hilly areas where the seeds may be carried down the slope by surface run-off when broadcast on burned areas, planting nursery-grown seedlings is preferred. Site preparation in this case is simply ring-clearing the specific spots where individual seedlings are to be planted. If there are standing miscellaenous trees, they sould be cut down to give full sunlight to the planted seedlings. The planted seedlings are individually ring-weeded for the first year.

In better terrain and when the amount of seed is not a problem, broadcast seeding is preferred. The whole plantation area is cleared of all vegetation and burned when dry. The seeds are then evenly broadcast on the fresh ash either by hand or aerially.

Intermediate treatments. Ipil-ipil is a slow starter. Weed control and fertilization accelerate establishment.

Ipil-ipil responds well to lime and phosphorus application in acidic and phosphorus-deficient soils. If the pH of the soil is below 5, lime is applied to bring up the pH. Applying 2–4 tons of lime per ha will bring the pH from 5.0 to 5.5. This is done before fertilizing with phosphorus, since the latter is likely to be fixed and rendered unavailable to plants when the soil is acidic.

Intermediate cutting is usually not conducted. The rotation, varying from 3 to 7 years depending on the end-use, is too short for any intermediate cuttings to be conducted economically or biologically justified.

Growth and yield. Ipil-ipil is a slow starter, like most legumes. Early in life the seedlings can be smothered by fast-growing weeds if not controlled. Volume yield during the first two years is usually low.

Once rapid growth begins, usually in the third year, the ipil-ipil forms a canopy of foliage that totally shades out weeds. Dry matter production in the leaves is large and soon improves the fertility and physical condition of the soil. Dried leaves have been estimated to contain 2 to 4.3% nitrogen and 1.3 to 4.0% potasium (Fig. 3). About 6 bags of dried leaves contain the same nitrogen that one bag of ammonium sulfate has (Table 4). Moreover, the tropical temperatures, moist soil, and the small size of the plant's leaflets encourage decay and within two weeks the fallen leaves rot to form humus. Compared with inorganic fertilizers the 'slow' release of nutrients from decaying vegetation and microorganisms allow the crop a better chance to absorb the nutrients as they leach through the soil. If release is too rapid, the tropical rainfall carries the nutrients beyond the root zone before they can be absorbed (2). Ipil-ipil has plenty of nodules on the roots and therefore it fixes nitrogen. It is believed that ipil-ipil increases nitrogen in the soil mainly through leaf fall. The beneficiary of this site improvement is the ipil-ipil itself and thus its growth is increased tremendously.

However, growth and yield varies greatly with the quality of the site. In the Philippines, from the North to the South, site indexes of 6 to 18 m based at age 5

Fig. 3. *Leucaena leucocephala* plantation in Batangas, Philippines. The heavy leaf litter is, thought to be the major source of nitrogen returned to the site.

Table 4. Fertilizer elements in dried *Leucaena leucocephala* leaves including fine twigs (2, 15).

Nutrient element	% of dry weight	Kg nutrient per ton of leaves
Nitrogen	2.09–4.30	20.9–43.0*
Phosphorus	0.15–0.40	1.5– 4.0
Potasium	1.34–4.00	13.4–40.0
Calcium	0.75–2.03	7.5–20.3
Magnesium	0.39–1.00	3.9–10.0

* About 6 bags of dried ipil-ipil leaves contain the same nitrogen as one 50-kg bag of ammonium sulfate.

years have been observed. Likewise, volume yield varies with spacing, more wood is produced at 2 m × 2 m than at closer spacings (Table 5).

Harvesting and regeneration. Harvesting ipil-ipil for fuelwood for domestic use by small scale land owners is by selection to a diameter limit which varies from 5 to 10 cm at the top and with a cutting cycle of 1 to 2 years. Every 1 to 2 years the owner goes back to the same area and cuts the stems that have grown larger than the diameter limit. The regeneration that replaces the harvested stems come from both natural seeding and coppice.

In industrial plantations of ipil-ipil where the wood is used as fuelwood for

Table 5. Volume yield in m^3 up to 5 cm diameter at the small end of *Leucaena leucocephala* (Hawaiian K-varieties) in the Philippines (16).

Age years	Spacing m	Site index in meters based at age 5 yrs. 6	8	10	12	14	16	18
2	1 × 1	8.4	10.4	12.9	16.0	19.8	24.4	30.3
	1 × 2	6.2	8.2	10.9	14.5	19.3	25.7	34.8
	2 × 2	5.3	7.3	10.0	13.8	19.1	26.3	36.3
4	1 × 1	33.5	45.8	62.7	85.7	117.2	160.3	219.2
	1 × 2	32.0	47.0	69.0	101.5	149.2	219.2	322.3
	2 × 2	31.2	47.6	72.5	110.4	168.3	256.4	390.7
6	1 × 1	53.1	75.1	106.2	150.1	212.2	300.1	–
	1 × 2	55.3	84.0	127.7	194.0	294.9	448.1	
	2 × 2	56.4	88.9	140.1	220.6	347.6	547.6	
8	1 × 1	66.9	96.1	138.2	198.6	285.6	–	
	1 × 2	72.8	112.4	173.7	268.4	414.6		
	2 × 2	75.9	121.6	194.7	311.9	499.6		
10	1 × 1	76.8	111.5	161.9	235.0	341.2		
	1 × 2	85.8	133.8	208.9	326.0	508.7		
	2 × 2	90.7	146.7	237.3	383.9	621.0		
12	1 × 1	84.2	123.0	179.9	262.9	–		
	1 × 2	95.7	150.4	236.2	371.1			
	2 × 2	102.1	166.2	270.7	440.9			

energy generation or for charcoal production, the plantation is usually clearcut at the end of the rotation and regenerated purely by coppice. Stumps of ipil-ipil trees of almost any age, and any variety, quickly resprout. Coppice growth is even more vigorous and grows more rapidly than seedlings because the sprouts are served by an already large root system. Coppice of Hawaii-type varieties have been known to grow 4 m tall and a diameter of 5 cm in one year while Salvador-type varieties have been known to produce sprouts that are even faster in growth than the Hawaiian-type, up to 6 m in one year. Ipil-ipil coppicing ability allows repeated harvested for many many years.

Agroforestry

Both falcata and ipil-ipil are planted in agro-forestry farms. Around the concession forest of the Paper Industries Corporation of the Philippines (PICOP) in Eastern Mindanao, within a radius of 100 km from its millsite, some 20,000 ha of private agricultural lands belonging to more than 3,500 farmers have been planted to falcata in combination with agricultural cash crops. About 10,000 ha more outside of this distance have been planted to falcata by about 500 farmers. The Development Bank of the Philippines gives out a loan to the farmer for the development of the farm while PICOP enters into a marketing agreement with the farmer where by PICOP guarantees to buy the falcata pulpwood. Twenty percent of the land is planted to agricultural crops and 80% to falcata. In addition, during the first two years when the falcata has not closed yet its canopy, crops are mixed with the falcata. Certain crops such as pineapple, coffee, and cocoa are mixed with falcata throughout the latter's rotation. Yields of these crops are known to be higher when grown in mixture with the falcata, obviously because of the nutrients added to the soil from nitrogen fixation and the leaf fall.

At the Manila Paper Mills in Northeastern Mindanao, the same agroforestry scheme is also being implemented with ipil-ipil as the tree crop.

CONCLUDING REMARKS

The use of biological nitrogen-fixation in Southeast Asian silviculture has been so far mostly passive. There is presently so much enthusiasm concerning the use of fast-growing hardwoods in industrial plantations as well as in the reforestation of denuded lands and many of these species happen to be nitrogen-fixers. There has been so far no specific active effort to integrate biological nitrogen fixation principles into the production of wood in the wood industry nor in the reclamation of denuded government lands.

The use of biological nitrogen fixation in Southeast Asia, however, may become more urgent in the near future. There are a number of reasons why there will be more need for this technology in the future, as follows:

1. The denuded lands are increasing rapidly and with time, they become more difficult to reforest without nutrient additions to the site.
2. Inorganic fertilizers are becoming more expensive as oil costs continue to increase.
3. The pulp and paper industry is now going into the development of second generation plantations which are expected to yield less than the first generation, requiring more nutrient inputs to maintain productivity.

These considerations, coupled with the increasing availability of more expertise in biological nitrogen-fixation research, have, in fact, started research. The next decade should see a rapid, active effort in research in Southeast Asia in this exciting new field.

RESEARCH WORKERS

1. Norami Ahmad, Forest Research Institute, Kapong, Selangor, Malaysia.
2. Director, Forest Research Institute, Borgar, Indonesia.
3. Ireneo L. Domingo, Paper Industries Corporation of the Philippines, Bislig, Surigao del Sur Philippines.
4. Mercedes U. Garcia, Department of Forest Biological Sciences, University of the Philippines at the Los Baños, College, Laguna, Philippines.

5. Enrique Pacardo, Program on Environment Science and Management, U.P. at Los Baños, College, Laguna, Philippines.
6. Sebastian S. Quiñones, Forest Research Institute, College, Laguna, Philippines.

REFERENCES

1. Hill GD: *Leucaena leucocephala* for pasture in the tropics. *Herbage Abstracts* 41: 111–119, 1971.
2. National Academy of Sciences: *Leucaena: promising forage and tree crop for the tropics*. Washington, DC, US National Academy of Sciences, 1977.
3. National Academy of Sciences: *Firewood crops. shrub and tree species for energy production*. Washington, DC, US National Academy of Sciences, 1980.
4. National Academy of Sciences: *Tropical legumes resources for the future*. Washington, DC, US National Academy of Sciences, 1979.
5. Haines SG, De Bell S: Use of nitrogen-fixing plants to improve and maintain productivity of forest soils. In: *Proceedings of symposium on impact of intensive harvesting on forest nutrient cycling*. Syracuse, New York, State University of New York College of Environmental Science and Forestry, 1979, p 279–303.
6. Parfitt RL: Shifting cultivation – how it affects the soil environment. *Harvest* 3: 63–66, 1976.
7. Zabala NQ: Interaction of *Anthocephalus chinensis* (Lamk.) Rich. ex Walp and *Albizia falcataria* (L.) Fosb. *The Pterocarpus* 1: 1–5, 1975.
8. Smith DM: *The practice of silviculture*, New York, John Wiley and Sons, 1962.
9. Chinte FO: Fast-growing pulpwood in plantations. *Philippine Forests* 5: 21–26, 1971.
10. Walters GA: A species that grew too fast, *Albizia falcataria*. *Journal of Forestry* 89: 168, 1971.
11. Suharlan A, Sumarma K, Sudiono Y: *Yield tables of ten industrial wood species*. Bogor, Indonesia, Forest Research Institute, 1975.
12. Sprinz PT: *Report on Albizia falcataria spacing and fertilization study*. Jakarta, Indonesia, Weyerhaeuser Company Tropical Forestry Research Center, 1977.
13. Revilla AV: Yield prediction in forest plantations. In: *Proceedings of the forest research symposium on industrial forest plantations*, Vergara NT, Bello, ED (eds), Manila, Philippine Forest Research Society, 1974, p 32–43.
14. Brewbaker JW: *Giant ipil-ipil: promising source of fertilizer, feed and energy for the Philippines*, Manila, USAID, 1975.
15. Yabes SI: *Ipil-ipil: the wonder tree*. Los Baños, Laguna Philippine Council for Agriculture and Resources Research, 1977.
16. Bawagan PV, Bonita ML, Domingo IL, Mendoza VB, Revilla AV: *Feasibility for an industrial tree plantation and wood-fired power plant at the Canluban Sugar Estate*. Laguna, Philippines, Consultancy Report, 1978.

12. Biological nitrogen fixation in forestry – research and practice in Australia and New Zealand

RUTH L. GADGIL

Forest Research Institute, New Zealand Forest Service, Private Bag, Rotorua, New Zealand

INTRODUCTION – CLIMATE AND FOREST RESOURCES

Australia and New Zealand are so frequently grouped together on a regional basis that outsiders may be led to assume similarities between the two countries that do not exist. Climate is related to land mass, and although New Zealand and much of the southern half of Australia lie in the same sub-tropical to temperate zone, the mainland Australian climate is far more extreme than that of the New Zealand islands. Because of the climate difference and the geographical separation, the flora and fauna of the two countries are quite distinct. The native New Zealand flora (except for the southeastern Canterbury Plains) is characterised by moist evergreen rainforest throughout, whereas forests in mainland Australia are of a sclerophyll type, ranging from wet to dry sclerophyll depending on rainfall. Small areas of rainforest exist in regions of high rainfall without a pronounced dry season. Forest and woodland occupy about 14% of the total land area in Australia and about 26% of the total land area of New Zealand (Table 1). The area occupied by exotic plantation forests is similar in both countries.

Table 1. Forested areas, 1978.

	Australia	New Zealand
	hectares (thousand)	
Total land area	768000	27000
Indigenous forest	43130	6250
Plantation forest	700	740

NITROGEN FIXATION IN INDIGENOUS FORESTS

In New Zealand the moist, maritime conditions have favoured the development of mixed kauri-podocarp-hardwood, podocarp-hardwood, podocarp-hardwood-beech (*Nothofagus* spp.) or pure beech forest. Seral forest of each of these broad types occurs locally, having been induced by storm damage, landslides or flooding in montane areas, or following the abandonment of farming on cleared lowlands. Most of the present indigenous forest, however, comprises apparent

Gordon, J.C. and Wheeler, C.T. (eds.). Biological nitrogen fixation in forest ecosystems: foundations and applications

ISBN 90-247-2849-5.
Printed in The Netherlands

climax associations with many individual trees (especially kauri and podocarps) 500–750 years old. In Australia, on the other hand, large areas exist in which the forest ecosystem (usually dominated by *Eucalyptus* spp.) has always been subject to wildfire. Here the tendency for the secondary succession to be continually renewed means that pioneer species are frequently encountered in the understorey and individual trees are less likely to survive for long periods. Well-grown *Eucalyptus* trees are usually resistant to fire and may survive for 200–300 years.

The nitrogen economy of the two forest types is very different. Undisturbed native forest in New Zealand has an understorey that is characteristically dense and rich in species, yet these include few nitrogen-fixing plants (1). Isolated studies in widely-differing forest areas suggest that small but regular nitrogen inputs can be expected in rainwater (3.5 kg ha^{-1} yr^{-}(2), 6.2 kg ha^{-1} yr^{-1} (3)); from fixation by epiphytic lichens (1–10 kg ha^{-1} yr^{-1} (4)) and from non-symbiotic fixation in litter (8 kg ha^{-1} yr^{-1} (5, 6)), soil (detectable but probably insignificant amounts (6)), the phyllosphere, and decaying wood (no specific New Zealand data). The 'nodules' or short roots which are characteristic of native conifer species in New Zealand are not analogous with the nodules of leguminous plants but result from the infection of lateral roots by endomycorrhizal fungi. Low rates of nitrogen fixation associated with podocarp roots have been reported but endophytic organisms are not thought to be involved (6). Rates of nitrogen fixation in the rhizosphere are about one tenth of those in litter layers and are not considered to have a significant effect on the amount of soil nitrogen in podocarp forests (6). Much of New Zealand's remaining native forest is no longer in its natural state. Since European settlement, intensive logging and the introduction of browsing animals have brought about profound changes with unknown effects on the nitrogen economy. Pioneer plants, native and exotic, are commonly found in disturbed bush. Their contribution to nitrogen input in the mixed forest community has not been determined, but the aggressive native shrub *Coriaria arborea* (tutu) is known to be capable of fixing 150 kg nitrogen ha^{-1} yr^{-1} in pure stands in the open (7).

Under the cool, moist conditions of Australia's island state of Tasmania the forests are less vulnerable than mainland forests to wildfire damage. Here nitrogen inputs from fixation in litter, soil, epiphytes and phyllosphere are likely to have more long-term effect than contributions from nitrogen-fixing plants (8).

Indigenous forests in mainland Australia tend to be dominated by *Eucalyptus* spp. and, probably as a result of frequent fire history, contain a wide range of pioneer nitrogen-fixing plants. Of these, the many species of *Acacia* and *Casuarina* are probably the best known and most widely distributed. During regeneration after fire, leguminous plants can fix 3–9 kg nitrogen ha^{-1} yr^{-1} in *Eucalyptus* stands (9). Estimates obtained for *Acacia* species in dry sclerophyll woodland in Victoria ranged from 0.005 kg nitrogen ha^{-1} yr^{-1} (*A. melanoxylon*) to 0.7 kg ha^{-1} yr^{-1} (*A. mearnsii*) (10). Annual nitrogen fixation by *Bossiaea laidlawiana*, a dominant understorey legume in karri forest (*Eucalyptus diversicolor*) varied from 6 kg ha^{-1} in a 7-year-old stand to 14 kg ha^{-1} in a mature karri stand with an 11-year-old understorey (11). Studies made on *Casuarina* suggest that although some species are able to fix nitrogen, others do not develop root nodules. It is unlikely that contributions of fixed nitrogen would exceed those of *Acacia* under natural conditions (8). The cycad *Macrozamia* is capable of fixing 19 kg nitrogen ha^{-1} yr^{-1} in the understorey of *Eucalyptus* forest (12) and can contribute up to 40 kg ha^{-1} annually after fire (13). Inputs from non-symbiotic fixation in the phyllosphere, litter and soil undoubtedly occur (13, 8) but are largely undocumented for Australian forest ecosystems. Hannon (14) has suggested that biological nitrogen accretion in indigenous forest soils may be limited by phosphorus level and governed by the rate of phosphorus liberation from parent material. In fires, more nitrogen than phosphorus would be lost, and counterbalancing gains might then be expected through biological nitrogen fixation in the developing post-fire ecosystem.

THE NEED FOR NITROGEN INPUTS IN MANAGED FORESTS

In recent years it has been widely acknowledged that forest management practices interfere with the nitrogen accumulation and recycling processes that develop in any undisturbed ecosystem. Very few of these practices increase the nitrogen supply, yet many have the potential for removing nitrogen from the forest site (15, 16, 17). Where reserves are large, the effects of net nitrogen loss may not be apparent in the short term. However, in New Zealand there are few managed forest sites where nitrogen is not a growth-limiting factor at some stage in the rotation (18). In Australia, most forest soils are deficient in phosphorus and nitrogen, often to a degree which is not experienced in New Zealand, Europe or North America (19). The need for nitrogen inputs to increase productivity is widely acknowledged (20, 21, 22) and Waring (17) has made an eloquent plea for a forest management policy aimed positively at increasing soil fertility.

For New Zealand it was estimated in 1978 that if all the responsive exotic forestry sites (State and privately-owned) were to be treated with fertiliser, 14000 tonnes of nitrogen element in fertiliser form would be required in 1985 (23). Often, for economic reasons, fertiliser application falls well short of the requirement, and it was apparent by 1980 that the projected level of fertiliser use was unlikely to be achieved (24). In Australia, Forest Service policy on fertiliser use varies from State to State. Nitrogenous fertiliser is applied in establishing *Pinus radiata* stands in South Australia, Western Australia and Tasmania but not in New South Wales and Victoria. Most private companies use nitrogenous fertiliser at the time of establishment (22). Fertiliser application in established stands is not yet an accepted practice in Australia.

THE POTENTIAL IMPORTANCE OF BIOLOGICAL NITROGEN FIXATION

In both Australia and New Zealand the scope for nitrogen application in forestry is already large and is likely to increase. The cost of nitrogenous fertiliser (urea) in relation to that of other fertilisers has increased in Australia throughout the period 1971–80 (25). In New Zealand the cost of urea may decrease when a new ammonia-urea plant at Kapuni comes into production in 1982–3.

Pastoral agriculture in both countries is almost independent of nitrogenous fertiliser (26, 27). This is the direct result of a system developed over the past 50 years in which nitrogen-fixing plants, usually clovers, are grown with grass and provide the bulk of the combined nitrogen required for stock rearing. Annual nitrogen input from *Trifolium repens* in the grass/clover mixture of developed lowland pasture is about 180 kg ha^{-1} in New Zealand (28). The system is usually dependent on regular applications of superphosphate fertiliser; often application of other nutrients is also necessary. This pastoral example of a non-exploitive system with low cost inputs and high potential productivity has not passed unnoticed in forest research.

EARLY FORESTRY TRIALS WITH NITROGEN FIXING PLANTS (1960–1970)

During the early 1960s considerable interest was aroused when Richards (29) reported that the introduction of exotic conifers in east Australian lateritic soils was associated with net nitrogen gains in the ecosystem (up to 50 kg ha^{-1} annually). Evidence of nitrogen fixation associated with the roots of *Pinus radiata* and *P. elliottii* var *elliottii* seedlings was obtained (30, 31), but doubt remained about the nature of the organisms involved. More recently Richards (32) demonstrated nitrogen-fixing activity in the rhizosphere of *P. elliottii* seedlings. There is no clear evidence to support the suggestion that mycorrhizal infection stimulates nitrogen fixation in the rhizosphere.

Trials investigating potential benefits from the introduction of nitrogen-fixing plants into forest stands date back to 1959. Using a fertiliser amendment (P, K, Ca, Cu, Zn, B and Mo), and a preliminary cover of *Trifolium repens*, Richards and Bevege (33) established a successful cover of *Lotononis bainesii, Desmodium uncinatum* and *Phaseolus atropurpureus* on a lateritic podsol in Queensland, Australia. The effect of these legumes on the growth of the indigenous conifers *Araucaria cunninghamii* and *Agathis robusta* and the exotic *Pinus elliotii* var *elliottii, P. taeda* and *P. caribaea* var *hondurensis* was studied for five years after planting. The legumes grew vigorously at first but declined during the third season, especially under the dense shade of the exotic pines. Growth of the indigenous conifers (normally inhibited by nitrogen deficiency) was stimulated by the legumes, but that of the exotic pines (normally reponsive to added nitrogen) was suppressed. Soil nitrogen content increased only where legumes were grown under indigenous conifers and it was assumed that the absence of soil nitrogen accumulation under *Pinus* resulted from suppression of the legumes by the trees.

In New South Wales, Australia, Waring (34) compared the abilities of a clover mixture and urea to supply nitrogen to *Pinus radiata* on a second rotation site where serious organic matter losses had occurred during the first rotation. A basal dressing of lime and superphosphate was used. During the first year after planting, tree growth in the clover plots was significantly retarded as a result of competition for water with the developing clover sward. In the second year, tree height growth in clover plots was equal to that in urea plots (45 kg nitrogen ha^{-1} applied annually) and significantly greater than that in control (no nitrogen) plots. After 3 years the increase in soil nitrogen due to clover was highly significant and twice as great as the increase due to urea application.

Studies in New Zealand began at Woodhill, one of the coastal sand dune forests of the North Island, in which marram grass (*Ammophila arenaria*) and perennial tree lupin (*Lupinus arboreus*) are used to stabilise the sand before *Pinus radiata* forest is established. The routine stabilisation procedure (35, 36) begins with the physical arrest of blowing sand on the foredune, using barrier fences. Marram grass is then planted behind the foredune and receives two applications of nitrogenous fertiliser annually (25 kg nitrogen element ha^{-1} yr^{-1}). After 1–2 years, lupin seed is sown and topdressing discontinued. Lupin plants usually provide sufficient cover for tree planting to commence within a further 2–4 years. Lupins and marram are tractor-crushed when the trees are planted, but after 6 months, vigorous regrowth of lupin has to be controlled, usually with aerially-applied hormone spray. Further releasing is unnecessary as the trees rapidly assume dominance. If tree canopy closure is permitted, the lupins and marram grass become completely suppressed, but, after thinning, the lupins germinate from buried seed and regenerate rapidly into a dense understorey. After 4–5 years the lupins are again suppressed, but quickly re-establish after a second thinning or clearfelling. Buried lupin seed can survive for long periods; the development of a continuous lupin ground cover has been observed after the clearfelling of a 36-year-old tree stand. Originally the planting of *P. radiata* on sand dunes was seen as a protection measure only. The potential for timber production was soon realised, however, and sizeable forest-based industries relying on sand forests have developed in the Auckland region and in the Manawatu (36). The total area of sand dune forests in New Zealand is about 60,000 ha.

Uncolonised coastal sand contains about 0.008% nitrogen, most of which is unavailable to plants (37). Except for a very localised area where copper deficiency was diagnosed (38), nitrogen is the only element known to limit tree growth. Because lupins are nitrogen-fixing plants and are a constant feature of the sand dune forest ecosystem, interest in their nutritional relationship with *P. radiata* began to develop.

In 1963 a trial was established at Woodhill Forest in which lupin regeneration, after the thinning of a 10-year-old *P. radiata* stand, was killed with hormone spray. Where lupin had been excluded, plots were treated with nitrogenous fertiliser at the following rates (kg ha^{-1}): 0,56 × 2, 112 × 2, and 213. Four years later tree basal area showed a small but significant response to nitrogen and to the presence of lupins (19). There was no differential response to fertiliser rate.

Glasshouse trials using coastal sand showed that growth and nitrogen nutrition of *P. radiata* seedlings were improved when lupin litter or seedling exudates were added (39). Stress in the form of shading, defoliation or drought was found to increase the rate of lupin root decomposition and this was shown to be a possible mechanism for increasing nitrogen transfer between lupins and *P. radiata* (40). It was also shown that lupin plants growing on the open dunes were capable of fixing at least 160 kg of nitrogen ha^{-1} yr^{-1} and that the lupin/marram association could accumulate 440 kg ha^{-1} of nitrogen in the biomass and litter in 5 years, an amount equivalent to that required by a first-rotation crop of *P. radiata* (37). Nitrogen accumulation was greatest in young lupin stands.

Attempts to assess the potential effects of nitrogen-fixing plants on tree growth in other soil types during this period met with varying degrees of success (41, 21). Most of these trials were conducted in the Nelson region (South Island, New Zealand) where tree growth is limited by nitrogen, phosphate and boron deficiencies. Attempts were made to establish a wide range of nitrogen-fixing plants, usually with the help of phosphate and boron fertiliser applications. The universal failure of *Trifolium repens*, *T. pratense*, *Lupinus luteus*, *L. arboreus*, *Lotononis* and *Acacia* on granitic soils was attributed to low soil pH. On the Moutere gravel soils (42), *Trifolium pratense*, *T. subterraneum*, *Lupinus luteus* and *L. angustifolius* all established well in a recently-cleared area, but were smothered by gorse (*Ulex europaeus*) within 2–3 years. *Lupinus polyphyllus* and 3 species of *Alnus* grew well in Tasman Forest and there was some evidence that alder caused a marginal improvement in height growth and foliar nitrogen levels in *Pinus radiata* after 3 years. *Lupinus arboreus* grew well among *P. radiata* regeneration at Golden Downs Forest and was responsible for a small increase in foliar nitrogen level and basal area increment 7–9 years later. A more definite basal area response (equivalent to that produced by 115 kg nitrogen ha^{-1} applied as urea) was observed where *L. arboreus* was established in a 10-year-old *P. radiata* stand in Tasman Forest. There was no replication of treatments in either of these lupin trials.

Some interest was expressed in the possibility of utilising nitrogen fixed by gorse, which is a spiny, aggressive and prolific weed in forests throughout New Zealand. Gorse is capable of fixing up to 200 kg nitrogen ha^{-1} annually (43). Several trials were initiated in the Nelson region, but little information exists about the methods and effectiveness of gorse control in exclusion plots. It is also doubtful whether the effects of gorse residues were ever entirely eliminated (41). Often the presence of live gorse suppressed the growth of young trees. Height growth increases were reported from some trials, but may have been the expression of competition for light in gorse plots. Basal area in one trial showed a greater response to gorse control than to the presence of gorse.

These early trials indicated very clearly that benefits to the tree crop in terms of nitrogen nutrition depend on the maintenance of a balance between successful establishment of the nitrogen-fixing plants on the one hand, and absence of undue interference with tree growth through competition for moisture, nutrients, light, etc., on the other. Responses appeared to be more obvious in established tree stands.

LUPINS IN THE NEW ZEALAND SAND DUNE FOREST ECOSYSTEM

Further investigation of the role of lupins in sand dune forestry continued at Woodhill Forest in the 1970s. Using the acetylene reduction assay method, it was shown that in the marram/lupin association of the open dunes, nitrogen fixation rates increased during the spring to an estimated 11.5 kg ha^{-1} $week^{-1}$ at time of lupin flowering (44). The development of fruiting heads was associated with a reduction in nitrogen fixation rate.

Studies in thinned forest stands provided no evidence that free-living organisms fixed significant amounts of nitrogen (45). Over a 13-week period in the spring, fixation rates in the lupin understorey were proportional to the amount of light penetration. In spite of considerable variability, the data indicated that nitrogen input from lupins in thinned forest stands was about one sixth of that in the open dunes (44). Extension of the survey to cover a full annual cycle showed a drop in fixation rate during the early summer and a slow rise in autumn and winter (46). During the dry summer months, infestation by kowhai moth larvae (*Uresiphita maorialis polygonalis*) caused repeated defoliation of lupin plants (a normal seasonal occurrence) and very little nitrogen-fixing activity was recorded.

Studies of nitrogen distribution in the biomass and litter of the developing forest ecosystem, showed that failure of lupin to regenerate after spray-releasing reduced the total above-ground nitrogen content of a 4-year-old *P. radiata* stand by 90 kg ha^{-1} (47). During the first 5 years of tree growth, marram grass and litter accumulated more nitrogen than either the lupins or the trees. There was no evidence of any overall nitrogen buildup in the mineral soil during this period and the storage role of the marram and undecomposed organic matter was clearly of the utmost importance for the future nitrogen nutrition of the trees (47, 48).

In a long-term lupin exclusion trial established at Woodhill in 1968, the onset of severe nitrogen deficiency in plots treated with hormone spray was not apparent until *P. radiata* was 5 years old. At first, tree diameter growth was depressed by the presence of lupins, particularly where fertiliser (a complete macronutrient supplement, including 50 kg nitrogen ha^{-1} twice annually for 10 years) was also applied (Table 2, D.S. Jackson, pers. comm.). Lupins in the understorey died out during 1973 and marram grass was suppressed by canopy closure in 1974. Although three experimental stocking reductions (2224 to 1483 stems ha^{-1} in 1970; 2224 to 741 stems ha^{-1} in 1972; 2224 to 1483 stems ha^{-1} in 1972 and to 371 stems ha^{-1} in 1976) were made, lupin regrowth was stimulated only by the 1976 reduction to 371 stems ha^{-1}. At the three denser stocking rates, lupins were therefore absent from the lupin plots from the end of 1973. Positive (but not significant) effects of lupins and of fertiliser on tree growth were first observed in 1973 and coincided with the first significant differences in foliar nitrogen levels. During 1974 the foliage of trees grown without either lupins or fertiliser became chlorotic, and foliar nitrogen levels dropped to below 1.0% in November 1974 and during each subsequent (southern) summer. In all other treatments nitrogen concentration in tree foliage has remained above 1.0%. The most recent tree growth measurements (Table 2) indicate that stocking reduction, lupins and fertiliser treatment have all had a significantly positive effect. The response to fertiliser treatment (discontinued after 1978) has been greater than the response to lupins, but trees grown with both fertiliser and lupins have achieved the fastest growth rate, especially in the heavily thinned plots. It is clear that the presence of lupins during the first 5–6 years improved tree growth later in the rotation but did not satisfy tree nitrogen requirements.

There is now a much better understanding of the need to manage the lupins as well as the trees in New Zealand sand dune forests, and it is recognised that large inputs of nitrogenous fertiliser will be required to maintain tree productivity if lupins do not establish successfully or if they do not regenerate after thinning. Normally there are few management problems. Inoculation of seed with *Rhizobium* is unnecessary; experimental inoculation with 38 strains of *R. lupini* showed little or no improvement over uninoculated seed in terms of dry matter and nitrogen yield of field-grown lupin plants (49). Generally speaking, lupins grow well on coastal sand without nutrient amendments. There is, however, one isolated area where lupin growth is limited by molybdenum, sulphur and boron deficiencies (50). In recent years the invasion of sand dune forests by pampas grass (*Cortaderia selloana* and *C. jubata*) has been seen as a threat to the effectiveness of lupins through physical competition and through uptake and immobilisation of fixed nitrogen. Attempts are being made to control the pampas grass with herbicides and grazing.

As in other parts of the world the consensus in Australia and New Zealand is that the potential for managing biological nitrogen fixation in forests does exist, but that much more research is required if the potential is to be realised. There is general agreement that symbiotic, rather than asymbiotic nitrogen fixation should be the basis for future research, partly because amounts of fixed nitrogen are larger in symbiotic systems and partly because the vigour of the non-fixing partner is usually a good indication of the amount of nitrogen being fixed.

Very few attempts have been made at establishing nitrogen-fixing species as tree crops. This is because *Eucalyptus* and exotic pines have proved to be so successful over a wide range of climate and soil types that milling technology and forest management are virtually committed to their monoculture. Some potential for nitrogen-fixing trees may exist if an end use other than that for conventional forest products could be justified. Examples might be the re-planting of nitrogen-deficient sites after mineworking (51), the growth of high-quality timber for specialised purposes (52), or the development of a fast-growing species for energy farming. In this context it is interesting to note the relatively high productivity and nitrogen accumulation rate which can be achieved in *Acacia* plantations on some sites (53).

Crane (22) has expressed the opinion that 'a great deal more research will be required before a satisfactory legume or nitrogen-fixing system is developed for *Pinus radiata*.' This statement cannot be disputed, particularly if an analogy with the clover/grass pastoral system is required. Whether such a broadly-based technology is attainable in conventional production forestry is very much an open question at present. The relatively low return on capital investment means that gains must be very convincing if understorey management is to be included in production forestry costs.

Turvey and Smethurst (25) were sceptical about the economic benefits to be derived from growing nitrogen-fixing plants in the understorey and considered that inorganic nitrogen amendments would be much cheaper if fertilisers, discing and herbicide treatments were needed to establish and control the legumes. Their calculations were based on a single input of 90 kg nitrogen ha^{-1} from either fertiliser or legumes and assumed that the legumes would be grown for two years before tree planting.

There are, of course, already instances where understorey management has to be taken into account. Land reclamation by afforestation nearly always requires the establishment of a cover crop and an input of nitrogen before trees can be planted, and nitrogen-fixing plants are often ideal for the purpose. The use of marram grass and lupins on coastal dunes is one example of this type of procedure. Other examples are the use of Rhodes grass (*Chloris gayana*) and *Acacia holosericea* during rehabilitation after surface mining in northern Australia (51) and the potential use of legumes in the reclamation of gold dredge tailings in the South Island of New Zealand (54).

Agroforestry is probably outside the scope of this review, but pasture development under trees (as opposed to tree planting in developed pasture) must be regarded as understorey management. There are signs in both Australia and New Zealand that enthusiasm for increased financial return from grazing in forests may outstrip the degree of understanding about the effects on ecosystem processes. In Western Australia, where farmland has been planted with *Pinus radiata* to reduce the salinity of the groundwater, trials have shown that *Trifolium subterraneum* pasture could be established under 13-year-old trees thinned to 143 and to 261 stems ha^{-1} if annual dressings of P, Cu, Zn and Mo were applied (55). After 2 years the increased shading caused a loss in pasture production, but a further thinning to 70 stems ha^{-1} gave a good pasture response. Total volume production by the trees was reduced by thinning, and after 5 years there was no evidence of increased nitrogen concentration in the needles (56). The effect on groundwater salinity has not been reported. In New Zealand, grazing to control forest weeds is being actively encouraged (57) yet little is known about the effects of livestock on tree growth and forest nutrient cycling.

Weed control is certain to become an increasing cost as the trend towards wider spacing and earlier thinning in *P. radiata* plantations continues (58). It is unlikely that noxious weeds will cease to be a problem in production forests unless they are carefully managed. Replacement with more easily controlled, less competitive understorey plants may prove to be an alternative to repeated herbicide treatment, particularly if access can be improved and fire risk lowered. The use of nitrogen-fixing plants in such situations offers the additional advantage of deliberate nitrogen addition to the ecosystem. Resulting increases in long-term and short-term productivity have yet to be evaluated, but are considered to be a distinct possibility in a region where responses to nitrogen application are so common.

CURRENT RESEARCH

In spite of their reservations about the economic benefits from growing nitrogen-fixing plants in the understorey, Turvey and Smethurst have established a legume screening trial in Gippsland, Victoria, and will be testing the performance of 7 indigenous and 8 exotic species in three soil types used for *Pinus radiata* plantations (25).

Annual lupins and subterranean clover are under investigation in western Victoria and South Australia as a source of nitrogen for *P. radiata* plantations (59). Losses of nitrogen through leaching are considerable in these sandy soils and nitrogen-fixing plants are seen as a potential alternative to fertiliser application providing that competition for moisture is not too severe. Results (reported for unnamed annual lupins only) showed no detrimental effect on the water relations of young pine trees. Foliar nitrogen concentration and tree height and diameter two years after planting were all greater where lupins were grown than where 30 g of nitrogen had been applied to each tree in fertiliser form.

The use of *Alnus viridis* and selected legumes (e.g., *Lotus* and *Lupinus*) in the establishment of protection forests on eroded mountain land is currently under investigation in the South Island of New Zealand. Preliminary trials have given encouraging results (60, 61) and the work is being extended to include intact but infertile mountain soils which may have potential for production forestry (62).

In the North Island of New Zealand there is interest in assessing the effect of *Lupinus arboreus* as an understorey plant in forests on the central pumice plateau. High-producing *P. radiata* stands on pumice soils have been shown to respond to nitrogenous fertilisers when thinned (63). *Lupinus arboreus*, which is known as a weed in these forests, may have some potential as a source of nitrogen if responses to sown lupins resemble those achieved at Woodhill (Table 2). Growth of this species is being compared with that of 35 other nitrogen-fixing plants (mostly forage legumes) in a spaced-plant trial at Whakarewarewa Forest, Rotorua. The trial, which also contains broadcast-sown plots of 8 species (in-

Table 2. Growth of *Pinus radiata* planted in 1968, Woodhill Forest New Zealand. Data from D.S. Jackson (pers. comm.).

Stocking	Treatment	Basal area of largest 371 stems ha^{-1}			
		1971	1973	1975	1981
		m^2 ha^{-1}			
2224 stems ha^{-1} throughout	Lupins excluded, no fertiliser	0.24	2.79	5.29	8.91
	Lupins only	0.28	3.29	6.05	11.06
	Fertiliser only	0.26	2.95	5.96	12.05
	Lupins + fertiliser	0.18	3.31	7.32	15.70
Thinned to 1483 stems ha^{-1} in September 1970	Lupins excluded, no fertiliser	0.18	2.62	5.01	9.80
	Lupins only	0.22	3.03	6.32	12.59
	Fertiliser only	0.23	3.20	7.01	15.23
	Lupins + fertiliser	0.20	3.12	7.75	16.80
Thinned to 741 stems ha^{-1} in September 1972	Lupins excluded, no fertiliser	0.26	2.94	6.11	12.08
	Lupins only	0.22	2.58	6.48	14.78
	Fertiliser only	0.22	3.31	7.79	18.20
	Lupins + fertiliser	0.19	3.00	8.17	19.26
Thinned to 1483 stems ha^{-1} in September 1972, then to 371 stems ha^{-1} in February 1976	Lupins excluded, no fertiliser	0.23	2.94	5.32	12.34
	Lupins only	0.30	3.39	6.97	17.75
	Fertiliser only	0.22	3.07	6.56	20.30
	Lupins + fertiliser	0.19	3.22	7.74	22.36

cluding *L. arboreus*), is located in a 10-year-old *P. radiata* stand grown to a sawlog regime and now at 300 stems ha^{-1}. Understorey vegetation was completely cleared and the soil disced before planting. No fertiliser will be applied. Species to be tested are all exotic except for the shrub *Coriaria arborea* which is also a weed in forests on the pumice plateau. Many forage legumes have been included because the effect of shade and litterfall on their growth performance is unknown. The trial will form the basis of a research programme investigating the potential of nitrogen-fixing plants for various options in the management of both the understorey and the trees.

RESEARCH WORKERS

The following is a list of workers currently engaged in research on biological nitrogen fixation in forestry in Australia and New Zealand:

1. Benecke U: Forest Research Institute, New Zealand Forest Service, P.O. Box 31–011, Christchurch, New Zealand.
2. Bergersen FJ: CSIRO Division of Plant Industry, Canberra, A.C.T., Australia.
3. Bevege DI: Department of Forestry, P.O. Box 5, Brisbane, Queensland 4000, Australia.
4. Charlton FJL: DSIR Grasslands Division, Private Bag, Palmerston North, New Zealand.
5. Crane WJB: Division of Forest Research, CSIRO, P.O. Box 4008, Canberra, ACT 2600, Australia.
6. Dalling MJ: School of Agriculture and Forestry, University of Melbourne, Parkville, Victoria 3052, Australia.
7. Davis MR: Forest Research Institute, New Zealand Forest Service, P.O. Box 31–011, Christchurch, New Zealand.
8. FitzGerald RE: Forest Research Institute, New Zealand Forest Service, P.O. Box 31011, Christchurch, New Zealand.
9. Gadgil RL: Forest Research Institute, New Zealand Forest Service, Private Bag, Rotorua, New Zealand.
10. Galbraith J: CSIRO Division of Land Resources Management, Wembley, Western Australia.
11. Green TGA: Department of Biological Sciences, University of Waikato, Private Bag, Hamilton, New Zealand.
12. Grove TS: CSIRO Division of Land Resources Management, Wembley, Western Australia.
13. Hingston FJ: CSIRO Division of Land Resources Management, Wembley, Western Australia.

14. Knowles RL: Forest Research Institute, New Zealand Forest Service, Private Bag, Rotorua, New Zealand.
15. Lamb D: Botany Department, University of Queensland, St. Lucia, Queensland 4067, Australia.
16. Langkamp PJ: School of Agriculture and Forestry, University of Melbourne, Parkville, Victoria 3052, Australia.
17. Lawrie AC: Botany Department, Monash University, Clayton, Victoria 3168, Australia.
18. Malajczuk N: CSIRO Division of Land Resources Management, Wembley, Western Australia.
19. Nambiar SEK: CSIRO Division of Forest Research, Mount Gambier, South Australia.
20. Nordmeyer AH: Forest Research Institute, New Zealand Forest Service, P.O. Box 31–011, Christchurch, New Zealand.
21. O'Connell AM: CSIRO Division of Land Resources Management, Wembley, Western Australia.
22. Richards BN: Department of Botany, University of New England, Armidale, New South Wales, Australia.
23. Shea SR: Forests Department of Western Australia, 54 Barrack St., Perth, Western Australia.
24. Silvester WB: Department of Biological Sciences, University of Waikato, Private Bag, Hamilton, New Zealand.
25. Smethurst PJ: APM Forests Pty Ltd., Box 37, Morwell, Victoria 3840, Australia.
26. Turvey ND: APM Forests Pty Ltd., Box 37 Morwell, Victoria 3840, Australia.
27. Waring HD: CSIRO Division of Forest Research, P.O. Box 4008, Canberra ACT 2600, Australia.

ACKNOWLEDGMENTS

I would like to thank the following people for their comments on the manuscript: Dr D. Lamb of the University of Queensland, Australia; Dr D.S. Jackson, Dr W.R.J. Sutton, Mr J.L. Nicholls, Dr G.M. Will and Mr R.L. Knowles of the Forest Research Institute, Rotorua, New Zealand.

REFERENCES

1. Greenwood RM: Rhizobia associated with indigenous legumes of New Zealand and Lord Howe Island. In: *Microbial Ecology*, Loutit MW, Miles JAR (eds). Berlin, Springer-Verlag, 1978, p 402–403.

2. Neary DG, Pearce AJ, O'Loughlin CL, Rowe LK: Management impacts on nutrient fluxes in beech-podocarp-hardwood forests. *NZ Journal of Ecology* 1: 19–26, 1978.
3. Bargh BJ: Output of water, suspended sediment and phosphorus and nitrogen forms from a small forested catchment. *NZ J For Sci* 7: 162–171, 1977.
4. Green TGA, Horstmann J, Bonnett H, Wilkins A, Silvester WB: Nitrogen fixation by members of the Stictaceae (Lichenes) of New Zealand. *New Phytol* 84: 339–348, 1980.
5. Silvester WB: Nitrogen fixation and mineralisation in kauri (*Agathis australis*) forest in New Zealand. In: *Microbial Ecology*, Loutit MW, Miles JAR (eds). Berlin, Springer-Verlag, 1978, p 138–143.
6. Silvester WB, Bennett KJ: Acetylene reduction by roots and associated soil of New Zealand conifers. *Soil Biol Biochem* 5: 171–179, 1973.
7. Silvester WB: Ecological and economic significance of the non-legume symbioses. In: *Nitrogen fixation*, Newton WE, Nyman CJ (eds). Washington State University Press, 1976, p 489–506.
8. Lamb D: Natural inputs of nitrogen to forest ecosystems. In: *Managing nitrogen economies of natural and man-made forest ecosystems*, Rummery RA, Hingston FJ (eds). Perth, Western Australia. *CSIRO Division of Land Resources Management*, 1981, p 1–25.
9. O'Connell AM, Grove TS, Malajczuk N: NitrogEn fixation in the litter layer of eucalypt forests. *Soil Biol Biochem* 11: 681–682, 1979.
10. Lawrie AC: Nitrogen fixation by native Australian legumes. *Aust J Bot* 29: 143–157, 1981.
11. Grove TS, Malajczuk N: Nitrogen inputs to *Eucalyptus marginata* and *E. diversicolor* forests. In: *Managing nitrogen economies of natural and man-made forest ecosystems*, Rummery RA, Hingston FJ (eds). Perth, Western Australia. *CSIRO Division of Land Resources Management*, 1981, p 199–204.
12. Halliday J, Pate JS: Symbiotic nitrogen fixation by coralloid roots of the cycad *Macrozamia riedlei:* physiological characteristics and ecological significance. *Aust J Plant Physiol* 3: 349–358, 1976.
13. Grove TS, O'Connell AM, Malajczuk N: Effects of fire on the growth, nutrient content and rate of nitrogen fixation of the cycad *Macrozamia riedlei. Aust J Bot* 28: 271–281, 1980.
14. Hannon NJ: The status of nitrogen in the Hawkesbury sandstone soils and their plant communities in the Sydney district III The sources of loss of nitrogen. *Proc Linn Soc NSW* 85: 207–216, 1961.
15. Will GM, Dyck WJ, Gadgil RL, Hunter IR, Madgwick HAI: Nitrogen in radiata pine forests in New Zealand. In: *Managing nitrogen economies of natural and man-made forest ecosystems*, Rummery RA, Hingston FJ (eds). Perth, Western Australia. *CSIRO Division of Land Resources Management*, 1981, p 146–152.
16. Raison RJ, Khanna PK, Woods PV, Godkin A: Loss of nitrogen from litter and understorey during prescribed burning in sub-alpine eucalypt forests. In: *Managing nitrogen economies of natural and man-made forest ecosystems*, Rummery RA, Hingston FJ (eds). Perth, Western Australia. *CSIRO Division of Land Resources Management*, 1981, p 291.
17. Waring HD: Silvicultural and management problems in supplying nitrogen for production forestry. In: *Managing nitrogen economies of natural and man-made forest ecosystems*, Rummery RA, Hingston FJ (eds). Perth, Western Australia. *CSIRO Division of Land Resources Management*, 1981, p 83–123.
18. Will GM: Nutrient deficiencies in *Pinus radiata* in New Zealand. NZ J For Sci 8: 4–14, 1978.
19. Waring HD: Forest fertilisation in Australia: early and late. In: *Australian Forest Nutrition Workshop, Productivity in Perpetuity*. Canberra, Australia, 1981, p 202–217.
20. Ballard R: Use of fertilisers at establishment of exotic forest plantations in New Zealand. *NZ J For Sci* 8: 70–104, 1978.
21. Mead DJ, Gadgil RL: Fertiliser use in established radiata pine stands in New Zealand. *NZ J For Sci* 8: 105–134, 1978.
22. Crane WJB: Nitrogen fertilisation and *Pinus radiata*. In: *Managing nitrogen economies of natural and man-made forest ecosystems*, Rummery RA, Hingston FJ (eds). Perth, Western Australia. *CSIRO Division of Land Resources Management*, 1981, p 327–339.
23. Ballard R, Will GM: Past and projected use of fertilisers in New Zealand forests. *NZ J For Sci* 8: 15–26, 1978.

24. Will GM: Use of fertilisers in New Zealand forestry operations 1980. *NZ J For Sci* 11: 191–198, 1981.
25. Turvey ND, Smethurst PJ: Biological and economic criteria for establishing a nitrogen fixing understorey in pine plantations. In: *Managing nitrogen economies of natural and man-made forest ecosystems*, Rummery RA, Hingston FJ (eds). Perth, Western Australia. *CSIRO Division of Land Resources Management*, 1981, p 124–145.
26. Evans LT: Nitrogen fixation – a symbiosis between industry and biology? In: *Current perspectives in nitrogen fixation*, Gibson AH, Newton WE (eds). Canberra, Australia. Australian *Academy of Science*, 1981.
27. Ball R, Brougham RW, Brock JL, Crush JR, Hoglund JH: Nitrogen fixation in pasture I. Introduction and general methods. *NZ Journal of Experimental Agriculture* 7: 1–5, 1979.
28. Hoglund JH, Crush JR, Brock JL, Ball R: Nitrogen fixation in pasture XII. General discussion. *NZ Journal of Experimental Agriculture* 7: 45–51, 1979.
29. Richards BN: Increased supply of nitrogen brought about by *Pinus*. *Ecology* 43: 538–541, 1962.
30. Richards BN: Fixation of atmospheric nitrogen in coniferous forests. *Aust For* 28: 68–74, 1964.
31. Richards BN, Voigt GK: Nitrogen accretion in coniferous forest ecosystems. In: *Forest-soil relationships in North America*, Youngberg CT (ed). Corvallis, USA. Oregon State University Press, 1965, p 105–116.
32. Richards BN: Nitrogen fixation in the rhizosphere of conifers. *Soil Biol Biochem* 5: 149–152, 1973.
33. Richards BN, Bevege DI: The productivity and nitrogen economy of artificial ecosystems comprising various combinations of perennial legumes and coniferous tree species. *Aust J Bot* 15: 467–480, 1967.
34. Waring HD: The nitrogen balance and soil fertility in pine plantations. *Aust Soils Conf Brisbane Proc* 5: 12–14, 1966.
35. Restall AA: Sand dune reclamation on Woodhill Forest. *NZ Jl For* 9: 154–161, 1964.
36. Wendelken WJ: New Zealand experience in stabilization and afforestation of coastal sands. *Biometeorology* 18: 145–158, 1974.
37. Gadgil RL: The nutritional role of *Lupinus arboreus* in coastal sand dune forestry 3. Nitrogen distribution in the ecosystem before tree planting. *Plant and Soil* 35: 113–126, 1971.
38. Will GM: Copper deficiency in radiata pine planted on sands at Mangawhai Forest. *NZ J For Sci* 2: 217–221, 1972.
39. Gadgil RL: The nutritional role of *Lupinus arboreus* in coastal sand dune forestry I. The potential influence of undamaged lupin plants on nitrogen uptake by *Pinus radiata*. *Plant and Soil* 34: 357–367, 1971.
40. Gadgil RL: The nutritional role of *Lupinus arboreus* in coastal sand dune forestry 2. The potential influence of damaged lupin plants on nitrogen uptake by *Pinus radiata*. *Plant and Soil* 34: 575–593, 1971.
41. Gadgil RL: Alternatives to the use of fertilisers. In: *Use of fertilisers in New Zealand forestry. Forest Research Institute Symposium No. 19*, Ballard R (ed). *New Zealand Forest Service*, 1977, p 83–94.
42. Stone EL, Will GM: Nitrogen deficiency of second generation radiata pine in New Zealand. In: *Forest-soil relationships in North America*, Youngberg CT (ed). Corvallis, USA. Oregon State University Press, 1965, p 117–139.
43. Egunjobi JK: Dry matter and nitrogen accumulation in secondary successions involving gorse (*Ulex europaeus* L.) and associated shrubs and trees. *NZ Jl Sci* 12: 175–193, 1969.
44. Sprent JI, Silvester WB: Nitrogen fixation by *Lupinus arboreus* grown in the open and under different aged stands of *Pinus radiata*. *New Phytol* 72: 991–1003, 1973.
45. Carter DA: Factors affecting nitrogenase activity of yellow lupin (*Lupinus arboreus*) over an annual cycle at Woodhill State Forest. MSc Thesis, University of Auckland, New Zealand, 1974.
46. Silvester WB, Carter DA, Sprent JI: Nitrogen input by *Lupinus* and *Coriaria* in *Pinus radiata* forest in New Zealand. In: *Symbiotic nitrogen fixation in the management of temperate forests*, Gordon JC, Wheeler CT, Perry DA (eds). Corvallis, Oregon, USA. Oregon State University, 1979, p 253–265.

47. Gadgil RL: Nitrogen distribution in stands of *Pinus radiata* with and without lupin in the understorey. *NZ J For Sci* 6: 33–39, 1976.
48. Gadgil RL: The nutritional role of *Lupinus arboreus* in coastal sand dune forestry 4. Nitrogen distribution in the ecosystem for the first 5 years after tree planting. *NZ J For Sci* 9: 324–336, 1979.
49. Caradus JR, Silvester WB: A comparative study of strains of *Rhizobium lupini* on *Lupinus arboreus* L. *NZ Jl Agric Res* 22: 329–34, 1979.
50. Gadgil RL, Knight PJ, Sandberg AM, Allen PJ: Molybdenum, sulphur and boron deficiency in lupin (*Lupinus arboreus* Sims) at Pouto Forest. *NZ J For Sci* 11: 114–127, 1981.
51. Langkamp PJ, Swinden LB, Dalling MJ: Nitrogen fixation (acetylene reduction) by *Acacia holosericea* on areas restored after mining at Groote Eylandt, Northern Territory. *Aust J Bot* 27: 353–61, 1979.
52. Nicholas ID: Tasmanian blackwood *(Acacia melanoxylon)*. What's New in Forest Research No 62, Forest Research Institute, Rotorua, New Zealand, 1978.
53. Frederick DJ, Madgwick HAI, Jurgensen MF, Oliver GR: Dry matter production and nutrient relations in 8-year-old *Eucalyptus regnans, Acacia dealbata* and *Pinus radiata*. In: *Australian Forest Nutrition Workshop, Productivity in Perpetuity*. Canberra, Australia, 1981, p 344.
54. FitzGerald RE: The potential role of legumes in the rehabilitation of levelled gold dredge tailings, Taramakau River, *N.Z. Proc NZ Grassld Ass* 42: 206–209, 1981.
55. Anderson GW, Batini FE: Clover and crop production under 13- to 15-year-old *Pinus radiata*. *Aust J Exp Agric Anim Husb* 19: 362–368, 1979.
56. Batini FE, Anderson GW: Agroforestry under 13- to 18-year-old *Pinus radiata*, Wellbucket, Western Australia. In: *Managing nitrogen economies of natural and man-made forest ecosystems*, Rummery RA, Hingston FJ (eds). Perth, Western Australia. *CSIRO Division of Land Resources Management*, 1981, p 346–353.
57. Knowles RL, Cutler TR: *Integration of forestry and pastures in New Zealand*. Paper prepared for the 11th Commonwealth Forestry Conference, September 1980. New Zealand Forest Service, Wellington, New Zealand, 1980.
58. Sutton WRJ: Comparison of alternative silvicultural regimes for radiata pine. *NZ J For Sci* 6: 350–356, 1976.
59. Nambiar SEK, Nethercott KH: Nitrogen supply to radiata pine through legumes. In: *Australian Forest Nutrition Workshop, Productivity in Perpetuity*. Canberra, Australia, 1981, p 344.
60. Benecke U: Nitrogen fixation by *Alnus viridis* (Chaix) DC. *Plant and Soil* 33: 30–48, 1970.
61. Nordmeyer AH: Legumes in protection forestry. *What's New in Forest Research* No 33, Forest Research Institute, Rotorua, New Zealand, 1976.
62. Davis MR: Legumes for infertile high-country soils. *What's New in Forest Research* No. 93, Forest Research Institute, Rotorua, New Zealand, 1981.
63. Woollons RC, Will GM: Increasing growth in high production radiata pine stands by nitrogen fertilisers. *NZ Jl For* 20: 243–253, 1975.

Index